Waste Reduction and Management in Food Security and Food Sustainability

Waste Reduction and Management in Food Security and Food Sustainability

Guest Editor

Ali Khoddami

MDPI

Basel • Beijing • Wuhan • Barcelona • Belgrade • Novi Sad • Cluj • Manchester

Guest Editor
Ali Khoddami
School of Life and
Environmental Sciences
The University of Sydney
Sydney
Australia

Editorial Office
MDPI AG
Grosspeteranlage 5
4052 Basel, Switzerland

This is a reprint of the Special Issue, published open access by the journal *Foods* (ISSN 2304-8158), freely accessible at: https://www.mdpi.com/journal/foods/special_issues/QZ24J7P552.

For citation purposes, cite each article independently as indicated on the article page online and as indicated below:

Lastname, A.A.; Lastname, B.B. Article Title. *Journal Name* **Year**, *Volume Number*, Page Range.

ISBN 978-3-7258-7240-4 (Hbk)
ISBN 978-3-7258-7241-1 (PDF)
https://doi.org/10.3390/books978-3-7258-7241-1

Contents

About the Editor

Ali Khoddami

Ali Khoddami is a food and agricultural researcher at the University of Sydney, where his research focuses on the intersection of food chemistry, sustainable production, and climate resilience. A leading expert in the nutritional and functional properties of crops, Dr. Khoddami has spearheaded several high-impact research initiatives, including the Australian Research Council (ARC) funded project investigating the food quality of Australian Indigenous Grains and the impacts of plant environments on their chemical profiles. Beyond native flora, Dr. Khoddami has significantly contributed to global food security through projects such as the GRDC-funded "Expanding Options for Sorghum," in which he explored the grain's potential as a food. His expertise in chemistry is evident in his collaborative work on sustainable poultry production, specifically investigating how heat-tolerant wheat and sorghum varieties respond to climate stressors such as drought. Dr. Khoddami's commitment to a circular economy is further demonstrated through his research on fruits and vegetables for sustainable diets, as well as his work on adding value to food waste, where he develops methods to reduce fresh produce waste and transform agricultural by-products into bioactive functional ingredients.

foods

MDPI

Editorial

Waste Reduction and Management in Food Security and Food Sustainability

Ali Khoddami

Sydney Institute of Agriculture, School of Life and Environmental Sciences, Faculty of Science, The University of Sydney, Sydney, NSW 2006, Australia; ali.khoddami@sydney.edu.au

Introduction

Climate change—one of the major global challenges—significantly affects agricultural productivity, the accessibility of water resources, and the stability food systems. Increases in temperature, extreme weather phenomena, and variable rainfall patterns led to a rise in pest and disease pressures and a reduction in crop yields [1,2], limiting the accessibility of rich food resources. Due to the increase in the global population, evolving food preferences, and lifestyle changes, food production is expected to be higher in the next 35 years compared to that of the entire history of mankind [3]. It is, therefore, necessary to provide food security for the growing world population while promoting the use of natural resources in a sustainable manner. In this regard, food security—as a multidimensional concept—refers to the economic, social, and physical access of all people to nutritious and sufficient food, that meets their dietary needs and preferences, at all times. Consequently, promoting a culture of avoiding food waste and managing food sustainably in different countries is essential to address environmental challenges and to ensure food security.

According to the Food and Agriculture Organization of the United Nations [4], about one-third of global food production meant for human consumption leads to food waste, and it occurs at different levels of the food supply chain. These levels include agricultural production, post-harvest transportation, processing, distribution, retail, and consumption. The application of advanced technologies, along with interventions in consumer behavior, can be considered efficient solutions for countering food waste [5]. Food waste harms the environment in a number of ways, including deforestation, water waste, greenhouse gas emissions from organic matter decomposition in landfills, and energy consumption for manufacturing and transportation [6,7]. On the other hand, the conversion of organic waste into high-value bioactive compounds is possible; this approach is considered a sustainable solution to reduce waste and increase the efficiency of resources. Vitamins are often found among food waste, despite the fact that they can be reused in different food- and medicine-related products [8]. These materials may also be used as secondary resources to create food items with added value. The significantly high levels of food waste can be considered an opportunity to address food insecurity, combat climate change and environmental damage, as well as to save energy [9].

Furthermore, the management of food waste at the consumer level—including the management of relevant behaviors—is considered imperative to decrease food waste and reduce its environmental impacts. Despite expressing an aversion to food waste, consumers often engage in behaviors that lead to food waste [10]. This is especially prevalent among the younger generation—a key consumer group that shapes the future population and plays an important role in reducing this environmental problem. The following is an attempt to briefly present research that has addressed various challenges, such as food security, food

 https://doi.org/10.3390/foods15040749

waste, and consumer behavior, as well as sustainable solutions to reduce environmental impacts, in order to reduce food waste and wastage from farms to households, and prevent environmental degradation and economic inefficiency.

In the article *"Global Comparison and Future Trends of Major Food Proteins: Can Shellfish Contribute to Sustainable Food Security?"* Contribution 1 (2025) examine the environmental quality and food security related to food production. The work also analyzes innovative and environmentally friendly protein sources, such as lab-grown foods, microorganisms, and insects, as well as vegetables. This study employs a combined approach to integrate quantitative and qualitative data from the literature and public databases. The effects of various protein resources on agricultural land use and greenhouse gas emissions were evaluated, and the possibility of carbon sequestration within the supply chain was investigated. In addition, future trends regarding global protein resources were reviewed, and shellfish were introduced as a significant, environmentally friendly and nutritional option. It was reported that this achievement is possible via shell biocalcification, which maintains low-impact production and contributes to the reduction in climate change. Moreover, in order to evaluate consumer, environmental, and nutritional challenges, all main protein resources like vegetal proteins, farmed seafood, and terrestrial livestock—along with alternative sources like cultured cells and insects—were considered. Bivalve aquaculture is a zero-emissions method that is dependent on natural phytoplankton and free of additives or antibiotics, which can have a minimal environmental footprint. The results of the study suggest that Bivalve aquaculture, including scallops, clams, mussels, and oysters, has great potential to increase food security and promote sustainable protein consumption by decreasing land use and assisting in mitigating climate change, environmental impacts, and atmospheric carbon. Since the need for global food is expected to increase in the future, seafood aquaculture is predicted to expand due to its benefits to human health and its lower environmental impact. The authors believe that challenges need to be addressed to fully exploit the potential of bivalve aquaculture as a sustainable economic activity.

The article *"What Drives Generation Z to Avoid Food Waste in China? An Empirical Investigation"* by Contribution 2 (2025) examines food waste avoidance behaviors among Generation Z in China from a psychosocial perspective. Food waste avoidance not only contributes to a reduction in environmental pressure and resource waste but also assists in achieving the objectives of sustainable development. Because of the features of the consumption environment and the consumers of this generation, the Theory of Planned Behavior (TPB) is the foundation for this research. After the collection of 417 valid responses via online survey and a questionnaire, structural equation modeling was employed to analyze the data. The findings indicate that Generation Z's intentions to prevent food waste are positively impacted by attitude, subjective norms, and perceived behavioral control. Additionally, attitudes and perceived behavioral control are positively and significantly impacted by moral self-identity. Furthermore, it is shown that scarcity mentality has a beneficial and moderating effect. The authors categorize the results of this paper into three areas (both theoretical and practical), including (I) the introduction of scarcity mindsets and moral identity leads to the extension of the TPB model and the deepening of our understanding about psychological mechanisms related to the avoidance of food waste; (II) experimental results indicate the TPB model's robustness, which demonstrates complex correlations among variables, indicating the significant moderating impacts of scarcity mindsets and moral identity, and demonstrating how social influences affect behavioral intentions in particular contexts; and (III) PBC is often influenced by environmental constraints, and environmental amenities can allow Generation Z to manage food more effectively. According to the authors, researchers should be expected to investigate the actual, real-world food waste behaviors of this generation, rather than simply their inten-

https://doi.org/10.3390/foods15040749

tions. Moreover, focusing on Generation Z leads to the lack of generalizability of the results. Furthermore, cross-national comparisons between Generation Z from various cultural origins may highlight notable differences in attitudes and behaviors and provide insightful information for the creation of successful international initiatives and policies.

The "*HRMS Characterization and Antioxidant Evaluation of Costa Rican Spent Coffee Grounds as a Source of Bioactive Polyphenolic Extracts*" by Contribution 3 (2025) presents a thorough analysis of Costa Rica's spent coffee grounds (SCGs), which serve as an abundant source of secondary metabolites possessing antioxidant qualities. The study used high-resolution mass spectrometry (HRMS) and ultra-performance liquid chromatography (UPLC-QTOF-ESI MS) to test aqueous extracts collected from various SCG samples. A total of twenty-one substances were identified based on the data, including four atractyligenin diterpenes, three caffeoylquinic lactones, and fourteen phenolic acids. Additionally, the study assessed antioxidant activity using four distinct assay techniques: the Trolox equivalent antioxidant capacity (TEAC) assay, the oxygen radical absorbance capacity (ORAC) assay, the potassium ferricyanide-reducing antioxidant power (PFRAP) assay, and the 2,2-diphenyl-1-picrylhydrazyl (DPPH) assay. The results revealed that the fine-grind sample outperformed the medium- and coarse-grind samples in antioxidant evaluations and had the greatest amounts of total polyphenols and phenolic acids. The findings highlight the importance of SCGs in metabolites that confer health benefits; therefore, there is interest in developing functional and therapeutic food innovations, given their potential health advantages.

The article "*Up-Cycling Broccoli Stalks into Fresh-Cut Sticks: Post-harvest Strategies for Quality and Shelf-Life Enhancement*" Contribution 4 (2025) examines the use of broccoli stems as a discarded agricultural by-product, which have less value in the context of fresh consumption and are only offered to the market as broccoli florets. In this research, broccoli stems—as a fresh cut product with added value—were evaluated by employing post-harvest storage strategies for 11 days at 5 °C to maintain their quality and increase the shelf life of the product. A mixed methodology was followed in this study to examine various storage and treatment approaches. In addition, Statgraphics Centurion XV software was used to perform ANOVA. Regarding various post-harvest treatments, hot water treatment (HWT) was reported to reduce browning and increase shelf life by 11 days. This result indicated the control of respiration and growth of microbes since responses related to natural defense in plant tissues are stimulated by heat, which leads to a rise in the resistance to microbial invasion. This feature led to the achievement of the highest sensory quality scores while being less efficient in the preservation of bioactive compounds. A combination of treatment with HWT + calcium ascorbate (CaAsc) was reported to be most effective in the optimization of quality and preservation of compounds that play a role in health promotion. The antioxidant effect of ascorbic acid in CaAsc was a potential reason for this synergistic effect. A combination of calcium and ascorbic acid that significantly preserves vitamin C. Trehalose (TREH) alone did not show a positive preservative effect and caused browning, increased respiration, and microbial proliferation. Moreover, basic hygiene and storage practices ensure microbiological safety during storage, such as a lack of foodborne pathogens—including E. coli and Salmonella spp. After the comparison of different treatment approaches, the combined method based on antioxidant and mild thermal treatments was reported to be promising for the preservation of broccoli stems. This indicates the applicability of these stems as a valorization of food waste in addition to satisfying the principles of a circular economy.

The paper "*Optimizing Farmers' and Intermediaries' Practices as Determinants of Food Waste Reduction Across the Supply Chain*" by Contribution 5 (2025) focused on the optimization of pre- and post-harvest practices, as well as the interaction between economic

https://doi.org/10.3390/foods15040749

and social traits of intermediaries and farmers for an increase in agri-food products' shelf life. Some approaches were reported to lead to a reduction in food losses and the highest levels of final returns for quality. Because grape producers in Egypt are vulnerable to changes in markets (at global levels)— due to the lack of land fragmentation and marketing information among small producers—this study examined data from 343 grape farmers and intermediaries in Egypt. The work used a combination of Ordinary Least Squares (OLS) and Fractional Regression Models (FRM) to analyze strong correlations between social and economic characteristics and stakeholder practices at different food supply chain stages. Key findings show that practices such as drip irrigation (13.9%), avoiding the use of ripening chemicals (24%), increasing farmers' cultivation area (6.1%), improved packaging (13.7%), and harvesting in the early morning hours (8.7%) can significantly reduce food waste. Moreover, for intermediaries, increasing the distance to market increased waste by 0.15%; whereas, employing suitable warehouses, participating in formal markets, and choosing the right transportation method reduced waste by 65.9%, 13.8% and 7.9%, respectively. Furthermore, food waste may be greatly decreased by combining these strategies. The authors additionally contend that post-harvest losses (PHL) are decreased by improving farmer knowledge and farmed area, and that formal agricultural markets help lower post-harvest losses by approximately 14% relative to informal markets. Furthermore, the decrease in PHL and the health of people and the environment were greatly enhanced by the lack of toxins. In order to lower PHL, the authors stress the need to integrate market types with suitable transportation and storage capabilities.

In the paper "*Moving Beyond Indices: A Systematic Approach Integrating Food System Performance and Characteristics for Comprehensive Food Security Assessment,*" the limits of food security indicators—such as possible ranking biases and insufficient understanding of food system dynamics—were studied by Contribution 6 (2025). Indicators of food security are used to make decisions and evaluate the ability of a nation to endure international economic and ecological crises. By using a methodology that incorporated two-step clustering and elastic-net regression-based feature selection, this study overcame these constraints and supported food system management by assisting decision-makers in implementing organized techniques for well-informed decision-making. This study used a new mathematical and operational approach using the Economist Intelligence Unit (EIU) dataset of 94 countries, selecting 15 indicators for the first model and 9 indicators for the second model from an initial set of 25 indicators. A two-stage clustering approach was then used to separate the countries into four categories according to a combination of income levels and food system features. This combined approach addressed the limitations of traditional input-based models, which place an emphasis on the availability of resources but do not link the composite indicator to food system outcomes. The nutritional condition of the population was taken into account as an essential output indicator in the input–output model, indicating the availability of daily food needs. According to the findings, countries like Switzerland and Germany that have high per capita incomes as well as integrated and industrialized food systems, which provide consistent food security for their people. However, food insecurity is observed more in traditional and rural food systems as well as in low-income countries. These systems could focus on policies that foster agro-technological innovation, increase access to agricultural finance, and enhance transport infrastructure. The suggested method cannot monitor changes or trends over time since it is only based on cross-sectional data.

The article "*Culinary Home Empowerment for Food Waste Prevention and Minimization: Feasibility and Efficacy Protocol*" by Contribution 7 (2024) outlines a protocol that assesses the feasibility, acceptability, and preliminary effectiveness of a household food-waste prevention and reduction program. The intervention features eight modules, each with

https://doi.org/10.3390/foods15040749

three to five short videos, as well as downloadable recipes and worksheets. The impact of this program would be examined in two distinct contexts: (1) a population-based setting in which home cooks voluntarily access the content and (2) a community-based setting in which the content is delivered intensively as part of a remote class. At the population level, the study processed outcomes such as feasibility, usage, acceptability, and satisfaction, as well as preliminary efficacy outcomes, including motivation, opportunity, and ability. It employs a single-arm quasi-experimental post-test-only design, in which the viewers are asked to complete a brief five-item post-video survey. Additionally, a separate recruited sample would undergo a more detailed assessment, including pre- and multiple post-test surveys. At the community level, the mixed-methods evaluation assesses feasibility, usage, acceptability, and satisfaction, as well as preliminary efficacy outcomes including motivation, opportunity, and ability. It also considers additional outcomes such as household food waste and sustainable dietary practices. The authors believe that the proposed protocol can translate findings into prevention lessons to minimize food waste at the consumer level. In keeping with a protocol article, this article is presented without results with statistical analysis prior to data collection. In order to guide policy and investment for program interventions, the authors hope that data analysis will reveal elements of the intervention that can prevent and reduce waste more broadly, informing policy and investment decisions.

In the article "*Solid-State Fermentation of Cereal Waste Improves the Bioavailability and Yield of Bacterial Cellulose Production by a Novacetimonas sp. Isolate*," the importance of cereal wastes like rice bran and cereal dust as by-products was examined by Contribution 8 (2024) using three mold isolates and a low-cost solid-state fermentation (SSF) pre-treatment. They also attempted to convert waste into a renewable, eco-friendly biomaterial that is crucial for the cost-effective and sustainable production of bacterial cellulose. The environmental effect of agro-industrial waste and the high cost of traditional culture media were demonstrated by substituting untreated rice bran or cereal dust for the culture medium. In contrast to the untreated control sample, the application of rice bran fermented with Rhizopus oligosporus produced a greater BC. In this work, when compared to Hestrin-Schramm (HS), the costs to produce culture medium decreased by up to 90%. The produced BC were characterized using derivative thermogravimetry (DTG), thermogravimetric analysis (TGA), X-ray diffraction (XRD), Fourier-transform infrared spectroscopy (FTIR), energy-dispersive X-ray spectroscopy (EDS), and scanning electron microscopy (SEM) techniques to gain insight into its chemical bonding, morphology, and internal structure. It was also found that there were comparable properties of biopolymer between the cellulose produced in waste-based and standard culture media, demonstrating that the SSF-based process can be effective in converting cereal waste to BC. The SEM technique demonstrated that the BC's nanostructure, fiber diameter, and shape were impacted by both the drying process and the selection of fungal strain for SSF. However, the obtained results from FTIR and XRD tests verified that the cellulose I structure remained intact in bacterial cellulose derived from treated substrates via solid-state fermentation. The research also found that SSF treatment enhanced the thermal stability of BC made from rice bran and cereal dust.

The article "*Tackling Food Waste: An Exploratory Case Study on Consumer Behavior in Romania*" by Contribution 9 (2024) conducted an analysis of the factors influencing household consumer behavior regarding food waste (FW) in Romania. Their aim was to understand the motivations to reduce FW, and investigate attitudes towards FW and food consumption behaviors in households. Data was collected using an online, self-administered questionnaire from a non-probability sample in Romania to examine consumer behaviors related to food waste avoidance. Moreover, a stepwise linear regression model and descriptive statistical analyses were performed to evaluate the Romanian consumers' behaviors to decrease FW as well as to specify the effect of various determinants on the motivation of consumers for avoiding FW. The findings point to important factors like the desire to lessen environmental effects; discussions about FW reduction in the home; established consumption patterns, particularly when cooking at home; the person's social circle, and personal, social, and ecological responsibilities; and involvement in awareness campaigns. In fact, the authors emphasize that FW avoidance behavior emerges from family environment, that discussions in the household are valuable, and that by socializing new generations with this spirit there will be a significant impact on FW reduction. The research points out the importance of educational and awareness-raising strategies, which are essential for changing consumer attitudes and behaviors in Romania. Furthermore, the discussion includes a comparative analysis with the results of studies documented by researchers in other countries, which highlights the differences in FW avoidance behavior of individuals in different cultural, social, and economic environments. Due to the limited number of participants, this study does not have appropriate generalizability to the entire Romanian population.

In the article "*Antioxidant Bio-Compounds from Chestnut Waste: A Value-Adding and Food Sustainability Strategy*" Contribution 10 (2025) concentrated on the antioxidant qualities of certain chestnut burr cultivars, to address the possibilities of a circular bioeconomy via the value-adding of agricultural leftovers and the dual issues of waste management and public health. According to the findings, almost one-third of food production is wasted, which increases environmental and human concerns. Utilizing the circular bioeconomy, biological waste may be converted into useful goods that can be used in sustainable food production. Based on their phenolic content, this study evaluates the hydroalcoholic extract content and antioxidant qualities of three chestnut burr cultivars from Mount Amiata that are certified under Indicazione Geografica Protetta (PGI)—Marroni (RM), Cecio (RC), and Bastarda Rossa (RB). In vitro assays were performed to assess their significance in combating oxidative stress. Because of their antioxidant properties, the investigations indicated that gallic and ellagic acids are the most prevalent components in the extracts. The results also demonstrated that the peel extracts of the Cecio and Bastarda Rossa cultivars exhibited antioxidant activity and much greater amounts of important bio-compounds, such as flavonoids, gallic acid, and ellagic acid. Moreover, a cytotoxicity test confirmed that all three extracts are safe for human cell lines. The in silico docking simulations that were performed in the study demonstrated that the chemicals may spontaneously bind to predicted targets, such as proteins associated with oxidative stress. This work advances functional foods and emphasizes the importance of chestnuts in promoting practices related to the circular bioeconomy, which focuses on the production of value-added products for biological applications. This approach would help to reduce the environmental pollution caused by the incineration of chestnut waste.

In the article "*Quality Properties of Bakery Products and Pasta Containing Spent Coffee Grounds (SCGs): A Review*" Contribution 11 (2024) examined spent coffee grounds (SCGs) as a type of waste that can be used as a material to create new foods. They found that incorporating different percentages of SCGs provided different quality properties in bakery and pasta products. This study used a literature review, with a comprehensive search of reputable databases and content analysis of published studies, to investigate the effects of adding different percentages of SCGs to different products. The study was conducted on baked products such as pasta, muffins, cakes, biscuits, bread, and cookies. The authors highlighted that, from a chemical point of view, SCGs contributed to increasing the nutritional value (protein, fat, fiber, ash, and antioxidant content) and health benefits of the baked products. They also emphasized that SCGs affected the physical properties of the products and that the effect depended on the concentration of SCGs, the coffee variety, and the processing method. The authors also investigated the effect of SCGs on the sensory

https://doi.org/10.3390/foods15040749

properties of the products. The incorporation of SCGs improved their sensory properties, adding a specific flavor and color to the baked goods. It is extremely important that the microbiological safety of pasta and bread, which guarantees the safety and quality of the product for human consumption, is not compromised when SCGs are added. Given their unique textures and structures, various kinds of baked products reacted to SCGs in different ways. SCGs enhanced the taste and fragrance of cookies, which are denser and less moist than bread or cakes. When assessing from an economic viewpoint, it is cost-effective to recycle SCGs, and this significantly assists in the minimization of the environmental footprint of the food industry and the reduction in waste. According to this work, the type of coffee was not specified in some studies—and some studies employed small sample sizes—which potentially limits the applicability and robustness of their findings.

Funding: This research received no external funding.

Institutional Review Board Statement: Not applicable.

Informed Consent Statement: Not applicable.

Data Availability Statement: Not applicable.

Conflicts of Interest: The author declares no conflicts of interest.

List of Contributions:

1. Tamburini, E.; Moore, D.; Castaldelli, G. Global comparison and future trends of major food proteins: Can shellfish contribute to sustainable food security? *Foods* **2025**, *14*, 2205.
2. Qi, X.; Li, M.; Chen, J.; Zhan, G.; Niu, L. What drives generation Z to avoid food waste in China? An empirical investigation. *Foods* **2025**, *14*, 323.
3. Navarro-Hoyos, M.; Vargas-Huertas, L.F.; Chacón-Vargas, J.D.; Leandro-Aguilar, V.; Alvarado-Corella, D.; Vega-Baudrit, J.R.; Romero-Esquivel, L.G.; Sánchez-Kopper, A.; Monge-Navarro, A.; Araya-Sibaja, A.M. HRMS characterization and antioxidant evaluation of Costa Rican spent coffee grounds as a source of bioactive polyphenolic extracts. *Foods* **2025**, *14*, 448.
4. García-Lorca, N.; Salas-Millán, J.Á.; Aguayo, E. Up-Cycling broccoli stalks into fresh-cut sticks: Post-harvest strategies for quality and shelf-life enhancement. *Foods* **2025**, *14*, 2476.
5. Ali, A.; Tan, Y.; Yang, S.; Xia, C.; Long, W. Optimizing farmers' and intermediaries' practices as determinants of food waste reduction across the supply chain. *Foods* **2025**, *14*, 2351.
6. Al-Ansari, M.; Nabeel, H.; Abdella, G.M.; El Mekkawy, T. Moving beyond indices: A systematic approach integrating food system performance and characteristics for comprehensive food security assessment. *Foods* **2025**, *14*, 1834.
7. Milliron, B.-J.; Neff, R.; Sherman, R.; Forde, D.; Miller, L.; Stott, D.; Mountford, A.; Deutsch, J.M. Culinary home empowerment for food waste prevention and minimization: Feasibility and efficacy protocol. *Foods* **2024**, *13*, 2529.
8. Henry, S.; Dhital, S.; Sumer, H.; Butardo, V., Jr. Solid-State fermentation of cereal waste improves the bioavailability and yield of bacterial cellulose production by a *Novacetimonas* sp. isolate. *Foods* **2024**, *13*, 3052
9. Danciu, C.-A.; Croitoru, A.; Antonie, I.; Tulbure, A.; Popescu, A.; Stanciu, C.; Sava, C.; Stanciu, M. Tackling food waste: An exploratory case study on consumer behavior in Romania. *Foods* **2024**, *13*, 3313.
10. Barletta, R.; Trezza, A.; Bernini, A.; Millucci, L.; Geminiani, M.; Santucci, A. Antioxidant biocompounds from chestnut waste: A value-adding and food sustainability strategy. *Foods* **2025**, *14*, 42.
11. Ahanchi, M.; Sugianto, E.C.; Chau, A.; Khoddami, A. Quality properties of bakery products and pasta containing spent coffee grounds (SCGs): A review. *Foods* **2024**, *13*, 3576.

https://doi.org/10.3390/foods15040749

References

1. Wiebe, K.; Robinson, S.; Cattaneo, A. Climate change, agriculture and food security: Impacts and the potential for adaptation and mitigation. In *Sustainable Food and Agriculture—An Integrated Approach*; Campanhola, C., Pandey, S., Eds.; The Food and Agricultural Organisation of the United Nations (FAO): Rome, Italy, 2019; pp. 55–74.
2. Eswaran, S.; Anand, A.; Lairenjam, G.; Mohan, G.; Sharma, N.; Khare, A.; Bhargavi, A. Climate change impacts on agricultural systems mitigation and adaptation strategies: A review. *J. Exp. Agric. Int.* **2024**, *46*, 1–12. [CrossRef]
3. Hemathilake, D.M.K.S.; Gunathilake, D.M.C.C. Agricultural productivity and food supply to meet increased demands. In *Future Foods*; Academic Press: New York, NY, USA, 2022; pp. 539–553. ISBN 978-0-323-91001-9.
4. Food and Agriculture Organization. Global food losses and food waste–Extent, causes and prevention. In *SAVE FOOD: An Initiative on Food Loss and Waste Reduction*; Food and Agriculture Organization: Rome, Italy, 2011; Volume 9.
5. Agya, B.A. Technological solutions and consumer behaviour in mitigating food waste: A global assessment across income levels. *Sustain. Prod. Consum.* **2025**, *55*, 242–256. [CrossRef]
6. Liu, Z.; Rousta, K.; Bolton, K.; Brancoli, P. Waste quantification and carbon footprint of street market food waste in Brazil. *Cogent Food Agric.* **2024**, *10*, 2305520. [CrossRef]
7. Zhao, P.; Zhang, S.; Wang, X.; Sun, H.; Guo, Y.; Wang, Q.; Sun, X. Research progress on anaerobic digestion of cellulose waste based on bibliometric analysis. *Sustainability* **2023**, *15*, 16060. [CrossRef]
8. Pandhi, S.; Mahato, D.K.; Kurian, A.; Pandey, S.; Thamarai, A.; Gaur, V.; Kamle, M.; Shukla, A.C.; Kumar, P. Transforming agri-food waste into value: Sustainable approaches for next-generation resistant starch production. *Bioresour. Technol.* **2025**, *437*, 133090. [CrossRef] [PubMed]
9. USEPA from Farm to Kitchen: The Environmental Impacts of U.S. Food Waste. Available online: https://www.epa.gov/land-research/farm-kitchen-environmental-impacts-us-food-waste (accessed on 1 August 2024).
10. Le Borgne, G.; Sirieix, L.; Valette-Florence, P.; Costa, S. Adopting waste-prevention routines: The role of consumer concern for food waste. *Appetite* **2021**, *163*, 105188. [CrossRef] [PubMed]

https://doi.org/10.3390/foods15040749

foods

MDPI

Article

Up-Cycling Broccoli Stalks into Fresh-Cut Sticks: Postharvest Strategies for Quality and Shelf-Life Enhancement

Nieves García-Lorca [1,2], José Ángel Salas-Millán [2] and Encarna Aguayo [1,2,*]

1 Postharvest and Refrigeration Group, Universidad Politécnica de Cartagena (UPCT), Paseo Alfonso XIII, 48, 30203 Cartagena, Spain; nieves.garcial@upct.es

2 Food Quality and Health Group, Institute of Plant Biotechnology (IBV-UPCT), Campus Muralla Del Mar, R&D Building, 30202 Cartagena, Spain; joseangel.salas@upct.es

* Correspondence: encarna.aguayo@upct.es

Abstract

Broccoli stalks are considered an agro-industrial by-product that, in the context of fresh consumption, is undervalued, as only broccoli florets are typically marketed. This study evaluated the up-cycling of broccoli stalks into a value-added fresh-cut product through postharvest preservation strategies. Stalks were peeled, cut into sticks (8 × 8 mm × 50–100 mm), sanitised, packaged under modified atmosphere conditions, and stored at 5 °C. Treatments included (a) calcium ascorbate (CaAsc, 1% *w*/*v*), (b) trehalose (TREH, 5% *w*/*v*), (c) hot water treatment (HWT, 55 °C, 1 min), and several combinations of them. HWT alone was highly effective in reducing browning, a key factor for achieving an extended shelf-life, controlling microbial growth and respiration, and obtaining the highest sensory scores (appearance = 7.3 on day 11). However, it was less effective in preserving bioactive compounds. The HWT + CaAsc treatment proved to be the most effective at optimising quality and retaining health-promoting compounds. It increased vitamin C retention by 78%, antioxidant capacity by 68%, and total phenolic content by 65% compared to the control on day 11. This synergistic effect was attributed to the antioxidant action of ascorbic acid in CaAsc. TREH alone showed no preservative effect, inducing browning, elevated respiration, and microbial proliferation. Overall, combining mild thermal and antioxidant treatments offers a promising strategy to valorise broccoli stalks as fresh-cut snacks. An 11-day shelf-life at 5 °C was achieved, with increased content of health-promoting bioactive compounds, while supporting circular economy principles and contributing to food loss mitigation.

Keywords: calcium ascorbate; hot water treatment; trehalose; glucosinolates; circular economy; revalorisation

1. Introduction

Broccoli (*Brassica oleracea* L.) is a cruciferous vegetable widely appreciated for its notable nutritional profile and health-promoting bioactive compounds, such as glucosinolates (GSLs), phenolic compounds, and vitamin C [1,2]. These phytochemicals exhibit antioxidant, anti-inflammatory, and chemopreventive effects, contributing to the recognition of broccoli as a functional food [3]. However, the commercial exploitation of broccoli is predominantly limited to its florets, which constitute only around 15% of the total biomass, whereas the stalks (21%), leaves (47%), and roots (17%) are commonly discarded, resulting in a significant underutilisation of valuable plant tissue [4,5].

Broccoli stalks are particularly rich in dietary fibre, GSLs, and residual sugars, positioning them as a promising raw material for the development of functional foods [6]. Nonetheless, their high-water activity and elevated respiration rates make them highly perishable, and they are prone to enzymatic browning, microbial spoilage, and nutrient degradation during storage [7,8]. These challenges are further exacerbated during minimal processing, where cutting, peeling, and washing lead to the disruption of cellular compartments, the activation of oxidative enzymes, and an increased microbial load [9]. In this context, transforming broccoli stalks into a fresh-cut, minimally processed, ready-to-eat product represents a novel strategy for the valorisation of agro-industrial by-products. Fresh-cut products are fresh fruits or vegetables that have been physically altered from their original form, such as by cutting, slicing, dicing, or peeling, while remaining in a fresh, raw state. These products are typically washed, minimally processed, refrigerated, and packaged to ensure safety and to extend shelf life while preserving their freshness [10]. This approach not only supports circular economy principles through waste reduction and resource efficiency but also offers a nutritious and practical alternative for consumers. Broccoli sticks could be marketed in a similar way to carrot or celery sticks, which are suitable for inclusion in salad bars and individual packs or as convenient ready-to-eat options for dipping. This innovation responds to the growing consumer demand for sustainable, health-oriented vegetable-based products, as reflected in the expansion of the global healthy snacks market, which was projected to reach USD 107 billion by 2024, with a CAGR of approximately 6.4% [11].

Of the different postharvest treatments available, antioxidant treatments, osmoprotectants, and mild thermal processes are among the most effective strategies for maintaining postharvest quality in fresh-cut produce. Calcium ascorbate (CaAsc), a salt formed by calcium and ascorbic acid, combines the antioxidant effect of ascorbic acid with the structural reinforcement conferred by calcium [12]. Ascorbic acid acts as a reducing agent that inhibits polyphenol oxidase (PPO)-mediated browning by reducing o-quinones to their precursor diphenols, thereby preventing the formation of melanins [13]. Simultaneously, Ca^{2+} ions reinforce the middle lamella and maintain firmness, slowing enzymatic degradation and limiting microbial infiltration [14,15]. CaAsc has demonstrated efficacy in preserving vitamin C, total phenolic content (TPC), and antioxidant capacity during cold storage in products such as melon [16], apple [17], and potato [18].

Trehalose (TREH), a non-reducing disaccharide, has attracted attention as an osmoprotectant and stress mitigator in postharvest physiology. It stabilises membranes and proteins under abiotic stress, which reduces oxidative damage and maintains tissue turgor [19,20]. Its application to fruits such as guava [21], cherry tomato [22], kiwifruit [23], and litchi [24] has delayed postharvest browning and improved membrane stability, antioxidant enzyme activity, and water retention. However, the effects of TREH are highly species- and matrix-dependent; thus, its effectiveness in broccoli cannot be assumed merely based on results from other crops. For instance, it may alter microbial dynamics in broccoli, due to its bioavailability as a carbon source for certain spoilage microorganisms [25], and may modulate PPO and peroxidase (POD) activity in unpredictable ways, potentially intensifying browning reactions [26]. Therefore, its effectiveness as a preservative must be validated for each target product.

Hot water treatment (HWT) is a traditional postharvest tool that is gaining relevance as a non-chemical preservation method that combines efficacy and consumer acceptance. HWT involves immersing the product in water at sublethal temperatures (typically 45–55 °C) for a short period of time (30 s to 2 min), thereby triggering the partial denaturation of degradative enzymes such as PPO, stimulating heat shock responses, and suppressing initial microbial loads [27,28]. HWT has been shown to delay senescence, reduce the respiration rate, and enhance the retention of antioxidant compounds during storage in fresh-cut apples, melons, and Chinese cabbage [12,29]. Moreover, HWT facilitates the uptake of solutes into plant tissues by increasing membrane permeability and promoting tissue porosity [30], which could enhance the efficacy of subsequent antioxidant dips such as CaAsc.

While these technologies have been widely studied in fresh-cut vegetables, the postharvest physiology and preservation of broccoli stalks remain unexplored. Their distinct anatomy and chemical composition, particularly the high concentrations of GSLs in outer tissues [31], may influence the efficacy of common preservation treatments. Previous studies have also shown that GSL profiles are sensitive to both enzymatic hydrolysis and thermal degradation [32,33], and thus require tailored strategies to minimise nutrient loss.

Therefore, the objective of the present study is to assess the individual and combined effects of CaAsc, TREH, and HWT, integrated with modified atmosphere packaging (MAP), on the shelf-life, microbial quality, sensory attributes, and bioactive compound retention in fresh-cut broccoli stalk sticks. This research seeks to establish an optimised postharvest protocol that enhances the valorisation of broccoli by-products and supports sustainable innovation in the fresh-cut vegetable industry.

2. Materials and Methods

2.1. Plant Material and Treatments

Fresh broccoli stalks (*Brassica oleracea* L. var. *Italica*, cv. Parthenon) are an undervalued by-product in the context of fresh consumption, as only the broccoli florets are typically marketed, while the stalks are largely discarded. Broccoli stalks were obtained from Levante Sur Coop. in the southeast of Spain (La Palma, Cartagena, Spain). Firstly, to obtain broccoli sticks, the broccoli stalks were washed with tap water to remove any dust and dirt. Subsequently, they were preliminarily disinfected by immersion in an 80 ppm peroxyacetic acid (PAA) solution for one minute. PAA was chosen because of its proven antimicrobial efficacy and its approval as a technological coadjuvant for washing fruits and vegetables, as recognised by the Spanish Food Safety Authority (AESAN) [34]. After being disinfected, the broccoli stalks were manually peeled with a knife and then cut into sticks measuring 8 × 8 mm in width and 50–100 mm in length with a vegetable stick cutter (model QTJ-J002X, VEVOR, Shanghai, China). Following the cutting process, the broccoli sticks underwent a second disinfection step in an 80 ppm peroxyacetic acid solution for one minute. Thereafter, the broccoli sticks were subjected to seven different treatments and also a control group that received no additional treatment beyond the disinfection procedure.

The treatments applied to the broccoli sticks were as follows: (a) CaAsc: the broccoli sticks were immersed in a cooled (5 °C) 1% (w/v) CaAsc solution for one min. (b) TREH: sticks were submerged in a cooled 5% (w/v) TREH solution for one min. (c) HWT: sticks were submerged for one min in water at 55 °C, followed by immediate cooling in a cold-water bath (5 °C). In addition to these three treatments, this study evaluated a further four combinations of the above treatments: (d) CaAsc + TREH: the sticks were sequentially immersed in a 1% CaAsc solution followed by a 5% TREH solution, each for one min and at 5 °C. (e) HWT + CaAsc: sticks were immersed in hot water at 55 °C for one min, then immediately cooled in a 1% CaAsc solution for a further min. (f) HWT + TREH: sticks were immersed in hot water at 55 °C for one min, followed by cooling in a 5% TREH solution for one min. (g) HWT + CaAsc + TREH: sticks were sequentially treated in hot water at 55 °C for one min, then cooled in both 1% CaAsc as well as 5% TREH solutions, each for one min. After the treatments, 120 g of broccoli sticks were packed in 500 mL trays and sealed with polypropylene film (microperforated Pplus, Amcor Plc, Zürich, Switzerland) for MAP, helping to maintain high relative humidity within the package. The trays were stored in a cold storage chamber (5 ± 1 °C), which was kept dark throughout the storage period (11 days), except for when the lights were briefly switched-on during sample collection, which was carried out on days 0, 4, 8, and 11. A minimum of three replicates (trays) per treatment were prepared.

2.2. Browning Index

Surface colour measurements were performed on 10 broccoli sticks per treatment and storage day using a Minolta CR-400 colorimeter (Minolta Inc., Tokyo, Japan). The device was calibrated with a white porcelain plate before use. For each broccoli stick, three random points were measured. Chromatic values were determined according to the CIELAB system and recorded as L^* (lightness), a^* (red–green component), and b^* (yellow–blue component). The browning index (*BI*) was calculated following the method described by Kasim and Kasim [35], using Equation (1).

$$BI = \frac{100\,(x - 0.31)}{0.17} \quad where\ x = \frac{(a^* + 1.75\,L^*)}{(5.645\,L^* + a^* - 0.012\,b^*)} \tag{1}$$

2.3. Microbiological Analysis

Microbiological analysis was conducted following standard methods to evaluate the microbial quality of the broccoli sticks. Samples were prepared by homogenising 10 g of stick sections from at least three different sticks per tray and treating in 90 mL of sterile peptone-buffered water for one min using a 400 Lab Stomacher (Seward Medical, London, UK). As described by Martínez-Hernández et al. [36], peptone-buffered water serves as an effective diluent for serial dilutions, ensuring accurate microbial enumeration. Serial dilutions were prepared and plated on selective media (Scharlau Chemie S.A., Barcelona, Spain) to quantify microbial groups under specific incubation conditions: mesophilic aerobic bacteria were plated on Plate Count Agar and incubated at 30 °C for 72 h; enterobacteria were plated on Violet Red Bile Dextrose Agar and incubated at 37 °C for 24 h; yeasts and moulds were plated on Rose Bengal Chloramphenicol Agar and incubated at 25 °C for 96 h; and pathogenic bacteria: *Salmonella* spp. was assessed using Xylose Lysine Deoxycholate Agar, while *Escherichia coli* was quantified using Tryptone Bile X-glucuronide Agar. All microbial counts were conducted in triplicate and expressed as log CFU/g where CFU (Colony-Forming Unit) represents the number of viable microorganisms capable of forming colonies under the given incubation conditions.

2.4. Gas Composition Analysis

The internal atmospheric composition in the trays was regularly monitored using a CheckPoint gas analyser (PCE Ibérica, S.L., Tobarra, Spain). For the analysis, a 25 mL gas sample was taken from each tray through a silicone septum, allowing the measurement of oxygen (O_2) and carbon dioxide (CO_2) levels. The device was calibrated between each set of measurements by sampling the ambient air in the room to ensure accuracy. Gas composition was measured in triplicate for each tray and treatment on specific days (0, 4, 8, and 11) of the product's shelf-life.

2.5. Sensory Analysis

The sensory evaluation was conducted by a trained panel of 10 individuals in a controlled sensory laboratory, where each panellist was seated in an individual booth to control for external factors and to ensure an objective sensory assessment. Water was provided for palate cleansing between samples, and each panellist was presented with three broccoli sticks per treatment, coded with random letters to prevent bias. Sensory analysis included evaluations of browning and dehydration using a 9-point scale, where

1 indicated very low quality (e.g., very brown or severely dehydrated) and 9 indicated optimal quality (no browning or fully hydrated). Appearance, firmness, crispness, and taste acceptability were assessed on a 9-point hedonic scale, in which 1 meant "dislike very much," 9 "like very much," and 5 represented the threshold for marketability. Finally, overall acceptability was rated by integrating all assessed attributes, also using the same 9-point scale, where scores $\geq$ 5 indicated potential consumer acceptance.

This trained sensory evaluation did not require an ethical statement, since it involved no invasive procedures or health-related interventions. Before starting the evaluation, the research team explained the scope and details of the project to the participants, including the purpose of the research, the identity of the researchers, data protection measures, privacy and data retention policies, the voluntary nature of participation, the right to withdraw at any time, and contact details for any questions that might arise. Finally, all participants signed a written informed consent form, confirming that they had read and understood the information provided, and that their questions had been answered.

2.6. *Chemical Analysis*

For the chemical analysis described below, the broccoli sticks were cooled, freeze-dried, ground into a fine powder, and stored at room temperature in vacuum-sealed bags until analysis.

2.6.1. Vitamin C Content

Vitamin C analysis was performed using High-Performance Liquid Chromatography (HPLC) as previously described by Rasines et al. [37] with slight modifications. A 0.5 g sample of freeze-dried broccoli stick powder was homogenised with 10 mL of cold extraction buffer (0.1 M citric acid, 0.05% ethylenediaminetetraacetic acid (EDTA), 4 mM sodium fluoride, and 5% methanol) using an Ultra-Turrax® homogeniser (IKA, Berlin, Germany). The mixture was kept on ice and in darkness. The homogenate was then filtered through a 4-layer gauze, and the pH of the extract was rapidly adjusted to 2.35–2.40 using 1 M HCl. Subsequently, 2 mL of the sample was centrifuged at 10,500× g for 5 min at 4 °C. The resulting supernatant was purified using Sep-Pak C18 cartridges (Waters, Dublin, Ireland), which were previously activated with 10 mL methanol, 10 mL ultrapure water, and 10 mL air. The purified extract was then filtered through a 0.45 μm nylon filter. A 750 μL aliquot of the filtered extract was derivatised with 250 μL 1,2-phenylenediamine (OPDA) and incubated at room temperature for 37 min. Afterwards, 20 μL of the derivatised extract was analysed by HPLC (1100 series, Agilent Technologies, Waldbronn, Germany), equipped with a G1322A degasser, G1311A quaternary pump, G1313A autosampler, G1316A column heater, and G1315B photodiode array detector. The chromatographic separation was carried out under isocratic conditions using a mobile phase consisting of 5 mM hexadecyltrimethylammonium bromide (CTAB) and 50 mM potassium dihydrogen phosphate in methanol/ultrapure water (5:95, v/v) at a flow rate of 1.6 mL min^{-1}. Chromatograms were recorded at 261 nm for ascorbic acid and 348 nm for dehydroascorbic acid (Merck-Sigma, Darmstadt, Germany). Each of the three replicates was analysed in triplicate. Vitamin C content was quantified as the sum of ascorbic acid and dehydroascorbic acid, using authentic standards for their identification, and the results were expressed as mg 100 g^{-1} dry weight (DW).

2.6.2. Antioxidant Capacity and TPC

Antioxidant capacity and TPC were determined in freeze-dried broccoli stick samples. For extraction, 0.2 g of the freeze-dried sample was mixed with 15 mL of methanol/water solution (70:30, v/v) and subjected to ultrasonication to enhance extraction efficiency. The sample was then vortexed for 30 s to ensure homogenisation. After centrifugation, the supernatant was filtered and collected. Three repetitions were performed and analysed for each treatment and storage day.

The antioxidant capacity was measured using the ABTS assay, following the method described by Re et al. [38] with slight modifications. The ABTS$\cdot^+$ radical was generated by dissolving 10 mg of ABTS reagent (2,2′-azinobis-(3-ethylbenzothiazoline-6-sulfonate)) in a potassium persulfate solution, followed by incubation in the dark for 12–24 h. Once the radical was formed, the ABTS$\cdot^+$ solution was diluted and adjusted to an absorbance of 0.7 ± 0.02 at 734 nm to establish a baseline. A stock solution of Trolox (Sigma Aldrich, St. Louis, MO, USA) was prepared in ethanol and diluted to different concentrations (5–100 μM) to generate a calibration curve. The absorbance was measured at 734 nm, and the curve was obtained by plotting absorbance values against Trolox concentrations. For the assay, 300 μL of the adjusted ABTS$\cdot^+$ solution was added to each well of a microplate, and an initial absorbance reading was taken. Then, 30 μL of water (for blanks), Trolox standard, or sample extract was added to the wells. After a 3 min incubation, the absorbance was measured again. The decrease in absorbance reflected the antioxidant capacity of the samples, which was expressed as milligrams of Trolox equivalents per gram of DW (mg TE g^{-1} DW), based on the calibration curve.

The TPC assay was determined using a multiscan plate reader (Tecan Infinite M200, Männedorf, Switzerland), following the method outlined by Salas-Millán et al. [39], with slight modifications. A volume of 20 μL from the previously extracted sample was pipetted into each well of a microplate, followed by 30 μL of Folin–Ciocalteu reagent (diluted 1:1 with ultra-distilled water) and 200 μL of a sodium hydroxide/sodium carbonate ($NaOH/Na_2CO_3$) mixture. The microplate was then incubated for 40 min at room temperature in the dark. Prior to measurement, the microplate was automatically shaken, and absorbance was recorded at 765 nm. TPC was expressed as milligrams of gallic acid equivalents per gram of DW (mg GAE g^{-1} DW).

2.6.3. Identification and Quantification of GSLs

The identification and quantification of GSLs were performed following the method described by Salas-Millán et al. [32], with slight modifications. Freeze-dried samples (0.1 g) were extracted using 5 mL of methanol/water (70:30, *v*/*v*). The extraction process involved incubation in a hot bath at 70 °C, followed by centrifugation and filtration. A 3 mL aliquot of the supernatant was then evaporated under nitrogen gas and reconstituted in 1 mL of mobile phase A. The mobile phase consisted of 0.1% formic acid in water (A) and 0.1% formic acid in acetonitrile (B). The analysis was performed in the negative ionisation mode using HPLC (Agilent 1200, Santa Clara, CA, USA), equipped with a G1311B quaternary pump, G1329B standard autosampler, and G1316A column heater, and coupled to a 6420 triple–quadrupole mass spectrometer (QqQ) with an electrospray ionisation (ESI) source. The autosampler was programmed to inject 10 μL of the eluted sample at a flow rate of 0.250 mL min^{-1}. The chromatographic gradient was set as follows: 5–30% B (14 min), 30–95% B (7 min), and 95–95% B (5 min), at 40 °C. For the identification of GSLs, the retention time, mass (MS), and MS/MS fragmentation spectra of the detected compounds were compared with a commercial standard or tentatively identified by comparing their fragmentation patterns with the available bibliographic data and the MassBank Europe, MassBank of North America, and PubChem databases. Calibration curves were generated using standards of each GSLs, which were used to quantify their concentrations. Finally, the results were expressed as mg 100 g^{-1} DW.

2.7. Statistical Analysis

Statistical analyses were conducted using Statgraphics Centurion XV, version 15.2.05 (Statgraphics Technologies, Inc., The Plains, VA, USA). A two-way analysis of variance (ANOVA) was conducted to evaluate the effect of postharvest treatments (T) and days of storage time (D) on each dependent variable ($p < 0.05$). Results were expressed as mean values ± standard error (SE). When significant differences were found ($p < 0.05$), Fisher's Least Significant Difference (LSD) test was used for multiple comparisons. A cluster correlation heatmap was generated using Ward's hierarchical clustering method and Euclidean distance. Ward's method was chosen as it minimises the total within-cluster variance, generating more compact and interpretable clusters for multivariate datasets such as those obtained in postharvest studies. Before the analysis, the data were normalised using the min–max scaling method. All analyses were performed using MetaboAnalyst 5.0 (https://www.metaboanalyst.ca (accessed on 5 November 2024)).

3. Results and Discussion

3.1. Browning Index

The evolution of BI in broccoli sticks across the different treatments during storage is shown in Figure 1. ANOVA revealed that both the storage time and the treatment had significant effects on the BI, with a strong interaction between these factors ($p < 0.05$). Notably, TREH, followed by CaAsc alone and the combination of both, exhibited the highest BI over time, even higher than the control. In contrast, HWT alone and its combinations led to only a slight increase in the BI, peaking at day 8, indicating that the treatment was effective in limiting browning compared to the control. The postharvest treatments proposed to mitigate this issue have distinct mechanisms and effects on enzymatic activity and subsequent browning. HWT at 55 °C for 1 min effectively reduced browning in the broccoli sticks, and this is likely due to the partial inactivation of PPO. Although PPO is considered to be thermostable in some plant species, moderate heat treatments can partially inactivate the enzyme, thereby slowing down enzymatic browning [26]. Similar results were observed in fresh-cut apples or melon peels, where HWT between 45 °C to 55 °C reduced browning and helped maintain the product [17,40]. Additionally, HWT is a non-chemical, easily applicable method for the fresh-cut industry that can effectively control enzymatic oxidation without negatively affecting sensory attributes.

CaAsc is a compound that combines calcium with ascorbic acid; the latter has well-known antioxidant properties. As mentioned in the introduction, ascorbic acid is widely used to inhibit enzymatic browning in fresh-cut vegetables due to its ability to reduce oxidised substrates back to their original phenolic forms, thereby preventing the formation of brown pigments. Ascorbic acid does not directly interact with the PPO enzyme but mitigates enzymatic browning by converting oxidised substrates back to their precursor diphenols [13]. Studies suggest that its anti-browning function is primarily attributed to its ability to reduce enzymatically formed o-quinones [41]. However, this effect may be less pronounced in certain fresh-cut fruits, such as pears [42], where ascorbic acid's protective role diminishes once it is fully oxidised to dehydroascorbic acid. At this stage, o-quinones can no longer be reduced, allowing melanin formation to proceed and leading to continued browning. Similarly, in our study, no significant reduction in browning was observed in the broccoli sticks. This is likely due to the limited capacity of ascorbic acid at the studied concentration to sustain its reducing effect. This contrasts with what has been observed in fresh-cut pears, where browning intensifies once ascorbic acid is fully oxidised. In line with these findings, Xiong et al. [18] observed that the browning in fresh-cut potatoes treated with CaAsc (0.5% w/v) was quite similar to that in untreated samples. However, ascorbic acid applied at the same concentration proved highly effective in delaying both browning and wound healing, reducing reactive oxygen species (ROS) levels, and enhancing antioxidant capacity. This suggests that CaAsc presents a lower efficacy in decreasing browning in certain fresh-cut fruits, probably due to limited reductive activity. On the other hand, TREH has been shown to stabilise membranes and protect plant tissues from stress-induced damage [19]. However, in our experiment, its application led to an unexpected increase in browning. This phenomenon could be attributed to several factors. Firstly, TREH may have interacted with PPO and POD, stabilising these enzymes and prolonging their activity, thereby enhancing the conversion of phenolic compounds into o-quinones. Secondly, TREH may have interacted with the resulting quinones, intensifying browning, as sugars have been reported to promote quinone polymerisation and the formation of darker pigments [25]. Additionally, TREH may have influenced the tissue's metabolic response, which may potentially have increased the respiration rate (as discussed in Section 3.3) and promoted the production of ROS, which could accelerate the oxidation of phenolic compounds. Unlike in litchi fruit (*Litchi chinensis*), where TREH (2.5 g/L) significantly reduced enzymatic browning by enhancing antioxidant enzyme activities and inhibiting PPO [24], in the broccoli sticks, it failed to prevent oxidative damage or delay browning, ultimately resulting in a more pronounced discolouration. These findings underscore the complexity of TREH's impact on fresh-cut products and highlight the need for product-specific evaluations before being applied as a preservative.

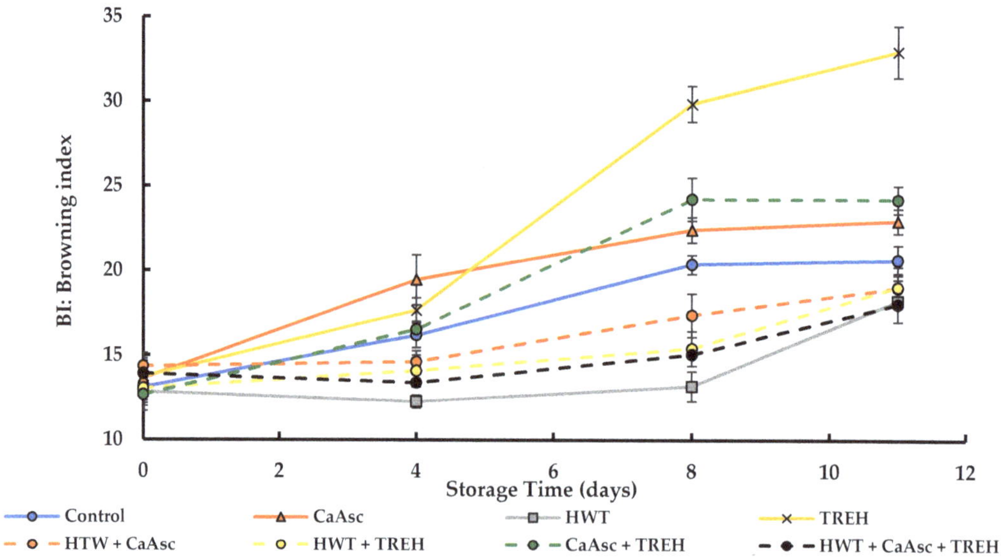

Figure 1. The browning index (BI) of the broccoli sticks during storage under different treatments. Data are presented as mean ± SEM (n = 10). ANOVA showed significant effects for main factors and their interaction. Fisher's LSD test: $LSD_{Days \times Treatment} = 2.81$, $p < 0.05$. CaAsc = 1% calcium ascorbate, TREH = 5% trehalose. HWT = hot water treatment at 55 °C for 1 min.

3.2. Microbiological Analysis

The evolution of total mesophilic aerobic bacteria and yeasts in the broccoli sticks under different treatments during storage is shown in Figure 2. ANOVA indicated that both the storage time and the treatment had significant effects on microbial growth ($p < 0.05$). From days 0 to 4, the mesophilic bacterial count remained below 4 log CFU/g in the CaAsc and HWT treatments. However, all the treatments, including the control, exhibited a progressive increase in bacterial populations, reaching 5.8 to 6.6 log CFU/g on day 11.

In contrast, the TREH treatment, alone or in combination with HWT, showed the highest bacterial counts over time, reaching 7.7 to 9.4 log CFU/g on day 11, which may be related to enhanced microbial survival or proliferation; nevertheless, further studies are needed to confirm a direct effect.

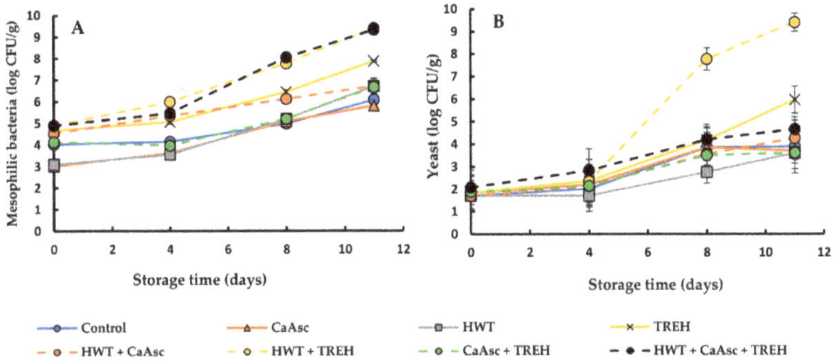

Figure 2. The growth of (**A**) mesophilic bacteria and (**B**) yeasts in the broccoli sticks during storage under different treatments. Data are presented as mean ± SEM (n = 3). ANOVA showed significant effects for main factors and their interaction. Fisher's LSD test: $LSD_{Days \times Treatment}$ = 0.36 for bacteria and 0.34 for yeast, $p < 0.05$. CaAsc = 1% calcium ascorbate, TREH = 5% trehalose. HWT = hot water treatment at 55 °C for 1 min.

Yeast counts remained stable during the initial 4 days but increased significantly with the storage time. From that day onwards, in particular, the HWT+TREH- and TREH-treated samples showed the highest yeast counts, reaching 9 log CFU/g and 6 log CFU/g, respectively, on day 11. Meanwhile, the rest of the treatments, the control included, showed lower yeast counts, ranging from 3.7 to 4.5 log CFU/g.

The combination of HWT + TREH proved to be the most favourable condition for microbial growth, leading to the highest counts of both mesophilic bacteria and yeasts. The enhanced microbial proliferation observed in the TREH-treated samples may be linked to the moisture-retaining capacity of TREH, which can modulate water activity in complex food matrices [25]. Additionally, certain bacteria and yeasts can metabolise TREH as an energy source, which could explain the increased microbial counts observed in those treatments. This effect was particularly pronounced when TREH was combined with HWT, as the heat may have enhanced nutrient dissolution and availability, further promoting microbial growth [43].

To explain the effect of CaAsc in the initial microbial control, it is important to consider the dual role of calcium and ascorbic acid in maintaining cell wall integrity and potentially influencing antimicrobial activity. Calcium reinforces cell walls, reducing pathogen penetration and enzymatic degradation, which can help limit microbial growth. Meanwhile, ascorbic acid may contribute to this effect by helping to maintain tissue integrity and the redox balance in the broccoli sticks, which could also play a role in microbial control [14,15,44]. On the other hand, HWT showed initial microbial control since heat may stimulate natural defence responses in plant tissues, enhancing resistance against microbial invasion [45] or it could delay enzymatic degradation by preserving cell wall integrity [28].

Pathogenic bacteria, including *Escherichia coli* and *Salmonella* spp., were analysed throughout the storage period, although no detectable counts were obtained (<2 log CFU/g). This underscores the effectiveness of the initial washing and disinfection steps in minimising microbial contamination and ensuring the microbiological safety of the broccoli sticks. Moreover, the absence of these pathogens meets the microbiological criteria established by Commission Regulation (EC) No 2073/2005 [46] for ready-to-eat foods, including minimally processed vegetables. Despite their absence, continued monitoring is crucial, as *E. coli* and *Salmonella* spp. remain key food safety indicators due to the potential risk they pose under favourable growth conditions. These findings are consistent with previous studies on fresh-cut produce, which highlight the role of rigorous hygiene practices and effective sanitation in controlling microbial proliferation [47].

3.3. Gas Composition Analysis

The treatments HWT + TREH, TREH, and HWT + CaAsc + TREH exhibited the highest CO_2 concentrations inside the packaging by the end of storage, with values exceeding 3% and reaching up to 6%. Concurrently, the oxygen levels in these samples decreased to between 15% and 17%, indicative of elevated respiration activity (Figure 3). This slightly enhanced metabolic activity suggests a moderate increase in oxidative demand, which may be associated with accelerated senescence and microbial proliferation. In fact, the highest mesophilic bacterial counts were found in these treatments (Section 3.2). This factor could partially explain the higher CO_2 levels, as microbial respiration contributes to CO_2 production during growth [48]. However, the HWT treatment, without TREH, resulted in the lowest CO_2 accumulation (<2%) and maintained higher O_2 concentrations (around 19–20%), indicating reduced respiration rates and a slower metabolic activity. This behaviour is closely associated with delayed senescence and extended shelf-life in fresh-cut produce.

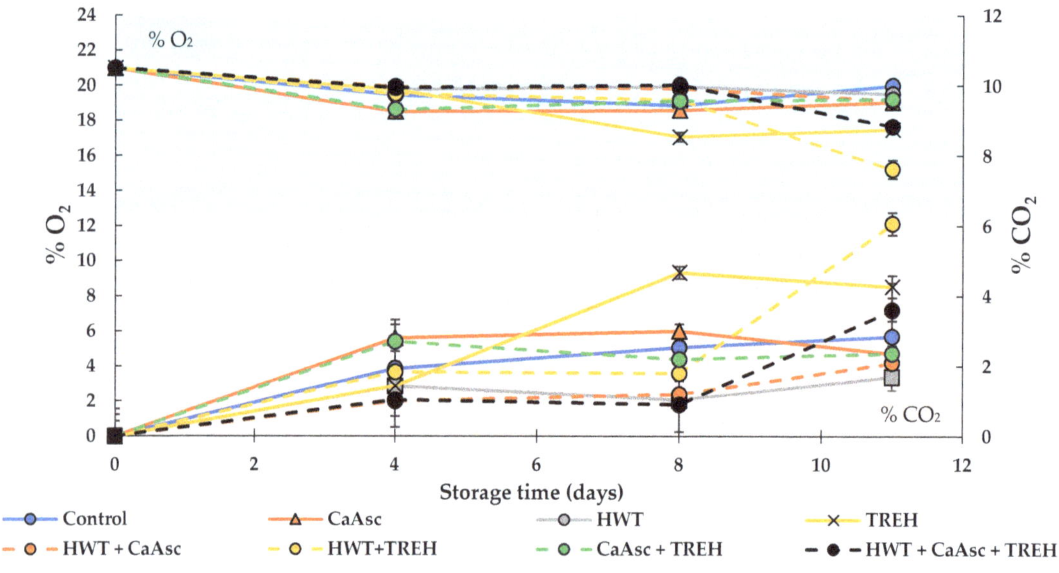

Figure 3. Changes in O_2 and CO_2 levels in the broccoli sticks during storage under different treatments. Data are presented as mean ± SEM (n = 3). ANOVA showed significant effects for the main factors and their interaction. Fisher's LSD test: $LSD_{Days \times Treatment}$ = 0.55 for O_2 and 0.66 for CO_2, $p < 0.05$. CaAsc = 1% calcium ascorbate, TREH = 5% trehalose. HWT = hot water treatment at 55 °C for 1 min.

These findings reinforce the effectiveness of mild thermal treatments to modulate plant tissue respiration, as previously reported in citrus fruits, where HWT initially increased respiration but subsequently stabilised it, preserving postharvest quality with no adverse effects [49].

Moreover, the results underline the efficacy of HWT as a non-chemical strategy to maintain freshness and reduce spoilage during cold storage. Interestingly, the effects of TREH appear to be product-specific. In whole peppers and apples, TREH has been shown to suppress weight loss and reduce respiration rates, resulting in improved quality and extended shelf-life [50,51], which would suggest that the effectiveness of TREH treatment may depend on the type of product and its metabolic characteristics.

3.4. Sensory Analysis

Figure 4 shows the sensory evaluation results of broccoli sticks on day 11 of storage under the different treatments. Appearance was strongly influenced by visual browning and dehydration. All the treatments, except for TREH, improved the visual appearance of broccoli sticks compared to the control. The most notable results were observed in the HWT treatment and its combinations, which consistently provided the lowest visual browning and dehydration. Regarding texture, crispness, and taste, the CaAsc and HWT treatments yielded the best results. In particular, the presence of calcium in the CaAsc treatment contributed to improved firmness and structural integrity, as previously demonstrated by other researchers [12,17]. The effect of HWT on maintaining firmness and improved appearance during storage has been reported for mandarin [49], fresh-cut apple, pear [52], and melon [16]. When considering overall sensory quality—integrating appearance, texture, and taste attributes—HWT emerged as the most effective treatment, extending the sensory shelf-life of the broccoli sticks to the 11-day limit.

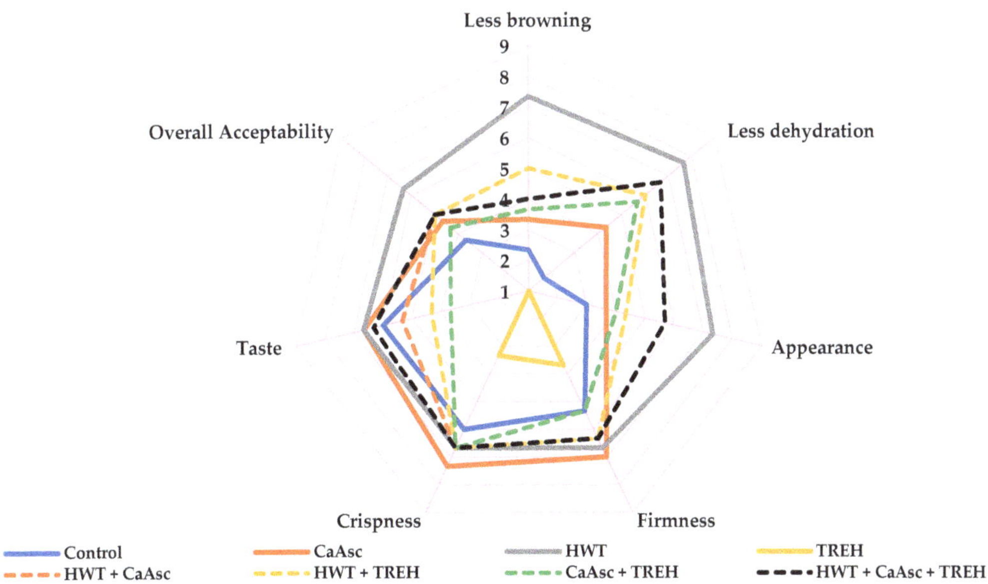

Figure 4. The sensory evaluation of the broccoli sticks on day 11 of storage under different treatments. Data are presented as mean values ($n = 10$). Data are based on a hedonic scale where 1 = unusable; 3 = poor; 5 = fair (limit of marketability); 7 = good; and 9 = excellent. ANOVA showed significant effects for the main factors and their interaction. Fisher's LSD test: $LSD_{Days \times Treatment} = 0.55$ for O_2 and 0.66 for CO_2, $p < 0.05$. CaAsc = 1% calcium ascorbate, TREH = 5% trehalose. HWT = hot water treatment at 55 °C for 1 min.

In contrast, the TREH treatment was the least effective, particularly in terms of browning and dehydration (Figure 5), where it received the lowest scores. TREH not only showed poor performance in these key attributes, which are crucial for consumer acceptance, but also in texture and appearance, highlighting its limitations as a standalone preservation method. These findings align with its inability to control microbial growth and prevent broccoli browning, as observed in this study. Interestingly, these results differ from previous research where TREH improved postharvest quality in other produce, such as peppers and guavas, by enhancing membrane stability, reducing water loss, and activating antioxidant enzymes such as peroxidase and catalase [21,51].

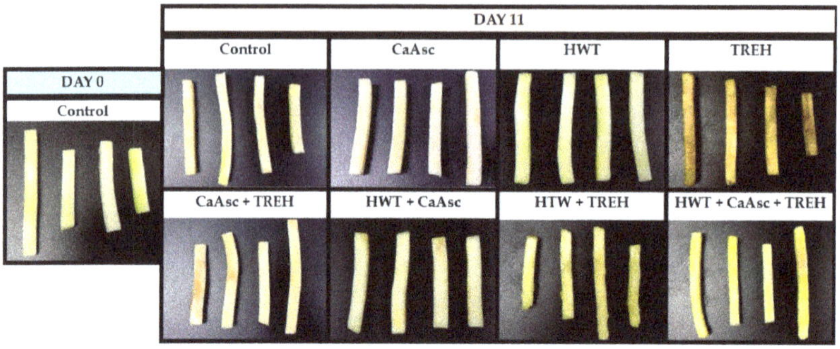

Figure 5. The visual appearance of the broccoli sticks: control sample on day 0 and samples under different treatments on day 11, included in the sensory analysis to assess visual changes over the storage period.

In line with these findings, TREH has been shown to improve firmness and water retention in cherry tomatoes [22] and kiwifruit [23]. The discrepancy in TREH's effectiveness between broccoli and other produce comes from differences in cell structure, water content, and metabolic responses, which limit the capacity to maintain sensory quality in the broccoli sticks. Consequently, our findings emphasize that TREH alone is not suitable for preserving the sensory or microbial quality of broccoli sticks during storage. However, combination treatments with TREH showed an enhanced sensory quality compared to individual treatments. Notably, the HWT + CaAsc + TREH treatment achieved some of the best overall scores, particularly in taste, suggesting a positive combination that may enhance certain sensory attributes and increase the appeal of broccoli sticks to consumers after extended storage. Sensory parameters were monitored throughout storage (days 0, 4, 8, and 11) to evaluate the progression of quality over time. Detailed results can be found in the Supplementary Section (Figure S1).

3.5. Chemical Analysis

3.5.1. Vitamin C Content

The changes in vitamin C content of the broccoli sticks are presented in Figure 6. At day 0, the HWT + CaAsc treatment exhibited the highest initial vitamin C content, reaching approximately 1500 mg/100 g DW, followed by the CaAsc treatment at around 1100 mg/100 g DW. As expected, immersion in a 1% CaAsc solution resulted in an increase in ascorbic acid content and, consequently, higher vitamin C. The enhanced response in HWT may be attributed to the thermal effect of HWT, which likely creates a gas expansion within the tissue that is followed by contraction during cooling, thereby promoting the absorption of the ascorbate solution (Figure S2). Furthermore, elevated temperatures may improve the diffusion rate of ascorbic acid and its solubility in both water and plant tissue. However, a gradual decline was observed with storage time in all the treatments that were combined with ascorbate. Interestingly, the CaAsc + TREH treatment showed a relatively stable vitamin C profile throughout storage, with minimal fluctuations. This suggests that the combination of CaAsc and TREH may contribute to ascorbic acid stabilisation, potentially due to TREH's known ability to protect cellular structures and reduce oxidative degradation [23].

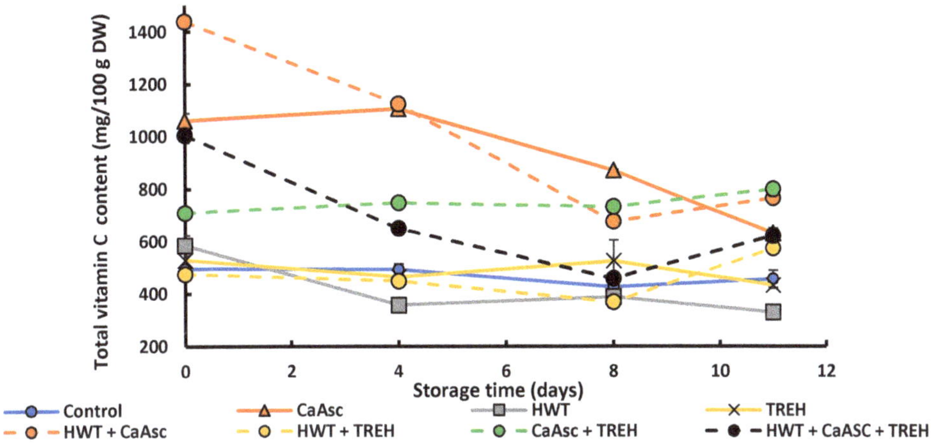

Figure 6. Total vitamin C content in the broccoli sticks during storage under different treatments. Data are presented as mean ± SEM (n = 3). ANOVA showed significant effects for main factors and their interaction. Fisher's LSD test: $LSD_{Days \times Treatment}$ = 69.95, $p < 0.05$. CaAsc = 1% calcium ascorbate, TREH = 5% trehalose. HWT = hot water treatment at 55 °C for 1 min.

In this context, it would seem that TREH acts as a supportive agent, enhancing the protective effect of CaAsc rather than providing direct stabilisation on its own. This interpretation aligns with findings by Baek et al. [53], who reported the limited efficacy of TREH alone in preserving vitamin C in fresh-cut apples, thus emphasizing the dominant role of ascorbate. Therefore, the improved vitamin C retention observed in the CaAsc + TREH treatment can be attributed to a synergistic effect, where ascorbate serves as the primary antioxidant while TREH supports cellular integrity and reduces degradation pathways.

In contrast, the control, HWT, TREH, and HWT + TREH treatments exhibited a steady decline in vitamin C content over time. By day 11, CaAsc + TREH and HWT + CaAsc were among the treatments with the highest vitamin C levels, even surpassing the CaAsc only treatment. These findings indicate that CaAsc is the main driver of vitamin C enhancement and retention, and that its effectiveness may be maintained or even improved when combined with other treatments, depending on their interactions. The variability observed across treatments highlights the importance of optimising combined preservation strategies to balance compound protection with potential degradation mechanisms.

3.5.2. Antioxidant Capacity and TPC

As shown in Figure 7, the treatments involving the application of CaAsc resulted in higher antioxidant capacity (A) and TPC (B) throughout storage, consistent with the trends observed for vitamin C levels (Figure 6).

The most pronounced effects were observed in the HWT + CaAsc treatment (27 mg TE/g and 17 mg GAE/g DW at day 0), followed by the CaAsc and HWT + CaAsc + TREH treatments, all of which maintained significantly higher antioxidant and phenolic values by the end of storage—even though HWT + CaAsc experienced a greater overall reduction of approximately 25%. The initial improvement in antioxidant preservation observed in treatments containing CaAsc is primarily attributed to the intrinsic activity of ascorbic acid and its protective role in stabilising phenolic compounds. Furthermore, when applied in combination with HWT, the uptake and efficacy of ascorbic acid appear to be enhanced, this is likely due to improved tissue permeability under mild thermal conditions, as previously discussed.

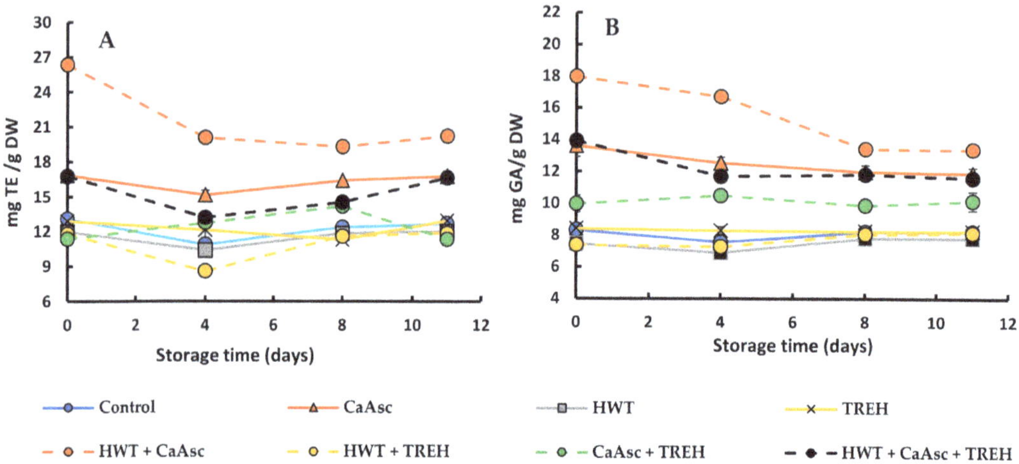

Figure 7. Changes in (**A**) antioxidant capacity and (**B**) TPC in the broccoli sticks during storage under different treatments. Data are presented as mean ± SEM (n = 3). ANOVA showed significant effects for the main factors and their interaction. Fisher's LSD test: $LSD_{Days \times Treatment}$ = 0.76 for antioxidant capacity and 0.89 for total phenolic content, p < 0.05. CaAsc = 1% calcium ascorbate, TREH = 5% trehalose. HWT = hot water treatment at 55 °C for 1 min.

The CaAsc + TREH treatment resulted in the moderate preservation of both antioxidant capacity and TPC, with values falling to between those of the CaAsc-based treatments and the remaining ones. In contrast, the control, TREH, HWT + TREH, and HWT treatments exhibited consistently lower levels of antioxidant capacity and TPC, ranging from 11.3 to 13.0 mg TE/g and 7.4 to 8.4 mg GAE/g, respectively. These values remained relatively stable over time but were substantially lower throughout the storage period. The diminished antioxidant potential and TPC observed in these treatments is probably due to the absence of ascorbic acid from CaAsc, highlighting the central role of ascorbic acid in enhancing and preserving antioxidant compounds in the broccoli sticks during cold storage.

Previous studies have also highlighted the effectiveness of combining thermal strategies in fresh-cut vegetables to enhance the retention of bioactive compounds during storage [27]. In addition, Xiong et al. [18] reported that, although CaAsc was less effective than ascorbic acid and sodium isoascorbate in controlling browning and wound responses in fresh-cut potatoes, it did contribute to quality maintenance during storage. In particular, CaAsc treatment led to a greater accumulation of total phenolic and flavonoid contents, as well as enhanced antioxidant capacity compared to the control. These effects have been associated with reduced oxidative stress and better preservation of membrane integrity, which could contribute to the stabilisation of antioxidant systems during storage. Similarly, López-Hernández et al. [29] demonstrated that short HWT (e.g., 53 °C for 3 s) delayed oxidative damage in fresh-cut Chinese cabbage, improving the retention of antioxidant activity and TPC over time. In broccoli, Perini et al. [28] found that HWT applied to the basal part of the stems reduced the postharvest losses of polyphenols and helped maintain antioxidant capacity, emphasizing its role in extending shelf-life and preserving nutritional quality.

3.5.3. Identification and Quantification of GSLs

A total of nine GSLs from the *Brassicaceae* family were detected and comparatively analysed across all treatments and storage times. These included aliphatic GSLs, including glucoraphanin, glucoiberverin, glucoerucin, and glucoberteroin; aromatic GSLs, such as glucotropaeolin and gluconasturtiin; and indole GSLs, including glucobrassicin, neoglucobrassicin, and 4-methoxyglucobrassicin. Their relative abundance patterns after normalisation are given in Figure 8A,B; they reveal differences associated with treatment and storage time. The corresponding absolute quantification data are available in Supplementary Table S1.

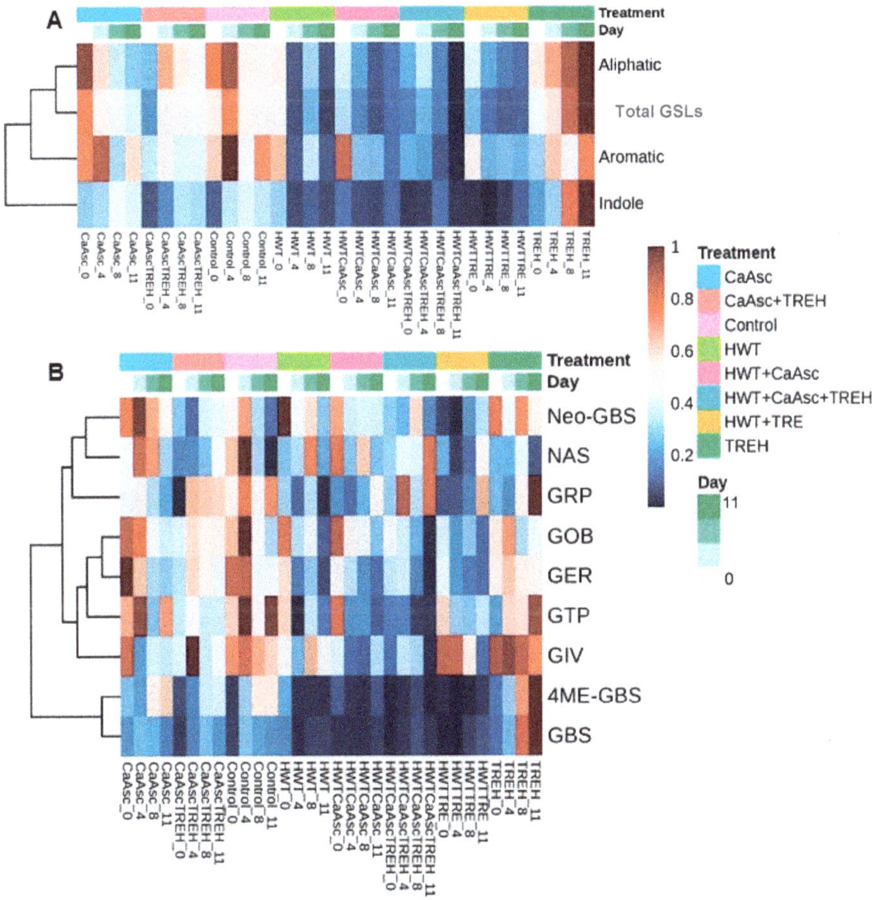

Figure 8. (**A**) Heatmap showing normalised levels of total, aliphatic, aromatic, and indole GSLs in the broccoli sticks stored at 5 ± 1 °C under different treatments (CaAsc = 1% calcium ascorbate, TREH = 5% trehalose. HWT = hot water treatment at 55 °C for 1 min). (**B**) Heatmap displaying normalised results of the quantification of individual GSLs: glucoraphanin (GRP), glucotropaeolin (GTP), glucoiberverin (GIV), glucoerucin (GER), glucobrassicin (GBS), neoglucobrassicin (NEO-GBS), glucoberteroin (GOB), 4-methoxyglucobrassicin (4ME-GBS), and gluconasturtiin (NAS) throughout storage. Values are shown on a colour scale from 0 to 1 after normalisation. Colour bars at the top indicate the treatments applied and the storage days (0, 4, 8, and 11). Data represent the mean of three independent replicates.

The distribution of polyphenols and GSLs within broccoli stalk is uneven, with higher concentrations in the bark compared to the core, as reported by Costa-Pérez et al. [31]. These differences are related to the distinct physiological functions of each tissue and may also result from the disruption of tissue integrity caused by cutting, which can activate hydrolytic enzymes and modify the composition of secondary metabolites.

In our study, the absolute total GSL concentrations ranged from 79.05 ± 3.93 to 176.94 ± 7.37 mg/100 g DW (Supplementary Table S1), with aliphatic ones generally being the most abundant group, mainly glucoerucin as the main important, followed by indole and aromatic GSLs. Aliphatic GSLs, which dominated the overall GSL profile, showed high normalised values at day 0 in treatments such as TREH, CaAsc, and their combination, whilst the HWT treatments, particularly HWT + CaAsc and HWT + CaAsc + TREH, exhibited notably lower initial levels. Over time, most treatments showed a decrease in total GSL content; this was especially evident in HWT-based combinations.

Indole GSLs followed a similar trend, with higher values in early storage which progressively diminished, although treatments such as TREH and CaAsc + TREH showed relative preservation. Aromatic GSLs remained low across all treatments and storage times, with glucotropaeolin ranging from 2.51 ± 0.08 to 3.67 ± 0.07 mg/100 g DW, and gluconasturtiin from 0.98 ± 0.01 to 2.87 ± 0.40 mg/100 g DW.

Overall, these findings suggest that thermal treatments may reduce GSL content, while formulations containing TREH or CaAsc alone may better preserve GSLs during storage. Interestingly, the relative stabilisation or late increase in these compounds observed in the TREH-based treatments on day 11 could be linked to postharvest stress responses and tissue degradation. As reported by Paulsen et al. [54], mechanical damage from minimal processing may initially trigger GSL biosynthesis, while progressive senescence facilitates enzymatic hydrolysis due to the loss of cellular compartmentalisation, potentially altering GSL dynamics throughout storage.

Glucoerucin was initially detected at a higher concentration in the control and CaAsc-treated samples, ranging from 80.32 ± 1.06 mg/100 g to 87.16 ± 1.06 mg/100 g DW. When evaluating the interaction between storage time and treatment, significant differences were only observed in the TREH-treated samples, which exhibited an increase in the concentration of glucoerucin and glucoraphanin over time. In the case of glucoerucin, this upward trend was not observed when TREH was combined with HWT; however, the reduction was less pronounced than in the other treatments. In the case of glucoraphanin and glucotropaeolin, the TREH-treated samples showed a notable increase in their relative levels by day 11.

As no treatment-specific mechanism has been identified to directly explain this phenomenon, the observed increase may be attributable to microbial activity. Microorganisms are known to degrade cell membranes and, in some cases, produce myrosinase-like enzymes that facilitate the breakdown of GSLs into bioactive compounds. This hypothesis is supported by the significantly higher counts of mesophilic bacteria and yeasts observed in the TREH-treated samples at the end of storage (Figure 2), suggesting increased microbial metabolism. These observations are consistent with previous findings by Salas-Millán et al. [39] and Wu et al. [55], who demonstrated that lactic acid bacteria can enhance the release and transformation of GSLs in broccoli by-products, even in the absence of endogenous plant myrosinase.

The initial concentration of glucoiberverin was significantly lower in the HWT + CaAsc and HWT + CaAsc + TREH treatments, approximately half that found in the other treatments. This is consistent with the heatmap results, where those two treatments exhibited the lowest normalised values. In contrast, the remaining treatments, particularly TREH, HWT + TREH, and the control, displayed higher initial concentrations, reflected by higher normalised values. No significant interaction effects were observed, as all treatments showed a decreasing trend in glucoiberverin concentration over time.

The most notable differences in gluconasturtiin concentration were observed in the treatments involving CaAsc and HWT, either alone or in combination. While the overall detected levels of gluconasturtiin were low, a decreasing trend was observed in the control throughout storage. In contrast, most treated samples maintained stable levels over time, and an increase was observed at the end of the storage period in samples treated with the HWT + CaAsc + TREH combination. This mirrors the trend observed for glucoraphanin, suggesting a possible synergistic effect.

Glucoberteroin was detected at the lowest concentration among all quantified GSLs, representing less than 1% of the total GSL content (Table S1). Interestingly, its initial levels were slightly higher in samples subjected to HWT, CaAsc, and their combination (HWT + CaAsc), although a consistent decrease was observed across all treatments over time. These findings align with observations by Mbudu et al. [56], who reported tissue-specific variability of glucosinolates in kohlrabi tissues.

For indole GSLs such as glucobrassicin, 4-methoxyglucobrassicin, and neoglucobrassicin, the initial concentrations were not markedly higher in any particular treatment, but a gradual decrease over time was observed across most treatments. Notably, treatments containing TREH—especially alone—showed a relative increase in glucobrassicin by day 11, which may be associated with microbial activity, as signs of spoilage (e.g., increased CFU/g counts) were evident from day 8 onwards. A similar association was reported by Salas-Millán et al. [32], where the spontaneous fermentation of broccoli stalks led to a notable rise in indole GSLs, including glucobrassicin and 4-methoxyglucobrassicin, in parallel with the growth of lactic acid bacteria and yeasts. Additionally, as highlighted by Li et al. [3], microbial metabolism can modulate the profile of bioactive compounds in broccoli, particularly under stress or degradation conditions. These findings suggest that, beyond the direct effect of the treatments applied, microbial interactions during storage may play a role in the late-stage accumulation or transformation of specific GSLs. These findings are consistent with the results reported by Núñez-Gómez et al. [5], who found that indole GSLs in broccoli stalks—such as glucobrassicin, 4-methoxyglucobrassicin, and neoglucobrassicin—were less stable during extraction, and were only detected in the extractable fraction, unlike more stable aliphatic GSLs such as glucoraphanin.

4. Conclusions

This study demonstrates the potential of up-cycling broccoli stalks, a commonly discarded by-product, into a high-value fresh-cut product through the application of targeted postharvest strategies. The main conclusions are as follows:

(a) Among the treatments evaluated, HWT alone was highly effective in reducing browning, a key factor in achieving an 11-day shelf-life, whilst also controlling microbial growth and respiration and yielding the highest scores in sensory quality. However, its ability to preserve bioactive compounds, particularly GSLs, was limited.

(b) To retain and enhance these health-promoting compounds, the HWT + CaAsc treatment was the most effective, primarily due to the antioxidant action of ascorbic acid present in CaAsc. This treatment significantly increased vitamin C retention (by 78%), antioxidant capacity (by 68%), and TPC (by 65%) compared to the control on day 11.

(c) Although the control treatment, based solely on peracetic acid disinfection, provided microbial reduction, the inclusion of antioxidant support in other combined treatments helped to maintain sensory quality and played a key role in stabilising vitamin C and reducing the degradation of both phenolic compounds and GSLs.

(d) Conversely, TREH applied alone did not contribute to extending the shelf-life. It was associated with enhanced tissue browning, elevated respiration rates, and increased microbial proliferation, factors potentially associated with membrane destabilisation and, consequently, the enhancement of GSL bioavailability. These results underscore the crop-specific nature of postharvest responses, as TREH, despite its reported benefits, did not perform consistently in this matrix. Future studies should explore the effect of different concentrations in cruciferous tissues.

(e) The absence of foodborne pathogens such as *Salmonella* spp. and *E. coli* confirms the efficacy of the initial sanitation and preservation procedures in ensuring microbiological safety throughout the storage time.

Overall, these findings highlight the value of integrated postharvest strategies, particularly those combining mild heat and ascorbate-based antioxidant dips, for preserving the quality and bioactive compounds in fresh-cut broccoli. The HWT + CaAsc treatment proved to be the most balanced option: it maintained sensory attributes and enhanced the retention of health-promoting compounds and achieved a shelf-life of 11 days at 5 °C under MAP. Given that both components are safe, commonly used, and cost-effective, this approach offers a scalable and sustainable solution for broccoli stalk valorisation in the fresh-cut industry. This approach not only supports food loss reduction and sustainable processing but also represents a viable up-cycling strategy aligned with circular economy principles. Future research should investigate the physiological mechanisms underlying these effects and evaluate their applicability to other vegetable by-products, contributing to loss and waste valorisation and the development of functional fresh-cut products.

Supplementary Materials: The following supporting information can be downloaded at: https://www.mdpi.com/article/10.3390/foods14142476/s1. Figure S1: Sensory evaluation of the broccoli sticks subjected to different treatments: (A) Control, (B) CaAsc, (C) HWT, (D) TREH, (E) CaAsc+TREH, (F) HWT+CaAsc, (G) HWT+TREH, and (H) HWT+CaAsc+TREH. Each radar chart shows the evolution of sensory attributes during storage (days 0, 4, 8, and 11). The evaluated parameters included overall acceptability, taste, crispness, firmness, appearance, less dehydration, and less browning. Data are expressed as mean values (n = 10), using a 9-point hedonic scale where 1 = unusable, 3 = poor, 5 = fair (limit of marketability), 7 = good, and 9 = excellent. CaAsc = 1% calcium ascorbate, TREH = 5% trehalose, HWT = hot water treatment at 55 °C for 1 min. Figure S2: Schematic representation of the proposed synergistic mechanisms of the combined hot water treatment (HWT, 55 °C for 1 min) and calcium ascorbate dip (CaAsc, 1%, 5 °C for 1 min) applied to fresh-cut broccoli stalks. HWT induces a thermal effect that creates gas expansion within the tissue; subsequent cooling in a chilled CaAsc solution causes tissue contraction; this sequence promotes the absorption of the CaAsc solution, enhancing antioxidant protection. Table S1: Glucosinolate contents (mg/100 g dry weight) in broccoli sticks during storage under different treatments.

Author Contributions: Methodology, validation, investigation, and data curation, N.G.-L., J.Á.S.-M. and E.A.; formal analysis and writing—original draft preparation, N.G.-L.; conceptualization and writing—review and editing, J.Á.S.-M. and E.A.; supervision, project administration, and funding acquisition, E.A. All authors have read and agreed to the published version of the manuscript.

Funding: This work was carried out within the framework of the AGROALNEXT programme and supported by the Ministry of Science, Innovation and Universities (MICIU), with funding from the European Union through NextGenerationEU (PRTR-C17.I1), and by Fundación Séneca, with funding from the Regional Government of Murcia (CARM). The authors also acknowledge the PRIMA Programme under Grant Agreement No. 2431 (FUSION: Comprehensive and sustainable solution to minimize food loss and waste and promote food security in the Mediterranean region). Nieves García-Lorca acknowledges the financial support received through the Investigo Programme, within the Recovery, Transformation and Resilience Plan (ID 2022-709 C23.I01.P03.S0020-0000217), funded by the European Union through NextGenerationEU, for predoctoral training.

Data Availability Statement: The original contributions presented in this study are included in the article. Further inquiries can be directed to the corresponding author.

Acknowledgments: The authors would like to express their gratitude to Levante Sur Coop. for providing the broccoli stalks used in this study. An AI-based tool was employed to enhance the clarity and translation of the manuscript. This tool was used exclusively for linguistic purposes; all scientific content, data interpretation, and conclusions are the sole responsibility and merit of the authors.

Conflicts of Interest: The authors declare no conflicts of interest.

References

1. Gawlik-Dziki, U.; Jezyna, M.; Świeca, M.; Dziki, D.; Baraniak, B.; Czyz, J. Effect of Bioaccessibility of Phenolic Compounds on in Vitro Anticancer Activity of Broccoli Sprouts. *Food Res. Int.* **2012**, *49*, 469–476. [CrossRef]
2. Ravikumar, I. Therapeutic Potential of *Brassica oleracea* (Broccoli)—A Review. *Int. J. Drug Dev. Res.* **2015**, *7*, 9–10.
3. Li, H.; Xia, Y.; Liu, H.Y.; Guo, H.; He, X.Q.; Liu, Y.; Wu, D.T.; Mai, Y.H.; Li, H.B.; Zou, L.; et al. Nutritional Values, Beneficial Effects, and Food Applications of Broccoli (*Brassica oleracea* Var. *italica* Plenck). *Trends Food Sci. Technol.* **2022**, *119*, 288–308. [CrossRef]
4. Sougrakpam, Y.; Deswal, R. Brassica Juncea Seeds and Seedlings Are Potential Functional Foods with Diverse Active Myrosinases. *Food Biosci.* **2023**, *56*, 103430. [CrossRef]
5. Núñez-Gómez, V.; González-Barrio, R.; Baenas, N.; Moreno, D.A.; Periago, M.J. Dietary-Fibre-Rich Fractions Isolated from Broccoli Stalks as a Potential Functional Ingredient with Phenolic Compounds and Glucosinolates. *Int. J. Mol. Sci.* **2022**, *23*, 13309. [CrossRef] [PubMed]
6. Salas-Millán, J.Á.; Aguayo, E. Fermentation for Revalorisation of Fruit and Vegetable By-Products: A Sustainable Approach Towards Minimising Food Loss and Waste. *Foods* **2024**, *13*, 3680. [CrossRef] [PubMed]
7. Helland, H.S.; Leufvén, A.; Bengtsson, G.B.; Skaret, J.; Lea, P.; Wold, A.B. Storage of Fresh-Cut Swede and Turnip in Modified Atmosphere: Effects on Vitamin C, Sugars, Glucosinolates and Sensory Attributes. *Postharvest Biol. Technol.* **2016**, *111*, 150–160. [CrossRef]
8. Villaño, D.; Fernández-Pan, I.; Arozarena, Í.; Ibañez, F.C.; Vírseda, P.; Beriain, M.J. Revalorisation of Broccoli Crop Surpluses and Field Residues: Novel Ingredients for Food Industry Uses. *Eur. Food Res. Technol.* **2023**, *249*, 3227–3237. [CrossRef]
9. Nasrin, T.A.A.; Yasmin, L.; Arfin, M.S.; Rahman, M.A.; Molla, M.M.; Sabuz, A.A.; Afroz, M. Preservation of Postharvest Quality of Fresh Cut Cauliflower through Simple and Easy Packaging Techniques. *Appl. Food Res.* **2022**, *2*, 100125. [CrossRef]
10. Silveira, A.C.; Aguayo, E.; Artés, F. Shelf-Life and Quality Attributes in Fresh-Cut Galia Melon Combined with Fruit Juices. *LWT—Food Sci. Technol.* **2013**, *50*, 343–348. [CrossRef]
11. Healthy Snacks Market Is Projected to Reach USD 187.7 Bn in 2033. Available online: https://dimensionmarketresearch.com/report/healthy-snacks-market/ (accessed on 30 June 2025).
12. Aguayo, E.; Requejo-Jackman, C.; Stanley, R.; Woolf, A. Effects of Calcium Ascorbate Treatments and Storage Atmosphere on Antioxidant Activity and Quality of Fresh-Cut Apple Slices. *Postharvest Biol. Technol.* **2010**, *57*, 52–60. [CrossRef]
13. Arias, E.; González, J.; Oria, R.; Lopez-Buesa, P. Ascorbic Acid and 4-Hexylresorcinol Effects on Pear PPO and PPO Catalyzed Browning Reaction. *J. Food Sci.* **2007**, *72*, C422–C429. [CrossRef] [PubMed]
14. Aguayo, E.; Escalona, V.H.; Artés, F. Effect of Hot Water Treatment and Various Calcium Salts on Quality of Fresh-Cut 'Amarillo' Melon. *Postharvest Biol. Technol.* **2008**, *47*, 397–406. [CrossRef]
15. Troyo, R.D.; Acedo, A.L. Effects of Calcium Ascorbate and Calcium Lactate on Quality of Fresh-Cut Pineapple (Ananas Comosus). *Int. J. Agric. For. Life Sci.* **2019**, *3*, 143–150.
16. Silveira, A.C.; Aguayo, E.; Chisari, M.; Artés, F. Calcium Salts and Heat Treatment for Quality Retention of Fresh-Cut 'Galia' Melon. *Postharvest Biol. Technol.* **2011**, *62*, 77–84. [CrossRef]
17. Aguayo, E.; Requejo-Jackman, C.; Stanley, R.; Woolf, A. Hot Water Treatment in Combination with Calcium Ascorbate Dips Increases Bioactive Compounds and Helps to Maintain Fresh-Cut Apple Quality. *Postharvest Biol. Technol.* **2015**, *110*, 158–165. [CrossRef]
18. Xiong, S.; Yun, J.; Zhang, C.; Li, W.; Zhou, F.; Tian, M.; Jiang, A. Assessment of the Effect of Ascorbic Acid, Sodium Isoascorbate and Calcium Ascorbate Treatments on the Browning and Wound Healing Process of Fresh-Cut Potatoes. *Food Chem.* **2025**, *463*, 141454. [CrossRef] [PubMed]
19. Colaço, C.A.L.S.; Roser, B. Trehalose—A Multifunctional Additive for Food Preservation. *Food Packag. Preserv.* **1994**, 123–140. [CrossRef]
20. Fernandez, O.; Béthencourt, L.; Quero, A.; Sangwan, R.S.; Clément Christophe, C. Trehalose and Plant Stress Responses: Friend or Foe? *Trends Plant Sci.* **2010**, *15*, 409–417. [CrossRef] [PubMed]
21. Vichaiya, T.; Faiyue, B.; Rotarayanont, S.; Uthaibutra, J.; Saengnil, K. Exogenous Trehalose Alleviates Chilling Injury of 'Kim Ju' Guava by Modulating Soluble Sugar and Energy Metabolisms. *Sci. Hortic.* **2022**, *301*, 111138. [CrossRef]

22. Fan, T.; Xia, M.; Cao, J.; Zhang, J.; Wang, T.; Cao, S. Trehalose Regulates the Quality and Antioxidant Capacity of Cherry Tomato during Postharvest Ripening. *Int. Food Res. J.* **2023**, *30*, 933–944. [CrossRef]
23. Chen, L.; Fan, K. Pulsed Vacuum Impregnated Trehalose to Improve the Physicochemical Quality of Frozen-Thawed Kiwifruit. *Int. J. Food Sci. Technol.* **2021**, *57*, 268–275. [CrossRef]
24. Liu, G.; Liu, S.; Liu, J.; Xiang, Y.; Zhu, L.; Xu, X.; Zhang, Z. Trehalose Delays Postharvest Browning of Litchi Fruit by Regulating Antioxidant Capacity, Anthocyanin Synthesis and Energy Status. *Postharvest Biol. Technol.* **2025**, *219*, 113249. [CrossRef]
25. Chen, A.; Tapia, H.; Goddard, J.M.; Gibney, P.A. Trehalose and Its Applications in the Food Industry. *Compr. Rev. Food Sci. Food Saf.* **2022**, *21*, 5004–5037. [CrossRef] [PubMed]
26. Tilley, A.; McHenry, M.P.; McHenry, J.A.; Solah, V.; Bayliss, K. Enzymatic Browning: The Role of Substrates in Polyphenol Oxidase Mediated Browning: Mechanisms of Enzymatic Browning. *Curr. Res. Food Sci.* **2023**, *7*, 100623. [CrossRef] [PubMed]
27. Grzegorzewska, M.; Badełek, E.; Szczech, M.; Kosson, R.; Wrzodak, A.; Kowalska, B.; Colelli, G.; Szwejda-Grzybowska, J.; Maciorowski, R. The Effect of Hot Water Treatment on the Storage Ability Improvement of Fresh-Cut Chinese Cabbage. *Sci. Hortic.* **2022**, *291*, 110551. [CrossRef]
28. Kabelitz, T.; Hassenberg, K. Control of Apple Surface Microflora for Fresh-Cut Produce by Post-Harvest Hot-Water Treatment. *LWT* **2018**, *98*, 492–499. [CrossRef]
29. López-Hernández, A.A.; Ortega-Villarreal, A.S.; Vázquez Rodríguez, J.A.; López-Cabanillas Lomelí, M.; González-Martínez, B.E. Application of Different Cooking Methods to Improve Nutritional Quality of Broccoli (*Brassica oleracea* Var. *Italica*) Regarding its Compounds Content with Antioxidant Activity. *Int. J. Gastron. Food Sci.* **2022**, *28*, 100510. [CrossRef]
30. Perini, M.A.; Sin, I.N.; Reyes Jara, A.M.; Gómez Lobato, M.E.; Civello, P.M.; Martínez, G.A. Hot Water Treatments Performed in the Base of the Broccoli Stem Reduce Postharvest Senescence of Broccoli (*Brassica oleracea* L. Var Italic) Heads Stored at 20 °C. *LWT* **2017**, *77*, 314–322. [CrossRef]
31. Costa-Pérez, A.; Moreno, D.A.; Periago, P.M.; García-Viguera, C.; Domínguez-Perles, R. A New Food Ingredient Rich in Bioaccessible (Poly)Phenols (and Glucosinolates) Obtained from Stabilized Broccoli Stalks. *Foods* **2022**, *11*, 1734. [CrossRef] [PubMed]
32. Salas-Millán, J.Á.; Aznar, A.; Conesa, E.; Conesa-Bueno, A.; Aguayo, E. Functional Food Obtained from Fermentation of Broccoli By-Products (Stalk): Metagenomics Profile and Glucosinolate and Phenolic Compounds Characterization by LC-ESI-QqQ-MS/MS. *LWT* **2022**, *169*, 113915. [CrossRef]
33. Yang, Q.; Luo, M.; Zhou, Q.; Chen, J.; Ji, S. Salicylic Acid Treatment Alleviates the Loss of Glucosinolates Accompanying Yellowing in Harvested Broccoli. *Postharvest Biol. Technol.* **2023**, *204*, 112466. [CrossRef]
34. Marín Sillué, S.; Berrada Ramdani, H.; Hernando Hernando, I.; López Rodríguez, R. Informe Del Comité Científico de La Agencia Española de Seguridad Alimentaria y Nutrición (AESAN) En Relación a La Seguridad Del Uso de Dos Soluciones Acuosas de Peróxido de Hidrógeno, Ácido Acético y Ácido Peracético Como Coadyuvantes Tecnológicos Para La Desinfección Bacteriana Del Agua de Lavado de Ciruelas, Cerezas y Peras. *Rev. Del Com. Científico De La AESAN* **2022**, *36*, 71–84.
35. Kasim, R.; Kasim, M.U. Biochemical Changes and Color Properties of Fresh-Cut Green Bean (*Phaseolus vulgaris* L. Cv.Gina) Treated with Calcium Chloride during Storage. *Food Sci. Technol.* **2015**, *35*, 266–272. [CrossRef]
36. Martínez-Hernández, G.B.; Amodio, M.L.; Colelli, G. Carvacrol-Loaded Chitosan Nanoparticles Maintain Quality of Fresh-Cut Carrots. *Innov. Food Sci. Emerg. Technol.* **2017**, *41*, 56–63. [CrossRef]
37. Rasines, L.; Castillejo, N.; San Miguel, G.; Aguayo, E. Can Household Storage Conditions Reduce Food Waste and Environmental Impact? A Broccoli Case Study. *Sci. Total Environ.* **2023**, *892*, 164779. [CrossRef] [PubMed]
38. Re, R.; Pellegrini, N.; Proteggente, A.; Pannala, A.; Yang, M.; Rice-Evans, C. Antioxidant Activity Applying an Improved ABTS Radical Cation Decolorization Assay. *Free Radic. Biol. Med.* **1999**, *26*, 1231–1237. [CrossRef] [PubMed]
39. Salas-Millán, J.Á.; Conesa-Bueno, A.; Aguayo, E. A Novel Antidiabetic Lactofermented Beverage from Agro-Industrial Waste (Broccoli Leaves): Process Optimisation, Phytochemical Characterisation, and Shelf-Life through Thermal Treatment and High Hydrostatic Pressure. *Food Biosci.* **2024**, *59*, 103999. [CrossRef]
40. Park, M.H.; Ko, D.Y.; Do, K.R.; Moon, H.W.; Malka, S.K.; Ku, K.M. Hot Water Treatment Alleviates Peel Browning in Oriental Melons through Cutin Biosynthesis: A Comprehensive Metabolomics Approach. *Postharvest Biol. Technol.* **2023**, *204*, 112451. [CrossRef]
41. Otwell, W.S.; Iyengar, R. Inhibition of Enzymatic Browning in Foods and Beverages. *Crit. Rev. Food Sci. Nutr.* **1992**, *32*, 253–273. [CrossRef] [PubMed]
42. Richard-Forget, F.C.; Goupy, P.M. Enzymatic Browning Reactions in Apple and Apple Products. *Crit. Rev. Food Sci. Nutr.* **1994**, *34*, 109–157. [CrossRef] [PubMed]
43. Newman, Y.M.; Ring, S.G.; Colaco, C. The Role of Trehalose and Other Carbohydrates in Biopreservation. *Biotechnol. Genet. Eng. Rev.* **1993**, *11*, 263–294. [CrossRef] [PubMed]
44. Allegra, A.; Inglese, P.; Guccione, E.; Farina, V.; Sortino, G. Calcium Ascorbate Coating Improves Postharvest Quality and Storability of Fresh-Cut Slices of Coscia and Abate Fétel Pears (*Pyrus communis* L.). *Horticulturae* **2022**, *8*, 227. [CrossRef]

45. Zhao, N.; Li, X.; Chen, W.; Shi, J. Effect of Hot Water Treatment on the Inhibition of Anthracnose, PG, PME Activity and PGIP Gene Expression in Harvested Papaya Fruits. *Acta Hortic.* **2013**, *975*, 487–494. [CrossRef]
46. European Commission. Commission Regulation (EC) No 2073/2005 on Microbiological Criteria for Foodstuffs. *Off. J. Eur. Union* **2005**, *L 338*, 1–26.
47. Althaus, D.; Hofer, E.; Corti, S.; Julmi, A.; Stephan, R. Bacteriological Survey of Ready-to-Eat Lettuce, Fresh-Cut Fruit, and Sprouts Collected from the Swiss Market. *J. Food Prot.* **2012**, *75*, 1338–1341. [CrossRef] [PubMed]
48. Jacxsens, L.; Devlieghere, F.; Debevere, J. Predictive Modelling for Packaging Design: Equilibrium Modified Atmosphere Packages of Fresh-Cut Vegetables Subjected to a Simulated Distribution Chain. *Int. J. Food Microbiol.* **2002**, *73*, 331–341. [CrossRef] [PubMed]
49. Hong, S.I.; Lee, H.H.; Kim, D. Effects of Hot Water Treatment on the Storage Stability of Satsuma Mandarin as a Postharvest Decay Control. *Postharvest Biol. Technol.* **2007**, *43*, 271–279. [CrossRef]
50. Li, C.; Sun, L.; Zhu, J.; Cheng, Y.; Huang, R.; Fan, Y.; Guo, M.; Ge, Y. Trehalose Maintains the Quality of Malus Domestica by Mediating Sucrose and Respiratory Metabolism. *Sci. Hortic.* **2022**, *295*, 110857. [CrossRef]
51. Ding, F.; Wang, R. Amelioration of Postharvest Chilling Stress by Trehalose in Pepper. *Sci. Hortic.* **2018**, *232*, 52–56. [CrossRef]
52. Abreu, M.; Beirão-da-Costa, S.; Gonçalves, E.M.; Beirão-da-Costa, M.L.; Moldão-Martins, M. Use of Mild Heat Pre-Treatments for Quality Retention of Fresh-Cut "Rocha" Pear. *Postharvest Biol. Technol.* **2003**, *30*, 153–160. [CrossRef]
53. Baek, S.M.; Ahn, S.I.; Lee, S.H.; Choi, J.M.; Hong, J.; Kim, Y.J.; Han, B.K. Prediction Model of Browning Inhibitor Concentration and Its Optimal Composition for Mass Processing of Ready-to-Eat Fresh-Cut 'Fuji' Apple (Malus Domestica Borkh.) Strains. *J. Food Sci.* **2024**, *89*, 4986–4996. [CrossRef] [PubMed]
54. Paulsen, E.; Moreno, D.A.; Martínez-Romero, D.; García-Viguera, C. Bioactive Compounds of Broccoli Florets as Affected by Packing Micro-Perforations and Storage Temperature. *Coatings* **2023**, *13*, 568. [CrossRef]
55. Wu, Q.; Liu, D.; Zhang, J.; Li, T.; Niu, H.; Xin, X.; Zhao, S.; He, C.; Zhang, C. Enhancing the Formation of Functional Glucosinolate Degradation Products in Fermented Broccoli Stalk By-Product with Lactic Acid Bacteria. *Food Chem.* **2025**, *464*, 141689. [CrossRef] [PubMed]
56. Mbudu, K.G.; Witzel, K.; Börnke, F.; Hanschen, F.S. Glucosinolate Profile and Specifier Protein Activity Determine the Glucosinolate Hydrolysis Product Formation in Kohlrabi (*Brassica oleracea* Var. *gongylodes*) in a Tissue-Specific Way. *Food Chem.* **2025**, *465*, 142032. [CrossRef] [PubMed]

foods

MDPI

Article

Optimizing Farmers' and Intermediaries' Practices as Determinants of Food Waste Reduction Across the Supply Chain

Abdelrahman Ali [1,2,*], Yanwen Tan [1,*], Shilong Yang [1], Chunping Xia [3] and Wenjun Long [4]

1 School of Economics and Management, South China Agricultural University, Guangzhou 510642, China; 20211161015@stu.scau.edu.cn
2 Agricultural Economics Department, Faculty of Agriculture, Fayoum University, Fayoum 63514, Egypt
3 College of Economics and Management, Huazhong Agricultural University, Wuhan 430070, China; xcp@mail.hzau.edu.cn
4 Research Center for Rural Economy in Ministry of Agriculture and Rural Affairs, Beijing 100125, China; 87280931@163.com
* Correspondence: aaa31@fayoum.edu.eg (A.A.); tanyw@scau.edu.cn (Y.T.); Tel.: +86-2085286242 (Y.T.)

Abstract

Improper stakeholder practices are considered a primary driver of food loss. This study aims to investigate the consequences of pre- and post-harvest practices on extending the shelf life of agro-food products, identifying which practices yield the highest marginal returns for quality. Using Fractional Regression Models (FRM) and Ordinary Least Squares (OLS), the research analyzed data from 343 Egyptian grape farmers and intermediaries. Key findings at the farmer level include significant food loss reductions through drip irrigation (13.9%), avoiding maturity-accelerating chemicals (24%), increased farmer-cultivated area (6.1%), early morning harvesting (8.7%), and improved packing (13.7%), but delayed harvesting increased losses (21.6%). For intermediaries, longer distances to market increased losses by 0.15%, while using proper storage, marketing in the formal markets, and using an appropriate transportation mode reduced losses by 65.9%, 13.8%, and 7.9%, respectively. Furthermore, the interaction between these practices significantly reduced the share of losses. The study emphasizes the need for increased public–private partnerships in agro-food logistics and improved knowledge dissemination through agricultural extension services and agri-cooperatives to achieve sustainable food production and consumption. This framework ensures robust, policy-actionable insights into how stakeholders' behaviors influence postharvest losses (PHL). The findings can inform policymakers and agribusiness managers in designing cost-efficient strategies for reducing PHL and promoting sustainable food systems.

Keywords: food handling practices; stakeholders' training needs; horticultural postharvest loss; good agricultural practices; grapes supply chain; shelf life

1. Introduction

Identifying and addressing the food loss drivers, including improper stakeholder practices, is essential to sustain and maintain the shelf life of agro-food products across the supply chain (SC) for as long as possible [1–3]. The Boston Consulting Group (BCG) (Boston Consulting Group https://www.bcg.com/publications/2018/tackling-1.6-billion-ton-food-loss-and-waste-crisis, accessed on 30 April 2025) estimates that by 2030, annual food losses and waste (FLW) (according to the FAO, food loss refers to the share of food lost across the food supply chain before it reaches the final consumer, while food waste represents the share of food wasted by the consumers) will hit 2.1 billion tons, worth USD 1.5 trillion, and decreasing it is crucial to achieve sustainable food system transformation, particularly in developing countries (which import most of their food) [4]. Meanwhile, improper handling and hygiene practices result in food contamination, constantly increasing the percentage of loss across the supply chain [5]. Furthermore, measuring the interaction between socioeconomic factors and stakeholder practices toward reducing the rate of loss across the food supply chain (FSC) plays an essential role in designing an appropriate solution to minimizing the share of lost food at different nodes across the supply chain [6,7]. The marginal benefits of preventing losses should exceed the marginal costs to make the intervention more attractive to relevant stakeholders [8]. Adopting good agricultural practices, including cultivation, harvesting, handling, and postharvest practices, dramatically improves the farmers' net profit and enables the sustainable use of limited resources in developing countries by reducing the percentage of FLW [9]. Generally, pre-sorting on farms can remove fruits with unacceptable shapes, sizes, and colors during harvesting, grading, and sorting at the farm gate. However, these fruits have the same nutritional value and could be used either fresh or processed to feed humans. Due to their appearance, they are often left in the field [3,10].

The handling practices of agro-food products (such as harvesting, pre-cooling, sorting and grading, packaging, and storing) play a substantial role in preserving quality and extending shelf life [11]. Improper practices or neglecting one of the aforementioned operations lead to quality degradation, especially for perishable and sensitive fruits like grapes. For example, poor harvesting methods, incorrect cooling, and poor storage contribute to the loss of products. Because grapes are very sensitive to temperature, they cannot be kept as a product for long periods. However, the substantial role of agricultural education and extension in disseminating the best farming practices, including postharvest operations for perishable products, is weak in most developing countries [12]. This highlights the urgent need to reform the public agricultural extension institutes to raise small stakeholders' knowledge, improve their practices and skills, and change their behaviors and attitudes to use limited resources more sustainably [13]. Considering the negative effects of spraying harmful chemicals to promote earlier fruit maturity (color, size, and solidity) or sustain their freshness after harvest, this practice can decrease the fruits' shelf life and result in food safety problems and negative environmental impacts [14].

Fruits provide essential vitamins and offer high profit margins for small-scale farmers and middlemen. Still, the share of postharvest fruit loss is very high due to a lack of agro-processing facilities and agro-logistic services, especially in developing countries. Varying perishability between different fruit types is a key factor affecting all aspects of this industry; growers must maintain minimum product quality for consumers, which impacts pricing and the percentage of losses [3]. However, farmers and middlemen remain hesitant to adopt the recommended postharvest practices. Efforts to disseminate and encourage stakeholders to adopt these practices can significantly reduce the share of postharvest loss (PHL) in grapes and other fresh fruit and food products (PHL refers to the share of lost or wasted food across the food supply chain, starting from harvesting and extending until the product reaches the final consumer).

Egypt is ranked fourth in the world for table grape production quantity (~1.7 million tons), and growth in this production has significantly increased Egyptian grape exports over the past 15 years [2]. The FAO estimated the share of postharvest loss to be around 18.6%, 5.3%, and 6.7% for the farmers, wholesalers, and retailers in North Egypt, highlighting the critical points of losses, including the harvest, wholesale, and retailer market levels, respectively [2]. Egyptian grape producers and value chain actors are increasingly vulnerable to shifts in global markets due to a lack of marketing information and land fragmentation among small-scale producers. They might delay mature crop harvesting to decrease transportation costs, which significantly increases the loss volume. Small-scale producers lacking export access need new strategies to upgrade the table grape value chain.

The interlink between the grapes supply chain and required post-harvest operations depends on the primary destination market (domestic or export). Furthermore, neglecting these operations (e.g., sorting, packing, grading, etc.) at the farm could lead to an increase in the percentage of food loss at the following downstream stages of FSC [3,15]. Stakeholders' practices and access to logistics significantly influence the proportion of loss across FSC, and this share varies depending on FSC efficiency, logistics service availability, and stakeholder practices [3,16]. For example, cold transportation and storage services mainly contribute to preserving the product for long periods [2]. Adopting good handling techniques, using appropriate equipment or technology, building capacity, enforcing thorough regulations, and improving marketing can avoid many problems [17]. FLW reduction interventions have garnered more global attention and investment following the global food crisis in 2008, when many countries sought to increase local food self-sufficiency through various strategies, including food loss and waste reduction, particularly in developing countries [2].

Individuals' preferences are based on the costs, benefits, and risks of the choices in line with their socioeconomic characteristics, so they will only act if the benefit outweighs the costs [18]. In the same way, we assumed (1) the stakeholders will invest to decrease FLW when the value of the saved food outweighs the cost of the intervention (practices or technology) [19]. (2) The interaction between the different proper practices could result in a greater reduction of losses across the supply chain. However, the importance of the correlation between the farmer and middleman characteristics and their practices is emphasized. The previous studies did not analyze the different impacts of the interaction between the stakeholders' socioeconomic characteristics and their practices toward food loss and waste, which could give an overview of the technical training needed to enhance the capacity of the small farmers and intermediaries to adopt the good agricultural practices, including the best postharvest operations. The current study aims to evaluate and analyze the interaction between stakeholders' socioeconomic characteristics and their impacts on their practices at the different FSC stages, to determine the appropriate intervention strategies for reducing these losses. The results of this study could help policymakers, agribusiness owners, and international funding donors design sustainable intervention reduction strategies for these losses, achieve a sustainable supply chain, attain environmental preservation, and ensure efficient use of limited resources available.

The recent literature indicates a scarcity of evidence on the impact of stakeholders' practices as a food loss driver [20]. Furthermore, the majority of earlier FLW research concentrated on calculating the proportion of loss [3,21,22], few studies have investigated the food loss driver, particularly horticultural crops. Furthermore, most of these studies used the macro-secondary data (e.g., food balance sheet, FAO) to estimate the FLW, which neglected the heterogeneity between the stakeholders and the significant role of those small stakeholders in FLW reduction strategies, and they provided only high-level insights about the nature and causes of these losses [23]. Despite the substantial investment in reducing losses, the effectiveness of these interventions remains limited due to a lack of coordination among different FCS actors, the failure to improve infrastructure, and inadequate support for capacity building [24]. Therefore, we aim to measure the interaction between socioeconomic characteristics and stakeholders' practices, which represents an important step in designing appropriate interventions to minimize this share at different nodes across the FSC.

The novelty of the current work could be highlighted through its theoretical and empirical contribution to a better understanding of the root causes and drivers of FLW at the small stakeholder level. This includes assessing the impacts of socioeconomic characteristics heterogeneity among stakeholders on access to logistics services and its effects on FLW reduction. This study presents evidence-based findings on how socioeconomic characteristics and their interaction with other technological factors influence the percentage of FLW across FSC in developing countries. To analyze the FLW drivers, we started with farmers' practices during farm operations, then the middlemen handling the product to reach the end users. The following sections include the methodology, data analysis, results, discussion, and conclusion.

2. Materials and Methods

2.1. Theoretical Framework

The reduction of food loss in agricultural supply chains is fundamentally dependent on the intricate interaction of social and economic behaviors demonstrated by farmers and intermediaries. These participants function within a network of incentives, limitations, and relational dynamics that significantly affect post-harvest losses. Grasping this relationship is essential for creating effective interventions [24]. For example, trust between farmers and intermediaries promotes an exchange of market information, where intermediaries disseminate market standards and demand signals, and farmers communicate production difficulties. This openness fosters improved planning and coordination to synchronize harvests with market requirements and minimize mismatch losses [19]. Furthermore, trust and strong networks lower transaction costs, enable collective investment in loss-reducing infrastructure, improve access to credit and information, and strengthen bargaining power, making economically viable solutions more accessible. Long-term relations encourage collaboration, where the intermediaries might provide credit for agro-inputs that mitigate losses (such as cold storage and transportation) or provide technical support; farmers tend to prioritize working with trusted partners, which may improve quality control.

The target-measure-act framework has been employed to analyze and measure the impacts of stakeholders' or individuals' practices as a food loss driver. This approach has been approved as a recognized approach for developing a rapid result for FLW reduction [25,26]. This study intends to evaluate the interaction between socioeconomic characteristics and the stakeholders' practices as food loss drivers across FSC to provide a better understanding of how these factors interact. We target the most important actors across FSC who are responsible for the highest share of FLW (farmers and middlemen). Also, the grapes have been considered the main fruit crop in the investigated area. We measured the impacts of these interactions on the stakeholders by employing the rational choice theory, where the stakeholders' decisions are based on the interaction between the availability, access, and utilization of the facilities and technologies that reduce FLW, and the socioeconomic characteristics of those stakeholders. Finally, the appropriate interventions have been discussed based on the study's results, as shown in Figure 1.

Rational choice theory (RCT), also known as a rational action or choice theory, is a framework for recognizing and often formally modelling social and economic behaviour [18]. It assumes that individuals' and agents' preferences are based on the benefit/cost ratio, where stakeholders will invest in stopping FLW when the value of the saved food outweighs the price of the intervention (practices or technology). Moreover, people use the current resources to maximize rewards, and agents or individuals can incorporate all relevant factors into their preference rankings for the overall alternative. Furthermore, social scientists can indirectly understand agents' desires through their disclosed decisions. Researchers extrapolate from observable behavior to rebuild the preference hierarchy that governs a rational agent's decisions.

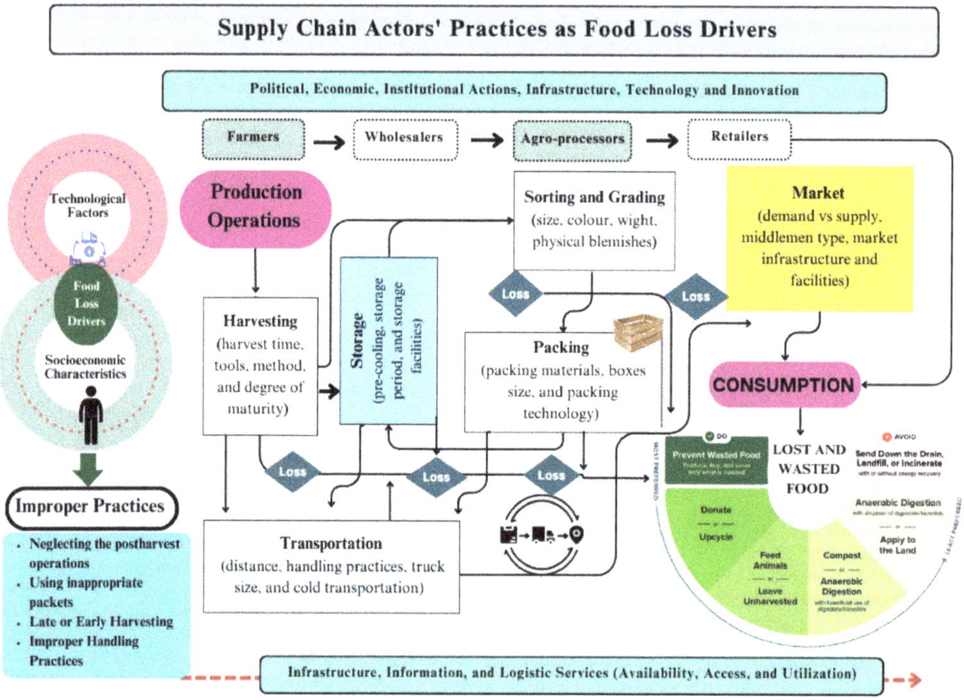

Figure 1. Supply Chain Actors' Practices as Food Loss Drivers.

RCT is used in microeconomics to model human or individual decision-making, where it assists economists in better understanding the behaviour of a society "in terms of individual actions as explained through rationality". This theory is widely used in economic studies and agricultural policy analysis, which explains human decision-making [27]. Recently, the fundamentals of this theory have been used in many studies to determine the appropriate intervention to reduce food and post-harvest losses [28,29]. The farmer is challenged with a choice of whether or not to adopt post-harvest handling practices to reduce loss and the extent of adoption. So, a farmer faces a tradeoff between incurring additional costs to mitigate loss and risking getting loss. For example, the farmer could risk late harvesting his crop because they wishes the market price might increase or the harvest labor rent might decrease. A rational farmer and middleman will invest in alleviating loss if there is an economic motivation to do so. Additionally, reducing loss increases the amount of the crop available for sale and consumption. Hence, the quantity of the crop saved by reducing loss and the cost of mitigating the losses play a crucial role in the farmer's decision [29].

Improper practices are mainly a result of a complex interaction across the supply chain between the stakeholders' socioeconomic characteristics and access to agricultural services. Where the value of crop saved by reducing loss and the cost of mitigating the losses play a crucial role in the farmers' and intermediaries' decisions.

2.2. Study Area and Sampling Strategy

A multi-stage sampling design has been adopted to select grape farmers and middlemen. According to Yamane, 384 observations are required to justify an infinite population [30]. We used the following equation to determine the sample size.

$$n = \frac{N}{1 + N * (e)^2} \tag{1}$$

where n is the proposed sample size, N is the total population size, N = (453) for the current study, and e is the level of precision (0.05).

In the current study, we only considered small farmers; hence, the proposed sample size would be 213 growers. The sample was chosen from the main grape production governorate in Egypt, according to the cultivated area (*El Minya*, located south of Cairo), as shown in Figure A1. We randomly selected the three main districts (*Bani Mazar, Mattai, and Samalout*), which represent more than 84% of the total cultivated area in *El Minya* governorate. Then, two villages were selected in each district (based on the grape cultivated area). The respondents' numbers in each village were determined according to the geometric mean of the number of farmers and the area planted with grapes in each village. Finally, 213 Egyptian grape farmers were randomly selected in these villages. Similarly, we interviewed 130 middlemen involved in the fruit and vegetable trade. We excluded the questionnaires of 13 farmers and 10 middlemen due to incomplete answers.

In this study, we used a structured questionnaire to collect data from the stakeholders through face-to-face interviews during the 2020 season. We developed the two questionnaires (one for farmers and the other for the middlemen). We pretested them, modified them based on the pretest results, and finally, we collected the data. The proposed questionnaires were prepared based on related literature and have been tested through informal group discussions (grape experts and agro-economists) and a pilot study. Table A2 provides more details about the socioeconomic characteristics of the selected farmers and middlemen.

2.3. Model Specification

Degradation in quantity and/or quality of agricultural products depends on both farmers' and middlemen's practices, which include preharvest and postharvest operations, as follows:

PHL (Y) = F (Pre-harvest practices, Post-harvest practices, Environmental factors, Institutional factors, Interaction effects), which could be presented in the following equation.

$$\text{PHL\_i} = \text{G}\beta 0 + \sum \beta \text{j Farmer\_Practices\_ji} + \sum \theta \text{k Middlemen\_Practices\_ki} + \sum \gamma \text{mControls\_mi} + \varepsilon_\text{I} \tag{2}$$

where (PHL_percent) is the dependent or outcome variable, Xi's predictor or independent variables, including the pre-harvest and postharvest variables, as shown in Table A1. β, $\gamma, \theta, \propto$, δ are coefficients for each predictor and ε_I is the error term. We include each practice as a separate regressor to identify which specific practices have an impact. For example, pre-harvest practices are the farm operations such as (irrigation system, pest control (including spraying the chemicals to speed up the maturity or improve the appearance of the final product), timing of harvest, stage of fruit maturity, initial sorting and grading at the farm gate, which the farmers mostly do. While post-harvest practices include (storage conditions, handling, packing, transportation time (cold-chain, distance to the market), etc.), are done by the middlemen, as shown in Table 1. The control variables (Z) include access to information, membership in an agricultural cooperative, farm size, education, etc.

Table 1. Descriptive analysis of the pre- and post-harvest practices for grapes (farmers' and middlemen).

Stage	Items	Description	Farmers % Frequencies N = 200	Middlemen % Frequencies N = 120
Farmoperations	Irrigation system	Drip	45	-
		Conventional	55	-
	Spray chemicals to speed up the maturity	Yes	27.5	-
		No	72.5	-
Harvesting	Harvesting time of the day	Early in the morning	87	-
		Any time of the day	13	-
	Stage of maturity	Fully ripe	79	-
		Partially ripe	21	-
	Delay harvesting (after optimum maturity)	Yes	56.5	-
		No	43.5	-
	The reason	To catch a high price	50.0	-
		To avoid the low-price	40.5	-
		None	9.5	-
	Method of harvesting	Scissors/knife	49.5	-
		Hand	50.5	-
		Machinery	-	-
Sorting and grading	Grade/pre-sort your produce immediately	Yes	58.5	70
		No	41.5	30
	Quality indicators for grading	Color	29.5	2.5
		Size	3.0	-
		Weight	19.0	-
		Physical blemishes	38.0	52.3
		All the above	10.5	39.2
Storage	Access to proper storage facilities	Yes	50.5	88.3
		No	49.5	11.7
	Storage problems with your produce?	Colour changing	18.5	-
		Wilting and shrinking	6.0	-
		Weight losses	6.0	2.5
		Insects and rotting infestations	7.0	45
		All the above	62.5	52
	Storage place	At farm	19	-
		At home	3.6	-
		In the market	77.4	100
	Number of storage days, farmers (N = 128), For middlemen (N = 107)	One Day	81.5	38.3
		Two days or more	18.5	61.7
Packing	Packaging material	Wood Box	-	-
		Crates of palm leaf	27.0	72
		Carton Box	30.0	-
		Plastic Box (improved packing)	43.0	28

Table 1. *Cont.*

Stage	Items	Description	Farmers % Frequencies N = 200	Middlemen % Frequencies N = 120
Transporting	**Product coverage during transportation?**	Yes	28.5	-
		No	31.5	52.5
		Sometimes	40.0	47.5
	Distance to the nearest market (Km)	50–100	-	80.8
		101–150	1.5	4.2
		151–200	21.0	11.7
		<200	77.5	3.3
	Transportation time	Early in the morning	56.5	47
		Any time of the day	43.5	53
	Transportation mode	Own transportation	12.3	23
		Hired transportation	87.7	77
Marketing	**Type of market**	Formal market	23	56
		Informal market	77	44
Dealing with losses during the different stages?		Donate to charity	11.5	-
		Feed animals	33.0	31.7
		Compost	5.0	-
		Landfill (left in the field)	22.0	-
		Garbage	28.5	68.3
Do you need facilitation to target the preferred marketing channel?		Cold-Transportion	15.5	1.8
		Equipped storage places	19	14.5
		Microfinance	14.0	50.9
		Market information	45.5	13.6
		None	6	19.2
Are you satisfied with your current main marketing channel?		Yes	16	47.5
		No	84	52.5
What are the constraints to target the preferred marketing channel?		Lack of marketing information	70.1	-
		High cost	19.2	-
		Needs big quantity	10.7	-

2.3.1. A. Fractional Regression Model (FRM)

Since PHL is a proportional/fractional outcome (e.g., 5%), a fractional regression model is ideal [31]. It uses a logistic or probit link function to ensure predictions stay within [0, 1] (or [0, 100%] after rescaling). Due to the nature of the data, we employed the fractional probit model in the current study.

$$E(\text{PHL\_i} \mid \text{Xi}) = G\left(\beta 0 + + \sum \beta j \text{ Farmer\_Practices\_ji} + \sum \theta k \text{ Middlemen\_Practices\_ki} + \sum \gamma m\text{Controls\_}i\right) \quad (3)$$

G (·): Logistic or probit function (e.g., G (z) = $\frac{e^z}{1+e^z}$).

PHL_i = Percentage loss (e.g., 0–100%). The average of percentage losses for each selected stakeholders has been presented in Table A2.

2.3.2. B. Addressing Interaction Effects:

We evaluated the interaction effects to capture synergies or trade-offs between stakeholders' practices, where interactions between the different practices might influence the percentage of PHL.

$$E(PHL\_i \mid Xi) = G(\beta 0 + \beta 1\ Practice\_i + \theta 1\ Practice\_k + \delta(Practice\_i * Practice\_k)i + \gamma Controls\_i \tag{4}$$

A negative δ implies that combined practices reduce losses more than their individual effects.

For robustness checks, we compared the results of FRM with Ordinary Least Squares (OLS) {with PHL logit-transformed; $ln\left(\frac{PHL}{100-PHL}\right)$}.

The variables were investigated to determine whether they affected the PHL. Then, the factors with an effect were used in the model to determine whether they had a negative or positive effect on the PHL. At that time, some of the affected variables were examined as dummy variables, assuming that the stakeholders adopting and applying the best agricultural practices are (1) and otherwise (0). We checked the autocorrelation and multivariate collinearity, revealing no extreme multivariate collinearity in the data.

3. Results and Discussion

Adopting the best agricultural and handling practices will extend the shelf life of agrofood products and consequently decrease the percentage of loss, highlighting the importance of agricultural extension in disseminating information among farmers and other actors across FSC [32]. The following section presents and discusses the results of the practices of grape farmers and middlemen.

3.1. Descriptive Analysis of Grape Farmers' and Middlemen's Practices as Food Loss Drivers

The self-report method in socioeconomic research is vulnerable to social desirability bias, which can impact the validity and reliability of the results. To overcome (over- or underestimations), a mixed method of qualitative and quantitative methods, along with real observations, was used during data collection to ensure accuracy and avoid bias.

Socioeconomic characteristics, including age, education, gender, and cultivated area, for the selected respondents are presented in Table A2. These demographics may influence the respondents' actions and practices, as young people with higher education can adopt best practices more quickly than older individuals. Likewise, the number of years of experience also plays a crucial role, as highly experienced farmers can determine the appropriate input quantities and types, irrigation times and water quantities, and the optimum stage of maturity for harvesting. For asset ownership, farmers and middlemen who own assets such as cars, tractors, and tricycles reported a lower loss percentage due to their ability to transport their production to or from the market, especially for small-scale farmers. At the same time, the organization's membership proved its importance in disseminating new agricultural operations and marketing information through training, facilitating the joining of the contract farming program, and providing access to microfinance, which must be considered during the design and implementation of intervention reduction strategies.

Postharvest operations have been divided between farmers and middlemen. The farmers mainly do the upstream supply chain operations. Still, middlemen are responsible for collecting and redistributing agro-food products. Both practices determine the cumulative share of lost and wasted food downstream of the supply chain [33]. The respondents reported that during harvesting, pre-grading, and sorting, they are excluding unacceptable fruits (shapes, sizes, and colours) at the farm gate, which is consistent with the recent findings reported that the final consumers will not accept these fruits due to their shape or size and should be removed from the beginning of supply chain [10]. Furthermore, around half of the farmers (45%) used drip irrigation. They harvested the fully ripening fruits (79%) in the early morning (87%), where the drip irrigation could increase the fruit's solidity and subsequently decrease the share of lost food across FSC. Additionally, around 72.5% of them don't spray chemicals to accelerate fruit maturity in normal cases (although they sometimes use these chemicals to achieve a higher price, especially at the beginning of the harvest season). An overview of the main selected practices (factors) that determine the share of grape loss is presented in Table 1.

Recent studies revealed that primary inspection and sorting play a significant role in reducing losses downstream FSC [34]. Likewise, inappropriate packing boxes, such as palm crates (common for Egyptian farmers), increase the percentage of injured fruits during handling, packing, and transporting [8]. The quality indicators for grading were considered by around 67.5% of the farmers' respondents, who were mainly based on the colour and physical blemishes of the product. Still, most of the middlemen were grading not only based on the physical blemishes but also considering (solidity, size, and weight). Furthermore, farmers and intermediaries reported that they store their production in the market (sometimes without cold storage services) before selling it for one or more days under normal temperature conditions, which increases the rejection of fruits due to color changes, wilting, insect infestation, rotting, and weight loss. Although open trucks are a common transportation mode for agricultural products in Egypt, most of the respondents reported that they don't cover the products during transportation from the farm to the wholesale market or retailers, which increases the losses and decreases the product quality, especially, during summer and for the long-distance transportation (sometimes more than 12 h).

Most respondents reported being dissatisfied with their current marketing channel, and the primary constraint to targeting the preferred channel was a lack of market information. Access, affordability, and availability of facilitation to target the preferred marketing channel significantly impact the loss, as farmers reported an urgent need for market information and microfinance, while small intermediaries require microfinance. Most farmers either leave the injured fruits in the field or use them to feed their animals. Whereas 50% leave the injured fruit in the field or send it to the garbage, and only 33% use it for animal feeding. This appears clearly during the season or when the market price is low, the farmers prefer to leave the yield in the field or feed the animals with the unacceptable fruits [35]. However, the intermediaries mainly send them to the garbage (especially in urban and semi-urban regions), which causes multiple health and environmental impacts. This highlights the importance of minimizing losses in the upstream supply chain to reduce the share of wasted food in subsequent nodes or even at the consumer level.

3.2. *Fractional Regression Models (FRM) Compared with OLS Results for Grapes (Farmers and Middlemen)*

Improper food supply chain practices increase the amount of damaged fruit and consequently the PHL ratio at various stages (e.g., over-maturity, improper handling, overloading boxes or trucks, and improper storage) [36]. Furthermore, infrastructure availability and agro-food processing facilities are crucial in enhancing the marketing systems to decrease the PHL [37]. Hence, controlling these factors along with FSC has a tremendous and immediate impact on reducing the percentage of PHL, which could be a way to enhance the welfare of small stakeholders (farmers and intermediaries) in developing countries.

3.2.1. FRM Compared with Ordinary Least Squares (OLS) Results for Grape Farmers

Identifying the optimal harvest practices to minimize losses is crucial in designing an intervention strategy. In Table 2, the results of the FRM and OLS show that the education of grape farmers positively reduces the PHL, with a 0.3% reduction in the share of lost food for FRM and a 4.9% reduction for OLS, on average, for educated farmers compared to illiterate farmers. Our current findings support the previous study that revealed the significant role of agricultural education in reducing PHL in many developing countries, such as the substantial role in the global initiative (Postharvest Education Foundation). In the past two decades, this initiative has highlighted the requirement for continuous training, policy support, and accessible technologies to sustain the impact of the PHL reduction interventions [38].

Table 2. Fractional probit regression results (grapes farmers).

Farmers	**Fractional Probit Regression**			**OLS Estimations**	
PHL	**Coef.**	**Std. Err.**	**Marginal Effect (dy/dx)**	**Coef.**	**Std. Err.**
Cons.	−1.9172	0.30926 ***	-	−3.07161 ***	0.51874
Age	−0.000896	0.0018351	−0.00308	−0.00258	0.00431
Education	−0.02017 *	0.011534	−0.00308	−0.04894	0.02938
Experience	−0.0006893	0.003843	−0.0001	−0.00040	0.00938
Total Cultivated area	−0.02872 **	0.011849	−0.00438	−0.06116 **	0.02971
Grapes Cultivated	−0.008235	0.015477	−0.00125	−0.013729	0.04151
Information Source	−0.0004830	0.008711	−0.00073	−0.01184	0.02157
Cooperative membership	−0.0374406	0.03239	−0.00571	−0.08225	0.07927
Access to microfinance	−0.0235572	0.03289	−0.00359	−0.08205	0.07880
Contract farming	−0.0385765	0.03430	−0.00588	−0.07118	0.08804
Irrigation system	−0.05994 *	0.03458	−0.00915	−0.13909 **	0.07874
Non-Chemical to speed up maturity	−0.12198 ***	0.03352	−0.01862	−0.24078 ***	0.08370
Harvest Method	−0.01706	0.03375	−0.0026	−0.00731	0.07672
Harvest time	−0.04546 **	0.02052	−0.00694	−0.08752 **	0.04600
Late harvesting	0.10239 ***	0.03033	0.01563	0.21671 ***	0.07474
Stage maturity	−0.04359	0.041401	−0.0066	−0.07258	0.09656
Presort and Grade immediately	−0.02042	0.03386	−0.00311	−0.04036	0.07695
Propre Storage harvest	−0.06058 *	0.031385	−0.00924	−0.1173	0.07637
Packaging material	−0.06801 ***	0.018199	−0.01038	−0.13785 ***	0.04685
Cover Product	−0.02572	0.02074	−0.00392	−0.05611	0.04795
Distance to market	0.0004	0.000262	0.0006	0.00003	0.00061
Farm price	−0.00851	0.038608	−0.00130	−0.00355	0.09433
Interaction effect					
Irrigation * Nonchemical	−0.0978 ***	0.01770	−0.01498	−0.19834 ***	0.04349
Irrigation * Harvest time	−0.09933 ***	0.02145	−0.01521	−0.19248 ***	0.05509
Harvest time * packing material	−0.02140 **	0.01087	−0.00327	−0.03803 **	0.02328
Harvest time * Non-late harvest	−0.04080 ***	0.01650	−0.00625	−0.08877 ***	0.04167
	N = 200 Wald chi2 (21) = 106.69 Prob > chi2 = 0.0000 Pseudo R^2 = 0.0077			F(21, 178) = 2.39 Prob > F = 0.0011 R-squared = 0.2201 Adj R-squared = 0.1281	

*, **, and *** are refer that $p < 0.1$, <0.05, and <0.01.

Moreover, non-use of chemicals has a strong positive effect on decreasing the PHL. It leads to an average decrease in the share of losses by 0.19% and 24% for FRM and OLS, in addition to reducing their negative health and environmental impacts. This highlights the essential role of agricultural extension services in enhancing the knowledge and skills of small farmers by training them on the best agricultural practices, including pre- and post-harvest operations [39]. In addition to the role of technical agricultural education, which significantly reduces pesticide application among vegetable farmers in China [40].

Furthermore, the irrigation system used during the production is one of the most essential factors in determining the share of PHL, where the production under drip irrigation could significantly decrease the share of lost food by 0.9% and 13.9% on average for FRM and OLS on average compared with the conventional (flood) irrigation. This is due to the low moisture and water content of grape fruits produced under drip irrigation, which deters pathogens, increases their solidity, and extends their shelf-life, subsequently decreasing losses. This agrees with a recent study that revealed that the total production under drip irrigation is higher in quantity (45%) and quality than the conventional or flood ones [41].

The results revealed that increasing the total cultivated area for the grape farmer significantly decrease the share of PHL at the grape farm level by 0.4% and 6% on average for FRM and OLS, highlighting that the increasing the cultivated land is associated with a decrease in PHL where the big farmers can benefit from (large-scale production) by lowering per-unit production and transportation costs, especially for long distances to the market. This is in line with the economic theory, where land fragmentation decreases economic efficiency and could force small farmers to delay harvesting their production to increase the quantity and reduce per-unit transportation costs [42]. This appears clearly in developing countries, but the collective marketing and cooperatives could open opportunities for small-scale grape producers to export their production or to attract local traders who secure agricultural micro-finance for them and then purchase their harvest [43]. At the same time, the large farms help farmers utilize advanced production technologies and improved varieties with a long shelf life, subsequently decreasing the share of losses.

The results of FRM revealed that harvesting early in the morning could decrease the losses by 0.7% and 8.7% on average for FRM and OLS, respectively, and it is statistically significant at 1%. Meanwhile, delaying harvest after the optimal maturity increases the share of PHL by 1.5% and 21.6% on average for FRM and OLS, respectively, compared to the optimal harvesting. Our results support the FAO findings, which revealed that the delays in harvesting due to climate, peak labor demand, and competition of Egypt's exporters can significantly impact export volumes, diverting production to local markets and increasing the losses [2]. This highlights the role of governmental support in finding new strategies for upgrading the value chain of small-scale table grape actors to access the export market. Furthermore, the grape farmers reported that they sometimes store the harvested fruits at the farm, which increases the share of losses by 0.9% and 11.7% on average for FRM and OLS, respectively, compared to farmers who sell their production directly. In the case of storing products, previous studies have highlighted the role of using edible coatings, such as gum Arabic, in delaying ripening and maintaining fruit quality during storage [44]. For packing materials, the results revealed that the losses increase by around 0.1% and 13.7% on average for FRM and OLS, respectively, when using the crates of palm leaf compared to the carton and plastic boxes, highlighting the importance of using appropriate packing materials for slowing respiration and keeping the fruits from mechanical damage. However, there is a significant correlation between using plastic crates and lower food contamination [5]. In Egypt, the familiarity with using plastic crates is low compared to that of traditional palm crates for most small fruit and vegetable farmers. Furthermore, the packinghouse companies or stations collect grapes from the farms and then repack them according to the destination market (local supermarkets, hotels, exports). They provide these services to exporters or traders. Sometimes, they buy from grape farmers and resell the packed product to their clients, which can result in more PHL due to the repeated loading and reloading of the product [2].

The interaction between the different farmers' practices significantly reduced the PHL. For example, combining drip irrigation and non-spraying chemicals has a negative and significant coefficient, which could decrease the PHL by an additional 1.5% and 19.8% on average for FRM and OLS, respectively. Likewise, combining the optimal harvesting time with drip irrigation results in a further reduction of 1.5% and 19.2% on average for FRM and OLS, respectively. Furthermore, using the appropriate packing and harvesting early in the morning could reduce the PHL by an additional 0.3% and 3.8% on average for FRM and OLS, respectively. Additionally, combining optimal harvesting maturity with early morning harvesting results in an average reduction of 0.6% and 8.8% for FRM and OLS, respectively. This highlights the importance of using holistic and integrated interventions to reduce the percentage of PHL, considering the cost and availability for the small farmers and stakeholders [3].

To compare the results of the FRM and OLS models, we observed that most variables align in the same direction (positive or negative) across the two models. However, significance levels vary due to methodological differences, as the Probit model assumes non-linearity and the OLS model assumes linearity. Examining the model fit, the FRM has a low pseudo R^2 (0.0077) but a significant Wald chi^2 (p = 0.0000), indicating that the model explains limited variance despite its overall significance. In contrast, the OLS model with interactions has a higher R-squared value and significant F-statistics, which suggests that linear relationships dominate the data. Where the OLS model explains ~22.1% of the log (PHL) variance, highlighting its effectiveness and suggesting better explanatory power. We expected this value to be lower for the farmers than the middlemen due to the nature of the included variables (postharvest operations done mainly by the middlemen rather than farmers), so we expect the middlemen to have a higher explanation of variance in the dependent variable (R-squared).

However, covering the products during transportation from the farm to the wholesale or local market was statistically insignificant. However, it remains one of the primary factors contributing to the increase in the PHL. However, the farm price is also statistically insignificant. Still, its fluctuations from one season to another and during the same year contribute directly and indirectly to increasing the PHL by delaying the harvest time or spraying some chemicals (sometimes) to speed up the maturity, negatively affecting product quality and increasing losses. When the farm gate price is high, the farmers try to avoid postharvest losses and keep their production as fresh as possible by covering the product or following the appropriate operations during harvesting and handling. That highlights the role of contract farming to reduce the price risk and achieve benefits for farmers and consumers; it can create linkages between them through adopting the short supply chain and using E-commerce [45].

The village's location near the main cities plays a significant role in allowing farmers to receive training through international projects, which increases their knowledge of best practices for farm operations and post-harvest handling. This is consistent with the findings that revealed the significant role of agricultural training and extension in reducing the PHL share [39,46]. Meanwhile, increasing farmers' experience could decrease the share of losses, especially for fruits, where farmers are familiar with water requirements, pest control, postharvest operations, and other related factors. Furthermore, we observed that new farms in the reclaimed areas produce lower yields than those in the old lands for the same varieties, possibly due to differences in soil fertility and the farmers' experience.

3.2.2. FRM Compared with OLS Results for Grape Middlemen

The main factors influencing the percentage of PHL for the grape's middlemen are presented in Table 3. These factors include methods of storage and transportation, whether in equipped vehicles or not, as well as long-distance transportation, which affects product spoilage, as presented below.

Table 3. Fractional probit regression results (grapes middlemen).

Middlemen	Fractional Probit Regression			OLS Estimations	
	Coef.	Std. Err.	Marginal Effect (dy/dx)	Coef.	Std. Err.
Cons.	−1.3841 ***	0.19093		−2.35058 ***	0.44873
Market type	−0.06916 ***	0.01320	−0.00775	−0.13881 ***	0.0363
Gender	0.05423 **	0.02522	0.00620	0.10136 *	0.06671
Age	−0.00059	0.00117	−0.000067	−0.00024	0.00295
Education	−0.01619	0.01349	−0.00183	−0.03823	0.03126
Experience	−0.02731	0.02203	−0.00310	−0.06235	0.06735
cooperative	−0.06501	0.04692	−0.00738	−0.14608	0.09549
Access to microfinance	−0.03843	0.04139	−0.00436	−0.07905	0.08440
Information source	−0.01274	0.021062	−0.001447	−0.0313615	0.04021
Proper Store harvest	−0.30398 ***	0.04929	−0.03468	−0.6593 ***	0.10466
Transportation time	0.00702	0.01234	0.00079	0.00547	0.04684
Pre-sort and -grade immediately	−0.01341	0.04148	−0.00152	−0.02186	0.081292
Transportation mode	−0.03435 ***	0.01200	−0.00382	−0.07932 **	0.03548
Packaging material	−0.067853	0.03223	−0.00680	−0.00680	0.03209
Cover product	−0.01486	0.02400	−0.00168	−0.03810	0.05689
Distance to market	0.00072 ***	0.00027	0.00008	0.00158 ***	0.00046
Interaction effect					
Market type * proper storage	−0.05233 ***	0.01122	−0.00596	−0.10655 ***	0.030337
Transport mode * proper storage	−0.06623 ***	0.01312	−0.00754	−0.15159 ***	0.03641
	N = 120 Wald chi^2 (15) = 166.88 Prob > chi^2 = 0.0000 Pseudo R^2 = 0.0089			F (15, 104) = 7.69 Prob > F = 0.000 R-squared = 0.5258 Adj R-squared = 0.4574	

*, **, and *** are refer that $p < 0.1$, <0.05, and <0.01.

The most essential factor is the market type (formal and informal), where the rate of losses differs according to the product's customer and the location of the middlemen in the FSC. Trading in formal/structured markets reduces the share of lost food by ~0.7% and 13.88% on average for FRM and OLS, respectively, compared to the informal ones. This highlights the importance of a short supply chain to decrease losses and maximize marketing efficiency, which has been confirmed in the literature [47]. Regarding the gender impacts, the results showed that the PHL for male traders is ~1.8% higher than for females for FRM and 10% higher for OLS, respectively. Female traders mostly pre-sort and grade the products before reselling, compared to males. This aligns with the recent findings confirmed that female traders are more likely to follow the best handling and hygiene practices than male traders in Nigeria, contributing significantly to food safety and reducing losses [5]. Another interesting point is that we observed that most retailers and hawkers are females, and they are more likely to use plastic crates and sort products due to their dealing with small volumes, which reduces losses. Furthermore, proper storage significantly reduced losses, as evidenced by a negative and statistically significant coefficient, which is more pronounced in formal wholesale markets and supermarkets than in informal ones. Proper storage infrastructure (cold storage) is essential for reducing the PHL for perishable products like horticultural products, especially during summer [48,49].

The results revealed that the proper storage facilities could reduce the PHL by ~3.4% and 65.9% for FRM and OLS, respectively. A recent study from Nigeria showed that higher temperatures encourage microbial growth, and the number of horticultural products infested with contaminants doubled during the dry season [5]. Current results highlight significant heterogeneity of PHL in developing countries, where most losses occur between the farm gate and the consumer due to the heterogeneity of access to cold storage facilities across the FSC [50]. We observed during the market visits that the consumers (while searching and selecting the fruits) are more likely to touch and move the products, especially when they are unpackaged. This could increase food contamination and losses, especially for sensitive products like grapes, highlighting the importance of Modified Atmosphere Packaging (MAP) for such products [51]. Furthermore, the transportation mode (own or hired transportation) plays a substantial role in determining the share of PHL. The results revealed that the traders who use their own transportation can reduce the PHL by ~0.3% and 7.9% for FRM and OLS, respectively, compared with those who hire transportation. These findings suggest that the cooperatives could contribute to providing cold transportation services at subsidized fees to support small traders and farmers, thereby reducing their losses.

The longer distance to market results in a higher PHL, where the PHL increases 0.008% and 0.15% on average for an increase in the distance for the FRM and OLS, respectively. The increased transportation risks or delays lead to higher quantity and quality losses caused by the growth of fungi and bacteria. Due to the use of open trucks and the long distance between the production site and the wholesale or retail market, most farmers reported that they didn't cover the products during their journey to the markets. The sunshine caused massive damage to the grapes and increased the PHL, which is in line with the findings of a recent study [52]. Likewise, a recent study revealed that the main determinants of food losses in developing countries are the lack of infrastructure, inadequate marketing systems, and improper handling practices, highlighting the role of public-private investment [53]. The long distance and unavailability of cold transportation increase the possibility of losses, so the accessibility of refrigeration and a steady cold chain are necessary to prevent both qualitative and quantitative losses in fresh products, such as fruits and vegetables [54]. The efficiency of the cold chain depends on adjusting product temperature at each step and managing worker practices, ensuring a clean cold chain for a sustainable PHL reduction strategy [49]. Refrigerated vehicles are expensive for small farmers because they often cannot access microfinance, but they can access these services through cooperatives. In Egypt, exporters have identified the shortage of packaging, pre-cooling, and cold storage facilities as a key factor restricting trade. The shortage of cold-chain storage services on farms and in the market represents the primary challenge for small farmers and intermediaries across the horticultural supply chain in developing countries [53].

The percentage of stored products for the next day outside the cold storage causes a decrease in the quantity & quality, and increases PHL [55]. The production source for middlemen plays an essential role in the PHL percentage, where local traders sometimes spend a long time collecting produce from small farmers, which negatively affects product quality and increases food loss in subsequent stages. This is consistent with [5]. The total quantity that the trader deals with (buying and selling) is one of the essential factors that can lead to an increase in the PHL, and sometimes result in more losses of net profit for the different middlemen due to quality deterioration. Meanwhile, the losses increase as the time of loading and unloading the products during transportation and handling due to the lack of agricultural marketing facilities and information exchange, which are crucial in moving agricultural products from the farm gate to the final consumer; this will benefit all actors across the FSC [21,56,57].

The interaction between the different middlemen's practices significantly reduced the PHL. For example, combining the market type and proper storage facilities could decrease the PHL by an additional 0.5% and 10.6% for the FRM and OLS, respectively.

Likewise, combining access to adequate storage facilities with the use of the middleman's transportation results in an additional reduction of 0.7% and 15% on average for the FRM and OLS, respectively. This highlights the importance of using holistic and integrated interventions to reduce the percentage of PHL, considering the cost and availability for the small farmers and stakeholders.

Examining the model fit, the FRM has a low pseudo-R^2 (0.0089) but a significant Wald chi^2 ($p = 0.0000$), indicating that the model explains limited variance despite overall significance. In contrast, the OLS model with interactions has a higher R-squared value and significant F-statistics, which suggests that linear relationships dominate the data. Where the OLS model explains ~52.6% of the log (PHL) variance, highlighting its effectiveness and suggesting better explanatory power. R-squared for the middlemen is higher than that of the farmers (higher explanation of variance in the dependent variable (%PHL), because of the nature of the included variables (postharvest operations done mainly by the middlemen rather than farmers). Although some factors, such as education, access to finance, and experience, were statistically insignificant in this model, they still explain a significant portion of the variance between stakeholders in the PHL. This highlights the need for future studies to better understand the interaction between these factors as drivers of food losses at the middlemen level.

Overall, this study found a statistically significant correlation between improper pre- and post-harvest practices and PHL between Egyptian grape farmers and middlemen. The current results confirm the application of the fundamentals of the theory of rational choice to farmers and middlemen for investing in PHL reduction. This highlights the manuscript's theoretical contribution through the empirical estimations in this field. However, the adoption rate of best recommended postharvest practices remains low among farmers and middlemen. Efforts to disseminate and encourage stakeholders to adopt these best-practice recommendations can significantly reduce the share of PHL in grapes and other fresh fruit and food products. We concluded the percentage of the main significant variables of the current results in Figure 2.

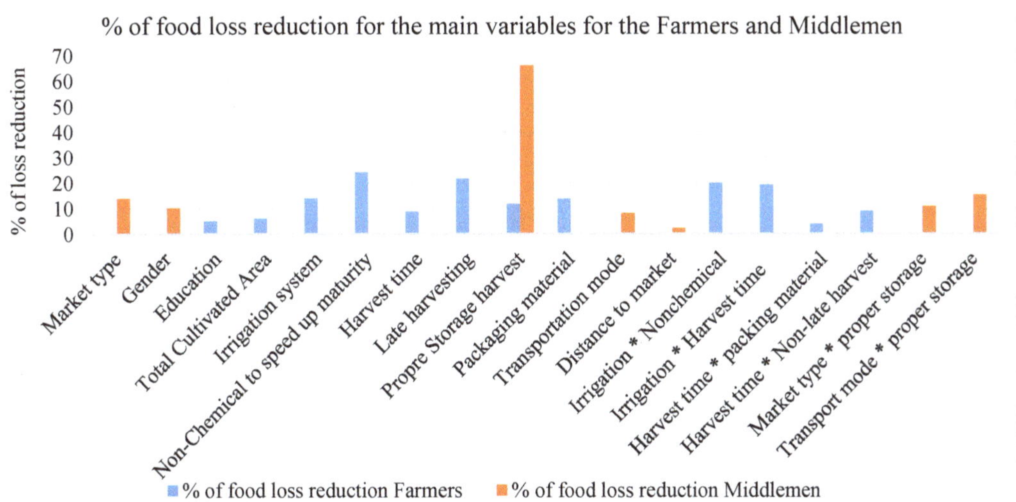

Figure 2. Percentage of loss reduction according to the results of the current study for the Farmers and Middlemen.

4. Conclusions

Poor postharvest practices, stemming from insufficient resources and knowledge, significantly contribute to losses throughout the food supply chain. Improving these practices is crucial for reducing FLW and enhancing the well-being of farmers and other stakeholders in developing nations. This study aims to investigate the impact of pre- and post-harvest practices on reducing food loss among Egyptian grape farmers and middlemen. Using a combination of the Fractional Regression Model and the Ordinary Least Squares model, the study found strong correlations between socioeconomic characteristics and stakeholders' practices at various stages of the food supply chain. The study also found that drip irrigation, delayed harvesting, and the use of appropriate packing materials can reduce losses at the farmer level. The study also found that farmers' education and cultivated area significantly influence PHL, with increased education and cultivated area leading to a decrease in PHL. Furthermore, the study emphasizes the importance of integrated interventions to minimize pesticide use in grape production, highlighting the need for further research to explore the economic feasibility of these combined approaches for small farmers.

For the agricultural markets, the study reveals that formal agricultural markets reduce post-harvest loss (PHL) by 13.88% compared to informal markets, underscoring the importance of efficient supply chains. Gender disparities and a lack of cold storage contribute to losses. Transportation mode and distance to markets also impact PHL. The study emphasizes the need for accessible refrigeration and a robust cold chain, including temperature control and hygienic practices at each stage. While refrigerated vehicles are expensive, microfinance through cooperatives could offer a solution. Inadequate packaging, pre-cooling, and cold storage facilities pose significant constraints, particularly for small-scale farmers and intermediaries. Finally, these findings underscore the importance of integrating market-type improvements with suitable storage facilities and transportation to reduce postharvest loss (PHL) by 10.6% and 15%, respectively. It emphasizes the need for holistic strategies that consider both cost and accessibility for all stakeholders. Proper storage adoption yields the highest marginal returns for quality, while subsidies for technologies like improved packing can mitigate losses. Agricultural cooperatives offer cost-effective cold storage and transportation services, while intensive training programs for farmers and intermediaries are also crucial. These findings can inform policymakers and agribusiness managers in designing effective, cost-efficient strategies to reduce PHL, boost smallholder profits, and foster sustainable food systems.

5. Limitations and Future Research Direction

Improved post-harvest handling, including harvesting, sorting, and the use of cold chain technologies, significantly reduces food losses in developing countries. Still, due to the lack of data, measuring the percentage of PHL and its impacts at different FSC nodes is challenging, highlighting the need for further investigations using primary data in the future to explore the leading causes according to different crops in various regions. Although we attempted to minimize the risk of social desirability bias in the self-reported methodology and due to heterogeneity among stakeholders, regions, crops, and the varying importance of different PHL causes, the current results may be limited in their generalizability to other crops in other areas. Investment in agrofood logistics, facilitated by public-private partnerships, is crucial for extending shelf life and enhancing the profitability of smallholder farmers. However, public-private coordination faces numerous challenges due to the weak institutional structures in developing countries, highlighting the need to investigate the impacts of public policy on the sustainability of agrofood systems. Public agricultural extension is vital in disseminating best practices among farmers.

Still, the number of agrarian facilitators decreases over the years in most developing countries (which could affect the estimation accuracy and the applicability of the suggested reduction interventions).

Further investigations are needed to assess the socioeconomic and environmental implications of integrating various farming practices with middlemen, thereby achieving the highest marginal returns for quality, minimizing losses, and enhancing supply chain efficiency. This framework could provide policymakers and agribusiness managers with more reliable, applicable, and economical insights for designing effective interventions. Future research should prioritize the role of the contract farming system in reducing post-harvest losses and securing additional resources for small farmers, which could help minimize the severity of market fluctuations. Innovation plays a significant role in securing low-cost logistics services for small-scale stakeholders, as well as proper handling and temperature control, which can reduce losses. This requires further investigation to ensure the applicability and economic viability of the proposed solutions.

Author Contributions: Conceptualization, A.A., Y.T., and C.X.; methodology, A.A., W.L., and Y.T.; software, A.A. and C.X.; validation, S.Y., W.L., Y.T. and A.A.; formal analysis, A.A.; investigation, A.A.; resources, W.L. and C.X.; data curation, A.A., W.L., and S.Y.; writing—original draft preparation, A.A., Y.T., W.L., and C.X.; writing—review and editing, C.X., S.Y., Y.T., W.L., and A.A.; visualization, S.Y. and W.L.; supervision, C.X.; project administration, Y.T.; funding acquisition, Y.T. All authors have read and agreed to the published version of the manuscript.

Funding: This work was supported by the National Philosophy and Social Science Foundation, China, grant number (23&ZD121).

Institutional Review Board Statement: Not applicable.

Informed Consent Statement: Not applicable.

Data Availability Statement: The original contributions presented in the study are included in the article, further inquiries can be directed to the corresponding authors.

Acknowledgments: We thank the anonymous reviewers and associate editor for their valuable help in improving the manuscript.

Conflicts of Interest: The authors declare no conflicts of interest.

Abbreviations

The following abbreviations are used in this manuscript:

PHL	Postharvest Losses
FSC	Food Supply Chain
FLW	Food Losses and Waste
RCT	Rational Choice Theory
OLS	Ordinary Least Squares
FRM	Fractional Regression Model

Appendix A

Table A1. Summarydescription of the main PHL explanatory variables.

Variable	Description
Stakeholders'Practices and Technological Factors	
Irrigation system (IRR)	(1) Drip (0) Conventional
Harvesting Time during the day (HT)	Early in the morning
	Any time across the day
Quality indicators for grading (QI)	Color
	Size
	Weight
	Physical blemishes
	All the above
Acess for proper storage facilities (SF)	Yes No
Storage problems of your produce? (SP)	color changing
	wilting and shrinking
	weight losses
	insects and rotting infestation
	All the above
Facilitation do you need to target the preferred marketing channel? (FMC)	Cold Transportion
	Equipped Storage places
	Microfinance
	Market information
	None
Dealing with losses during the different stages? (DPHL)	Donate for charity
	Feed animals
	Compost
	Landfill (left at the field)
	Garbage
Delay harvesting after optimal maturing? (DH)	Yes No
The reason (RED)	To get a high price
	To avoid the low-price
	None
Cover Product during transportation? (CP)	Yes No Sometimes
Harvest Method	(a) By knife cutting (b) by hand twisting (c) Machinery Harvesting
What stage of maturity of the fruit do you harvest?	(a) Fully matured (b) Partially matured
Grade/pre-sort your produce immediately	Yes No

Table A1. *Cont.*

Variable	Description
Number of storage days	Days
Spray chemicals to accelerate the maturity	Yes No
Packaging material	1. Wood Packaging Box 2. Crates of palm leaf 3. Carton Packaging Box 4. Plastic Packaging Box (improved packing)
Distance to market	Kilometers
Type of the market	Formal market
	Informal market
Transportation time	Early in the morning
	Any time of the day
Transportation mode	Own transportation
	Hired transportation
Stakeholders' Socioeconomic Characteristics	
Gender	Male Female
Age	Years
Education level	1 = university or college or equivalent 2 = secondary school (technical training) 3 = Primary/preparatory school 4 = Only read and write 5 = Illiteracy
Marital status	1. Married 2. Single 3. Divorced 4. Widower
Family members Number	Number
Experience Years	Years
Total Cultivated Area	Feddan (4200 m^2)
Cultivated area with study crop	Feddan
Source of information	1. Friends 2. Family members 3. Neighbors 4. Market agencies 5. Agricultural cooperation and extension 6. Companies and Inputs distributors
Member of an organization or cooperative	Yes No
Borrowed a loan (Access to finance)	Yes No
Contract to sell your produce	Yes No
Are postharvest losses a problem for you?	Yes No
Total_Production	Kg/feddan
Farm gate price per Kg	LE/kg
PHL_percent	**% (Dependent variable)**

Table A2. Sample description and descriptive statistics for the investigated respondents (grape farmers and Middlemen).

Stakeholders		Farmers N = 200		Middlemen N = 120									
				Village Trader N = 30		Wholesaler N = 25		Fruit Grocery N = 30		Supermarket N = 10		Hawker N = 25	
Items	**Description**	**F**	**%**	**F**	**%**	**F**	**%**	**F**	**%**	**F**	**%**	**F**	**%**
Gender	Male	196	98	27	90	25	100	6	12	14	70	31	62
	Female	4	2	3	10	-	-	44	88	6	30	19	38
Age	20 to <30 years old	35	17.5	3	10	-		8	16	4	20	6	12
	30 to <40	56	28	10	33.3	6	12	4	8	6	30	4	8
	40 to <50	60	30	14	46.7	24	48	22	44	5	25	26	52
	50 to <60	39	19.5	3	10	16	32	12	24	5	25	8	16
	>60	10	5	-	-	4	8	4	8	-	-	6	12
Education level	University	16	8	4	13.3	5	10	-	-	8	40	-	-
	Secondary school	28	14	4	13.3	9	18	5	10	11	55	4	8
	Primary/preparatory school	38	19	3	10	7	14	5	10	1	5	7	14
	Adult education/Read and write	61	30.5	11	36.7	19	38	30	60	-	-	9	18
	Illiterate	57	28.5	8	26.6	10	20	10	20	-	-	30	60
Marital status	Married	184	92	25	83.3	47	94	30	60	15	75		90
	Single	14	7	2	6.7	-	-	2	4	3	15	1	2
	Divorced/Widower	2	1	3	10	3	6	18	36	2	10	445	8
Number of Family members	2–4	36	18	9	30	15	60	12	41	7	70	7	28
	5–7	118	59	18	60	8	32	15	50	3	30	15	60
	>7	46	23	3	10	2	8	3	9	-	-	3	12
Items	**Description**	**F**	**%**	**F**	**%**	**F**	**%**	**F**	**%**	**F**	**%**	**F**	**%**
Experience Years	<5 years	14	7	-	-	-	-	5	16.7	3	30	7	23.3
	5 to <10	68	34	13	43.3	5	16.7	12	40	2	20	18	60
	10 to <15	76	38	12	40	7	23.3	10	33.3	4	40	5	16.7
	>15	42	21	5	16.7	18	60	3	10	1	10	-	-

Table A2. *Cont.*

Stakeholders		Farmers N = 200		Middlemen N = 120									
				Village Trader N = 30		Wholesaler N = 25		Fruit Grocery N = 30		Supermarket N = 10		Hawker N = 25	
Source of Information	Friends	60	30	7	23.3	15	60	10	33.3	-	-	13	52
	Family members	6	3	-	-	-	-	-	-	-	-	-	-
	Neighbors	14	7	10	33.3	5	20	-	-	-	-	7	28
	Market agencies	60	30	10	33.3	4	16	17	56.7	6	60	5	20
	Cooperation and agro-extension	13	6.5	3	10	1	4	-	-	-	-	-	-
	Companies and Inputs' distributors	48	24	-	-	-	-	3	10	4	40	-	-
Total Land Ownership	<1 *Feddan*	70	35	-	-	-	-	-	-	-	-	-	-
	1–3 *Feddan*	96	48	-	-	-	-	-	-	-	-	-	-
	More than 3 *Feddan*	34	17	-	-	-	-	-	-	-	-	-	-
Grapes Cultivated Area	<3 *Feddan*	116	58	-	-	-	-	-	-	-	-	-	-
	3–5 *Feddan*	52	26	-	-	-	-	-	-	-	-	-	-
	More than 5 *Feddan*	32	16	-	-	-	-	-	-	-	-	-	-
Member of an organization or cooperative	Yes	98	46	5	16.7	15	60	3	10	8	80	0	0
	No	112	54	25	83.3	10	40	27	90	2	20	25	100
Borrowed a loan (Access to finance)	Yes	118	59	8	26.6	20	80	4	13.3	6	60	2	8
	No	82	41	22	73.3	5	20	26	86.7	4	40	23	92
Contract to sell your produce	Yes	24	12	10	33.3	18	72	-	-	-	-	-	-
	No	176	88	20	66.7	7	28	-	-	-	-	-	-
Are post-harvest losses a problem for you?	Yes	174	87	30	100	19	76	10	33.3	10	100	7	28
	No	26	13	0	0	6	24	20	66.7	0	0	18	72
Total Production	Ton/feddan *	8.9 Ton/fed											
Farm gate price per Kg	LE/kg **	5.34		6.14		6.82		8.7		8.14		9.17	
PHL_percent	%	6.12%		1.65%		2.11%		2.7%		2.4%		1.9%	

* Feddan = 4200 square meters, it is the common unit of land area in Egypt. ** The average of annual exchange rate 2020. 1 $ = 15.7562 Egyptian pound. Central Bank of Egypt (CBE) https://www.cbe.org.eg/ (accessed on 30 April 2025).

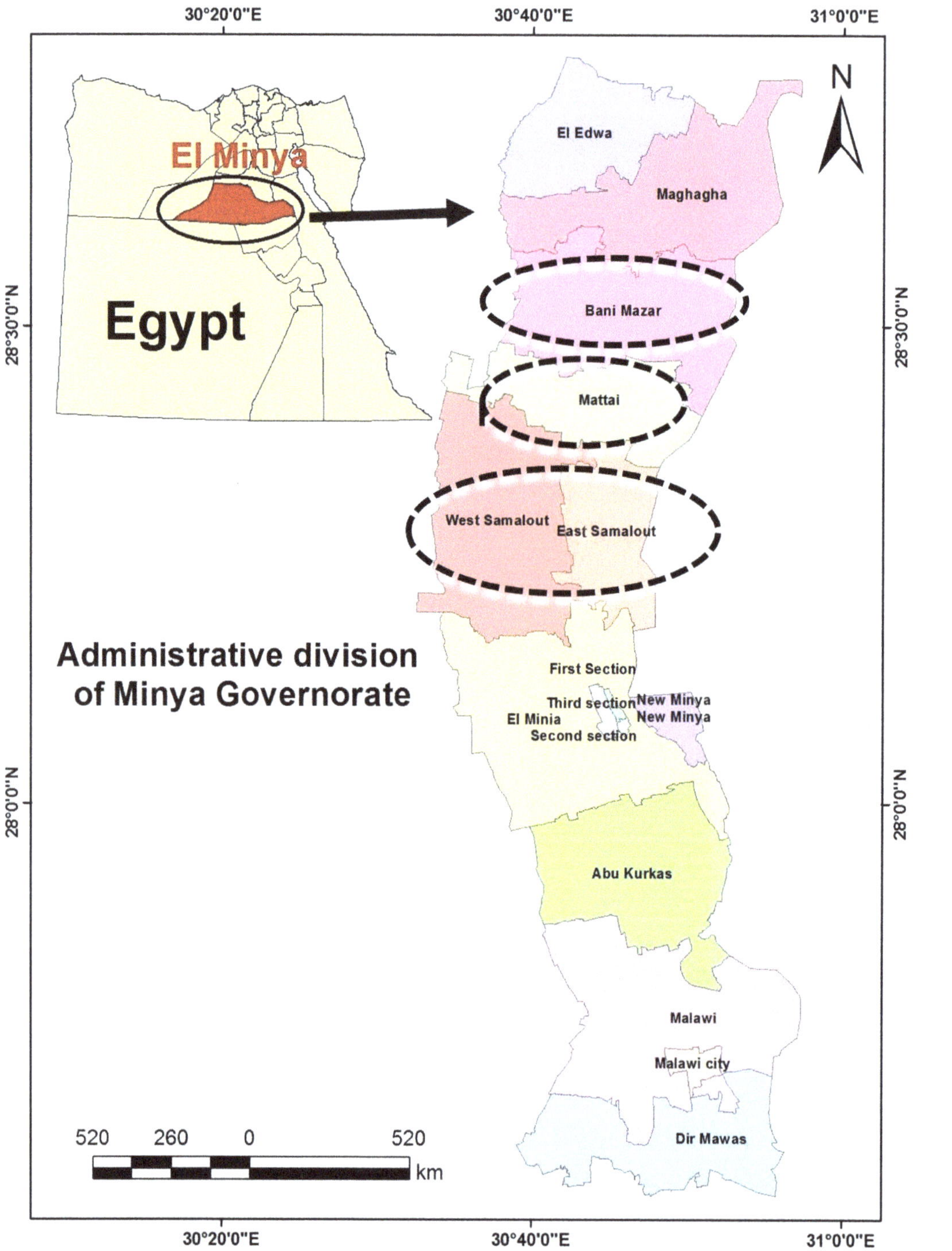

Figure A1. Study area map (*Minya* Governorate, Egypt). The left shows the location on the county map, whereas the right shows the detailed map of the three selected districts in the selected governorate (*Minya*).

References

1. Herzberg, R.; Trebbin, A.; Schneider, F. Product Specifications and Business Practices as Food Loss Drivers—A Case Study of a Retailer's Upstream Fruit and Vegetable Supply Chains. *J. Clean. Prod.* **2023**, *417*, 137940. [CrossRef]
2. FAO. *Food Loss Analysis for Grapes Value Chains in Egypt*; FAO: Cairo, Egypt, 2021.
3. Ali, A.; Tan, Y.; Medani, K.; Xia, C.; Abdullahi, N.M.; Mahmood, I.; Yang, S. Horticultural Postharvest Loss' and Its Socio-Economic and Environmental Impacts. *J. Environ. Manag.* **2025**, *373*, 123458. [CrossRef] [PubMed]
4. Davis, B.; Lipper, L.; Winters, P. Do Not Transform Food Systems on the Backs of the Rural Poor. *Food Secur.* **2022**, *14*, 729–740. [CrossRef]
5. Abatan, I.E.; Liverpool-Tasie, L.S.O.; Obadina, A.O. When Washing Is Not Enough: A Cross-Disciplinary Analysis of Hygiene and Handling Practices among Vegetable Traders in Nigeria. *Food Secur.* **2025**, *17*, 657–669. [CrossRef]
6. Parmar, A.; Hensel, O.; Sturm, B. Post-Harvest Handling Practices and Associated Food Losses and Limitations in the Sweetpotato Value Chain of Southern Ethiopia. *NJAS—Wagening. J. Life Sci.* **2017**, *80*, 65–74. [CrossRef]
7. Isack, M.E.; Lyimo, M. Effect of Postharvest Handling Practices on Physicochemical Composition of Tomato. *Int. J. Veg. Sci.* **2015**, *21*, 118–127. [CrossRef]
8. Anriquez, G.; Foster, W.; Ortega, J.; Santos Rocha, J. In Search of Economically Significant Food Losses: Evidence from Tunisia and Egypt. *Food Policy* **2021**, *98*, 101912. [CrossRef]
9. Abi Tarabay, P.; Chahine-Tsouvalakis, H.; Tohmé Tawk, S.; Nemer, N.; Habib, W. Reduction of Food Losses in Lebanese Apple through Good Harvesting and Postharvest Practices. *Ann. Agric. Sci.* **2018**, *63*, 207–213. [CrossRef]
10. Johnson, L.K.; Dunning, R.D.; Bloom, J.D.; Gunter, C.C.; Boyette, M.D.; Creamer, N.G. Estimating On-Farm Food Loss at the Field Level: A Methodology and Applied Case Study on a North Carolina Farm. *Resour. Conserv. Recycl.* **2018**, *137*, 243–250. [CrossRef]
11. Péra, T.G.; Rocha, F.V.d.; Caixeta Filho, J.V. Tracking Food Supply Chain Postharvest Losses on a Global Scale: The Development of the Postharvest Loss Information System. *Agriculture* **2023**, *13*, 1990. [CrossRef]
12. Farris, J.; Maredia, M.K.; Mason, N.M.; Ortega, D.L. Farmer Personality and Community-Based Extension Effectiveness in Tanzania. *World Dev.* **2024**, *173*, 106424. [CrossRef]
13. McNamara, P.; Tata, J. Principles of Designing and Implementing Agricultural Extension Programs for Reducing Post-Harvest Loss. *Agriculture* **2015**, *5*, 1035–1046. [CrossRef]
14. Muncke, J. Endocrine Disrupting Chemicals and Other Substances of Concern in Food Contact Materials: An Updated Review of Exposure, Effect and Risk Assessment. *J. Steroid Biochem. Mol. Biol.* **2011**, *127*, 118–127. [CrossRef] [PubMed]
15. Onwude, D.I.; Chen, G.; Eke-Emezie, N.; Kabutey, A.; Khaled, A.Y.; Sturm, B. Recent Advances in Reducing Food Losses in the Supply Chain of Fresh Agricultural Produce. *Processes* **2020**, *8*, 1431. [CrossRef]
16. Salamandane, C.; Fonseca, F.; Afonso, S.; Lobo, M.L.; Antunes, F.; Matos, O. Handling of Fresh Vegetables: Knowledge, Hygienic Behavior of Vendors, Public Health in Maputo Markets, Mozambique. *Int. J. Environ. Res. Public Health* **2020**, *17*, 6302. [CrossRef]
17. Stathers, T.; Holcroft, D.; Kitinoja, L.; Mvumi, B.M.; English, A.; Omotilewa, O.; Kocher, M.; Ault, J.; Torero, M. A Scoping Review of Interventions for Crop Postharvest Loss Reduction in Sub-Saharan Africa and South Asia. *Nat. Sustain.* **2020**, *3*, 821–835. [CrossRef]
18. Agnew, S. Rational Choice Theory. In *Environmental Criminology*, 2nd ed.; Routledge: London, UK, 2014; pp. 1–10. [CrossRef]
19. Surucu-Balci, E.; Tuna, O. Investigating Logistics-Related Food Loss Drivers: A Study on Fresh Fruit and Vegetable Supply Chain. *J. Clean. Prod.* **2021**, *318*, 128561. [CrossRef]
20. Verma, M.; Plaisier, C.; van Wagenberg, C.P.A.; Achterbosch, T. A Systems Approach to Food Loss and Solutions: Understanding Practices, Causes, and Indicators. *Sustainability* **2019**, *11*, 579. [CrossRef]
21. Ambler, K.; de Brauw, A.; Godlonton, S. Measuring Postharvest Losses at the Farm Level in Malawi. *Aust. J. Agric. Resour. Econ.* **2018**, *62*, 139–160. [CrossRef]
22. Chaboud, G.; Daviron, B. Food Losses and Waste: Navigating the Inconsistencies. *Glob. Food Secur.* **2017**, *12*, 1–7. [CrossRef]
23. Caldeira, C.; Corrado, S.; Goodwin, L.; Sala, S. Global Food Waste. In *Responsible Consumption and Production*; Springer International Publishing: Cham, Switzerland, 2019; pp. 1–12. ISBN 9783319710624.
24. Stöckli, S.; Niklaus, E.; Dorn, M. Resources, Conservation & Recycling Call for Testing Interventions to Prevent Consumer Food Waste. *Resour. Conserv. Recycl.* **2018**, *136*, 445–462. [CrossRef]
25. Hanson, C.; Flanagan, K.; Robertson, K.; Axmann, H.; Bos-brouwers, H.; Broeze, J.; Kneller, C.; Maier, D.; Mcgee, C.; Connor, C.O.; et al. *Reducing Food Loss and Waste: Ten Interventions to Scale Impact*; World Resources Institute: Washington, DC, USA, 2019.
26. Blanckenberg, A.; Opara, U.L.; Fawole, O.A. Postharvest Losses in Quantity and Quality of Table Grape (Cv. Crimson Seedless) along the Supply Chain and Associated Economic, Environmental and Resource Impacts. *Sustainability* **2021**, *13*, 4450. [CrossRef]
27. Nedergaard, P. The 2003 Reform of the Common Agricultural Policy: Against All Odds or Rational Explanations? *J. Eur. Integr.* **2006**, *28*, 203–223. [CrossRef]
28. Nash, J.; Peña, O.; Galford, G.; Gurwick, N.; Pirolli, G.; White, J.; Wollenberg, E. *Reducing Food Loss in Agricultural Development Projects Through Value Chain Efficiency*; CCAFS: Wageningen, The Netherlands, 2017.

29. Chegere, M.J. Post-Harvest Losses Reduction by Small-Scale Maize Farmers: The Role of Handling Practices. *Food Policy* **2018**, *77*, 103–115. [CrossRef]
30. Yamane, T. *Statistics, an Introductory Analysis*, 2nd, ed.; Harper and Row: New York, NY, USA, 1967; Volume 60.
31. Papke, L.E.; Wooldridge, J.M. Wooldridge Econometric Methods for Fractional Response Variables with an Application to 401(k) Plan Participation Rates. *J. Appl. Econom.* **1996**, *11*, 619–632. [CrossRef]
32. Prokopy, L.S.; Carlton, J.S.; Arbuckle, J.G.; Haigh, T.; Lemos, M.C.; Mase, A.S.; Babin, N.; Dunn, M.; Andresen, J.; Angel, J.; et al. Extension's Role in Disseminating Information about Climate Change to Agricultural Stakeholders in the United States. *Clim. Change* **2015**, *130*, 261–272. [CrossRef]
33. Arah, I.K.; Ahorbo, G.K.; Anku, E.K.; Kumah, E.K.; Amaglo, H. Postharvest Handling Practices and Treatment Methods for Tomato Handlers in Developing Countries: A Mini Review. *Adv. Agric.* **2016**, *2016*, 6436945. [CrossRef]
34. Liu, J.; Lundqvist, J.; Weinberg, J.; Gustafsson, J. Food Losses and Waste in China and Their Implication for Water and Land. *Environ. Sci. Technol.* **2013**, *47*, 10137–10144. [CrossRef]
35. Johnson, L.K.; Bloom, J.D.; Dunning, R.D.; Gunter, C.C.; Boyette, M.D.; Creamer, N.G. Farmer Harvest Decisions and Vegetable Loss in Primary Production. *Agric. Syst.* **2019**, *176*, 102672. [CrossRef]
36. Ali, A.; Xia, C.; Ouattara, N.B.; Mahmood, I.; Faisal, M. Economic and Environmental Consequences' of Postharvest Loss across Food Supply Chain in the Developing Countries. *J. Clean. Prod.* **2021**, *323*, 129146. [CrossRef]
37. Siam, G.; Abdelhakim, T. *Analysis of the Tomato Value Chain in Egypt and Establishment of an Action Plan to Increase Its Efficiency*; CIHEAM, Mediterranean Agronomic Institute of Montpellier: Montpellier, France, 2018.
38. Kitinoja, L.; Postharvest, T.; Foundation, E. Innovative Approaches to Food Loss and Waste Issues. 2016, 14. Available online: https://www.researchgate.net/profile/Lisa-Kitinoja/publication/303185499_Innovative_Approaches_to_Food_Loss_and_Waste_Issues/links/573881f308ae9f741b2bcbec/Innovative-Approaches-to-Food-Loss-and-Waste-Issues.pdf (accessed on 2025-07-01).
39. Kitinoja, L.; Barrett, D. Extension of Small-Scale Postharvest Horticulture Technologies—A Model Training and Services Center. *Agriculture* **2015**, *5*, 441–455. [CrossRef]
40. Tian, Q.; Yu, Y.; Xiang, Z.; Li, C. Agricultural Technical Education, Interpersonal Trust, and Pesticide Use by Vegetable Farmers in China. *J. Agric. Educ. Ext.* **2021**, *27*, 211–227. [CrossRef]
41. Ali, A.; Xia, C.; Jia, C.; Faisal, M. Investment Profitability and Economic Efficiency of the Drip Irrigation System: Evidence from Egypt. *Irrig. Drain.* **2020**, *69*, 1033–1050. [CrossRef]
42. Manjunatha, A.V.; Anik, A.R.; Speelman, S.; Nuppenau, E.A. Impact of Land Fragmentation, Farm Size, Land Ownership and Crop Diversity on Profit and Efficiency of Irrigated Farms in India. *Land Use Policy* **2013**, *31*, 397–405. [CrossRef]
43. Diab, Y.A.A.; Mousa, M.A.A.; Warnock, D.F.; Hahn, D.E. Opportunities for Producing Table Grapes in Egypt for the Export Market: A Decision Case Study. *Int. Food Agribus. Manag. Rev.* **2009**, *12*, 57–70.
44. Saleem, M.S.; Ejaz, S.; Anjum, M.A.; Nawaz, A.; Naz, S.; Hussain, S.; Ali, S.; Canan, İ. Postharvest Application of Gum Arabic Edible Coating Delays Ripening and Maintains Quality of Persimmon Fruits during Storage. *J. Food Process Preserv.* **2020**, *44*, e14583. [CrossRef]
45. Dsouza, A.; Mishra, A.K.; Webster, S. Vertical Coordination and Post-Harvest Losses: Implications on Food Loss. *Appl. Econ. Perspect. Policy* **2023**, *45*, 460–486. [CrossRef]
46. Wilson, C.L.; Mohammed, M.; Tokala, V.Y. *Postharvest Extension and Capacity Building for the Developing World*, 1st ed.; Mohammed, M., Tokala, V.Y., Eds.; CRC Pres, Taylor & Francis Group: Boca Raton, FL, USA, 2019; ISBN 9781138069282.
47. Zhang, X.; Qing, P.; Yu, X. Short Supply Chain Participation and Market Performance for Vegetable Farmers in China. *Aust. J. Agric. Resour. Econ.* **2019**, *63*, 282–306. [CrossRef]
48. Raut, R.D.; Gardas, B.B.; Kharat, M.; Narkhede, B. Modeling the Drivers of Post-Harvest Losses—MCDM Approach. *Comput. Electron. Agric.* **2018**, *154*, 426–433. [CrossRef]
49. Kitinoja, L. *Use of Cold Chains for Reducing Food Losses in Developing Countries*; The Postharvest Education Foundation (PEF) White Paper No. 13-03; The Postharvest Education Foundation (PEF): La Pine, OR, USA, 2013.
50. Gustavsson, J.; Cederberg, C.; Sonesson, U.; Van Otterdijk, R.; Meybeck, A. *Causes and Prevention of Food Losses and Waste*; FAO: Rome, Italy, 2011.
51. Matera, A.; Altieri, G.; Genovese, F.; Scarano, L.; Genovese, G.; Pinto, P.; Rashvand, M.; Elshafie, H.S.; Ippolito, A.; Mincuzzi, A.; et al. Impact of the Pre-Harvest Biocontrol Agent and Post-Harvest Massive Modified Atmosphere Packaging Application on Organic Table Grape (Cv. 'Allison') Quality during Storage. *Appl. Sci.* **2024**, *14*, 2871. [CrossRef]
52. Filimonau, V.; Ermolaev, V.A. Mitigation of Food Loss and Waste in Primary Production of a Transition Economy via Stakeholder Collaboration: A Perspective of Independent Farmers in Russia. *Sustain. Prod. Consum.* **2021**, *28*, 359–370. [CrossRef]
53. Ali, A.; Xia, C.; Ismaiel, M.; Ouattara, N.B.; Mahmood, I.; Anshiso, D. Analysis of Determinants to Mitigate Food Losses and Waste in the Developing Countries: Empirical Evidence from Egypt. *Mitig. Adapt. Strateg. Glob. Chang.* **2021**, *26*, 23. [CrossRef]

54. Al-Refaie, A.; Al-Tahat, M.; Lepkova, N. Modelling Relationships between Agility, Lean, Resilient, Green Practices in Cold Supply Chains Using Ism Approach. *Technol. Econ. Dev. Econ.* **2020**, *26*, 675–694. [CrossRef]
55. Ndiritu, S.W.; Ruhinduka, R.D. Climate Variability and Post-Harvest Food Loss Abatement Technologies: Evidence from Rural Tanzania. *Stud. Agric. Econ.* **2019**, *121*, 30–40. [CrossRef]
56. Chaboud, G.; Moustier, P. The Role of Diverse Distribution Channels in Reducing Food Loss and Waste: The Case of the Cali Tomato Supply Chain in Colombia. *Food Policy* **2021**, *98*, 101881. [CrossRef]
57. Pham, T.T.; Theuvsen, L.; Otter, V. Determinants of Smallholder Farmers' Marketing Channel Choice: Evidence from the Vietnamese Rice Sector. *Asian Econ. J.* **2019**, *33*, 281–300. [CrossRef]

foods

MDPI

Review

Global Comparison and Future Trends of Major Food Proteins: Can Shellfish Contribute to Sustainable Food Security?

Elena Tamburini [1,*], David Moore [2] and Giuseppe Castaldelli [1]

1 Department of Environmental and Prevention Sciences, University of Ferrara, 44121 Ferrara, Italy; ctg@unife.it
2 School of Biological Sciences, Faculty of Biology, Medicine and Health, University of Manchester, Manchester M13 9PT, UK; david@davidmoore.org.uk
* Correspondence: tme@unife.it

Abstract

Food security and environmental quality related to food production are global issues that need urgent solutions. Proteins are crucial for diets, and demand is growing for innovative and more environmentally sustainable sources of protein, like vegetables, microorganisms, and insects, and lab-grown food that can meet nutritional and environmental goals. This study analyzes a time series to assess the sustainability of different protein sources by evaluating their effects on emissions of greenhouse gases and the use of agricultural land while accounting for the carbon sink potential across the supply chain. The study also explores future trends in global protein sources, emphasizing shellfish as a key to achieving food security from both nutritional and environmental perspectives. By reviewing terrestrial livestock, farmed seafood, vegetal proteins, and alternative sources like insects and cultured cells, the study assesses sustainability, food security potential, and challenges from nutritional, environmental, and consumer viewpoints. We conclude that shellfish aquaculture, particularly oysters, mussels, clams, and scallops, has significant potential in enhancing food security, fostering sustainable protein consumption, reducing land use, and contributing to climate change mitigation by sequestering significant amounts of atmospheric carbon.

Keywords: food security; climate change; shellfish; bivalve mollusks; terrestrial livestock; alternative protein

1. Introduction

The latest United Nations report, which combines the mortality, fertility, and migration scenarios for 195 countries, indicates that the total world population, currently estimated to be 7.6 billion, is forecast to increase to 9.8 billion in 2050 and reach 11.2 billion in 2100 [1]. Moreover, the extent of the middle and rich classes, especially in Asian countries, has undergone rapid expansion, with a considerable impact on consumption patterns and for higher-value processed food [2]. With the rise in living standards, people, especially in urban areas, tend to move up in the food chain toward more energy-dense diets that include more animal products [3].

Aggregated demand for animal proteins is predicted to surpass that of all other foods by 2100, driven primarily by the South and South East Asia and Sub-Saharan Africa regions, whereas consumption in OECD (Organization for the Economic Cooperation and Development) countries will remain virtually unchanged, due to slow population growth and environmental concerns [4]. Similarly, the consumption of vegetal sources of protein, such as pulses, roots, tubers, and cereals, which continue to dominate the protein supply per capita, is expected to increase in the near future [5]. In fact, the animal-to-vegetable-protein ratio, which ranged from 0.29 in Africa to 1.08 in Europe in 2017 [6], is projected to nearly double in Africa by 2030 while remaining constant at 1.09 in Europe [7]. The market for so-called alternative proteins, such as plant-based, insect-based, cultured meat, and mycoprotein, is expected to grow in developed countries at a compounded average rate of 6% per year [8].

The existing challenges to world food production are predicted to be exacerbated by climate change, as increased atmospheric CO_2 concentrations diminish the nutritional value of grains and legumes, impacting especially essential nutrients such as zinc and iron [9]. Meanwhile, the expanding global population is becoming wealthier, and rising incomes are driving greater consumption of resource-intensive foods, particularly those that are sources of proteins [10]. According to published estimates, the world will need to produce more food in the next 35 years than it has in all of human history, given projected population growth and the continued evolution of dietary preferences [11]. However, current dietary protein production and consumption are depleting resources, degrading the environment, and fueling chronic diseases [12]. A critical challenge lies in redesigning the food system to ensure it is healthy, sustainable, and resilient to climate change, thereby supporting the achievement of both the Sustainable Development Goals and the Paris Agreement [13]. These human and environmental impacts light up intense debate on how to shift away from resource-intensive animal-based proteins, while food security remains a complex and pressing challenge [14]. Notably, the environmental impact of agricultural protein production has already surpassed established planetary sustainability thresholds, including those related to climate change, land use change, biogeochemical fluxes, and biodiversity [15]. Agriculture, including livestock, currently contributes approximately 20% of anthropogenic greenhouse gas emissions and accounts for roughly 70% of global freshwater withdrawals, with its long-term sustainability increasingly jeopardized by prevailing trends [16].

On the other hand, there is much concern about future fish consumption based on expected population growth as well as due to the large number of people still suffering from undernourishment. At the same time, there may be limits to the potential for expanded production from wild-catch fisheries and intensive aquaculture because they are both not immune to environmental concerns, like overfishing or the release of fish feed and pharmaceuticals into the water [17]. In this context, bivalve aquaculture (i.e., clam, oyster, mussel, and scallop) could represent a key option to meet food security objectives from both nutritional and environmental perspectives while preserving low-impacts production, supporting land use reduction, and contributing to climate change mitigation via shell biocalcification [18]. The production is rapidly expanding, already accounting for over 15% of the animal protein consumption of approximately 1.5 billion people [19].

A substantial body of literature examines the environmental impact of producing various animal protein sources using the life cycle assessment (LCA) framework, including analyses of their respective nutrient values [20]. Life cycle assessment (LCA) is a suitable methodology for evaluating the potential impacts of protein production across the entire production cycle. LCA includes the extraction and processing of raw materials, manufacturing, transportation and distribution, usage, reuse, and maintenance, as well as recycling and waste management [21].

This critical review aims to investigate future global trends and environmental impacts of the main food protein. We have extended the selection to cover all principal protein sources for which data are available, including terrestrial livestock, farmed seafood, vegetal proteins, and alternative protein sources, like insect protein and cultured meat, with a focus on bivalve mollusks.

2. Materials and Methods

2.1. Data Collection Method

This study is a traditional systematic and critical review based on a mixed approach, merging literature data and data collection from public databases [22].

Our study integrates well with the mixed method, which seeks a more complete understanding through the integration of qualitative and quantitative research. Based on a reasoned and informed review of published research since 2000, the key methodological features have been collected in order to increase understanding. We have extracted live weight mass and raw production data in the time interval of 1990–2020 from the FAO database, the Food and Agriculture Statistics of the FAO (FAOStat Trade Data) [23], and FishStaJ 4.02.06, which collects global fisheries and aquaculture production data, as well as the daily protein supply per capita—the current one and that expected in 2050. Using these official FAO data, we selected the first three species in terms of production mass at the global level for protein categories of terrestrial animal, vegetable, and marine aquaculture protein sources. Specifically, we have selected poultry, pork, and beef as animal meats; whole wheat flour, soybeans, and peas like vegetable crops; milkfish (*Chanos chanos*), Atlantic salmon (*Salmo salar)* and rainbow trout (*Oncorhynchus mykiss*) like farmed fish; and farmed crustaceans, mainly white leg shrimp (*Penaeus vannamei*). We have also included insects and cultured meat as examples of innovative and alternative protein sources. The FAO lists only 79 marine bivalve mollusks farmed out of the approximately 8000 species grouped into four main groups of clams, oysters, scallops, and mussels. The latest available data summary from the FAO (2024) indicates that the production of bivalve mollusks in global aquaculture represents approximately 17 million tons per year. In this study, oysters, clams, and mussels have been included.

A total of 15 sources of food protein were included in this study. World productions data by species, current and forecast, have been obtained from the FAO database, whereas the consumption forecast and nutritional information are from literature (Table 1). All protein content, byproduct/edible weight ratio, Feed Conversion Ratio (FCR), and Nutrient Rich Food ($NRF_{9.3}$) scores were expressed in live or fresh products. The 9.3 is a subscript that considers nine nutrients that are encouraged in the diet-protein (vitamins A, C, and D, calcium, fiber, potassium, protein, magnesium, and iron) and three nutrients that should be limited (added sugars, saturated fat, and sodium) [24,25]. FCR measures nutrient retention in terms of the weight of the feed fed to an animal over its lifetime divided by the live weight gained [26]. The standard deviation for protein content was indicated as that reported in the literature we accessed for each reported value.

Table 1. Production, consumption, and characterization of the principal proteins by source (FAO, 2020; [23]).

Protein Source	World Production (2020)	World Production Increasing Since 1990	World per Capita Consumption (2020)	World per Capita Consumption Increase by 2050	Protein Content per 100 g of Live or Raw Product	By-Products/Edible Weight	FCR	$NRF_{9.3}$	Source
	(Mtons * of live ** or fresh *** weight)	(%)	(g/day)	(%)	(g)	(%)	(kg feed: kg live or fresh weight)		
Poultry	133.4	260	109.2	66	20.9 ± 2.9	29	2.5:1	10.4	[24,27–30]
Pork	109.8	60	112.3	28	22.0 ± 3.6	52	5.0:1	7.0	[24,27,28,30,31]
Beef	72.1	56	68.0	56	20.4 ± 1.7	66	10:1	3.7	[24,27,28,30,32]
Whole wheat flour	752.0	200	180.0	395	14.0 ± 1.0	32	-	32.3	[24,30,33,34]
Soy	353.5	220	123	186	14.3 ± 4.4	22	-	21.0	[24,30,35,36]
Pea	14.6	0	9.3	NA	5.9 ± 1.2	30	-	23.1	[24,30,37,38]
Farmed salmon	2.7	200	35	4.3	21.2 ± 2.0	38	1.35:1	21.8	[24,30,38,39]
Farmed milkfish	1.3	200	NA	NA	21.0 ± 2.3	45	1.35:1	21.8	[24,26,30,40,41]
Farmed rainbow trout	0.8	170	25	NA	19.4 ± 1.7	45	1.35:1	21.8	[24,26,30,42,43]
Farmed crustacean	9.4	370	7.2	200	21.7 ± 2.5	62	2.4:1	3.6	[24,26,30,40]
Insect	0.2	0	NA	10,000	16.9 ± 3.9	0	1.2–2.2:1	2.15–13.14	[28,44,45]
Cultured meat	0.0	0	NA	NA	av. 20% < meat	0	1.2:1	NA	[32,46]
Clam **Mussel** **Oyster**	4.4 1.1 6.0	800 16 400	19.8	44.5	12.8 ± 2.7 11.9 ± 3.1 7.5 ± 1.9	56 35 90	- - -	46.9	[24,30,32]

* Mtons = 10^6 tons. ** live weight = the weight of an animal before it has been slaughtered and prepared as a carcass. *** fresh weight = the weight of vegetable food recently harvested or obtained from the plant, not preserved, is in its raw, unprocessed state.

2.2. Production-Related GHG Emissions, Land Use, and Carbon Capture Capacity

The data presented in Table 2 for the selected protein sources were derived from the scientific literature published since 2000. Several studies and reviews based on LCA were found, especially for terrestrial animal proteins and seafood. The main effort was to verify the system boundaries and functional units, as in some cases, the environmental impact calculations included a cradle-to-plate analysis of the CO_2 eq. emissions that were calculated on the average cooked portion. We focused on the cradle-to-gate studies, up to the end of production (the gate of the production site) and up to 1 kg of live weight for animals and 1 kg of fresh product for vegetables. Unfortunately, it was more difficult to find studies on insects and cultured meat. According to the LCA methodology, one of the impact categories resulting from the analysis is the use of land, which declares the impact of managing and occupying land on the environment for human purposes, including the direct use of land and all indirect contributions related to the product (i.e., fuels, chemicals, feed, and fertilizers) that require land for its production. Carbon sequestration capacity was also derived from the scientific literature. The actual role of shellfish in carbon sequestration has been somewhat debated since the early 2000s, but most of the recent scientific literature is clearly oriented toward considering the shellfish production process as net carbon sequestration [47–51].

Table 2. Cradle-to-gate carbon footprint (expressed as kg CO_2 eq. per kg of live weight or fresh product), land use (measured as m^2 of agricultural land per year and per kg of product), and carbon sequestration capacity for the protein sources selected for the present study.

Protein Source	Carbon Footprint (kg CO_2 eq. */kg live ** or Fresh *** Weight)	Land-Use (m^2/a·kg) ****	Carbon Sequestration Capacity (kg CO_2 */kg live ** or Fresh *** Weight)	Source
Poultry	2.0–6.8	5–8	-	[52,53]
Pork	4.8–11.0	8–15	-	[52]
Beef	24.5–42.6	43.5–420	-	[54–56]
Whole wheat	1.2–1.5	0.9–3.8	0.03–0.32	[57–59]
Soy	0.7–0.8	2.8–6.8	0.10–0.14	[60,61]
Pea	0.5	3.2–4.9	0.12–0.55	[62]
Farmed salmon	3.8–16.7	3.1	-	[63]
Farmed milkfish	3.6–4.7	2.6–4.1	-	[64–66]
Farmed rainbow trout	2.2–3.7	0.8–1.6	-	[66,67]
Farmed crustaceans	3.1–6.8	2.2	-	[66,68,69]
Insect	2.7–4.0	3.6	-	[70,71]
Cultured meat	6.6–25	7.7	-	[72]
Clam Mussel Oyster	0.02–0.7	5.0×10^{-5}	0.15–0.27	[47,73–75]

* kg CO_2 eq. = "kilogram of carbon dioxide equivalent" is a unit of measurement used to compare the global warming impact of different greenhouse gases, expressing them in terms of carbon dioxide. ** live weight = the weight of an animal before it has been slaughtered and prepared as a carcass. *** fresh weight = the weight of vegetable food recently harvested or obtained from the plant, not preserved, is in its raw, unprocessed state. **** (m^2/a·kg) = "square meters per year per kilogram" is a unit of measurement used to quantify the amount of land area (measured in square meters) required to produce a certain amount of a product (measured in kilograms) over a specific time period (measured in years).

3. Results

3.1. Worldwide Protein Consumption by Source

Using the World Bank's classification of countries by income levels [76], Figure 1 depicts the worldwide average daily protein supply per capita. Vegetable protein sources continue to dominate protein consumption (about 60%), with meat providing about 15% of available proteins worldwide, although its contribution to the diet varies widely among countries [77–79].

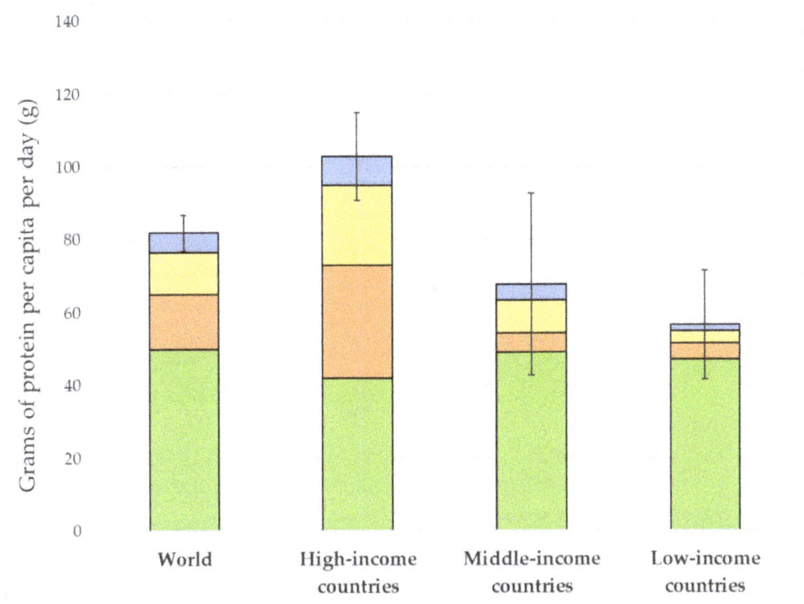

Figure 1. Protein consumption expressed in grams per capita per day by source: plant proteins (green bars), terrestrial animal meat (orange bars), eggs and diary proteins (yellow bars), and seafood (blue bars) (source FAO [23]; reference year, 2020). Based on the World Bank classification, countries have been grouped into high-income (with a gross national income per capita (GNI) above USD 12,375); middle-income (with a GNI between USD 1036 and USD 12,375); and low-income (a GNI of USD 1035 or less).

In countries with a high standard of living, vegetables are reduced approximately to 35% of daily protein, eggs and dairy products contribute to 28%, and animal protein (terrestrial and seafood) cover up to 40% of the overall intake. From low-income to low-middle-income countries, the amount of animal protein rises, gradually closing the gap with high-income countries [80], confirming the strong correlation between per capita income and meat consumption. Seafood currently provides approximately 7% of total proteins consumed by the global population, including capture fisheries, and aquaculture [81].

Although global protein production has dramatically increased significantly since 1990 (Table 1), by approximately 60% for pork and beef and by more than 200% for fish and plants, the forecast for 2050 indicates only a modest increase for all sources, except wheat. In contrast, alternative proteins, like mycoprotein, insect protein, cultured meat, and, as we shall see, potentially shellfish protein, have tremendous growth potential for the next few decades.

In 2030, the per capita consumption of insects and insect-based products (meals and powders) is expected to grow by about 10,000% worldwide, reaching about 3 mln tons in

2030, whereas their consumption as a source of nutrition is almost exclusively restricted to Africa, South America, and South-East Asia at present [82]. Recent estimates indicate that over 2000 insect species are now considered edible, primarily crickets, mealworms, beetles, and caterpillars [83]. Similarly, cultured meat is now at an early stage of development, but it is expected to reach 0.4–2.1 mtons of production by 2030 and gradually replace approximately 30–40% of traditional meat consumption [84].

As shown in Table 1, all animal sources have similar protein content but very different by-products: edible weight ratios (from 29% of poultry to 66% of beef). Beef also has the worst FCR (10%) and the lowest feed nutritional score, expressed as $NRF_{9.3}$ = 3.7. Meanwhile, insects and cultured meat have the highest FCR and zero waste but a lower nutritional score than plant sources.

The dietary protein content of raw, unprocessed beef, pork, and poultry products examined in this review ranged from 20.4 ± 1.7 g/100 g to 22.0 ± 3.6 g/100 g. Similarly, raw farmed fish products exhibited comparable protein levels, with rainbow trout providing 19.4 ± 1.7 g/100 g and salmon offering 21.2 ± 2.0 g/100 g. Insects (16.9 ± 3.9 g/100 g) and cultured meat demonstrated protein contents nearly equivalent to that of conventional meat and seafood products. In contrast, plant-based sources displayed significantly lower protein concentrations, ranging from peas (5.9 ± 1.2 g/100 g) to soy and wheat flour (14.3 ± 4.4 g/100 g and 14.0 ± 1.0 g/100 g, respectively). Moreover, from a nutritional standpoint, plant proteins often struggle to provide the diverse range of essential amino acids found in meat products [85]. Shellfish have a protein content comparable with that of plant proteins, the highest values of by-products (primarily shells), and the highest $NRF_{9.3}$ compared with other proteins. A higher $NRF_{9.3}$ corresponds to low energy-dense and healthier foods. Shellfish and seafood are lower in saturated fat than red meat, pork, and poultry, but they may not contain as many mineral nutrients as red meat, like iron and zinc. The unique dietary value of molluscan shellfish is them being low in calories but high in nutrients, especially minerals and vitamins that are not always found in other meats [86]. Moreover, they are rich in non-cholesterol sterols, low in saturated fatty acids, and rich in omega-3-fatty acids, which are essential for maintaining overall human health from reducing blood pressure to lowering risks of depression and anxiety [87]. In terms of efficiency, seafood species are very efficient converters of feed into protein, far more efficient than most terrestrial livestock systems. For instance, poultry converts about 18% of their consumed food and pigs convert about 13%, as compared with 30% in the case of fish [88] and up to 50% for shellfish [62]. As an example, based on the $NRF_{9.3}$ values and average protein content of meats, it is estimated that 100 g of beef is required to provide the same nutritional score as 21 g of oyster meat, 13 g of mussel meat, or 12.5 g of clam meat.

3.2. Carbon Footprint of Protein Sources and Land Use

According to the selected LCA studies, referring to 1 kg of live/fresh weight and cradle-to-gate boundaries, the carbon footprint and land use of the different protein sources are presented in Table 2, which indicates that beef has the largest impact, both in terms of carbon dioxide equivalent (CO_2 eq) emissions per 1 kg of live weight (24.5–42.6 CO_2 eq/kg live weight) due to rumen fermentation and land use (43.5–420 $m^2/a{\cdot}kg$). Regardless of production systems, which range from highly intensive to highly extensive, or the choice between localized organic supply chains and global distribution, beef production requires substantial land use. Some studies estimate land consumption at up to 420 m^2 per year per kilogram of live beef [89]. Additionally, it has been suggested that producing feed for Europe's livestock demands cropland equivalent to seven times the size of the European Union [90]. Our data demonstrate a wide range of product variation. Pork and poultry have a medium-to-high carbon footprint and land consumption (4.8–11.0 CO_2 eq/kg live weight and 8–15 $m^2/a \cdot kg$, respectively). In both cases, the variety may be the result of the availability of conventional production systems or free-range animals, which significantly impact both categories. For pigs, a significant portion of the emissions originate from nitrous oxide emissions released during feed production [91]. For farmed seafood, the CO_2 eq. values are primarily driven by aquaculture methods and equipment and the distance to aquaculture sites, which in turn influence fuel use and overall energy consumption.

The allocation of terrestrial land for seafood production has been quantified as agricultural land utilized for cultivating the plant-based components of feed.

Intensive aquaculture practices for farmed salmon, rainbow trout, and milkfish require high-protein nutrients, additives, antibiotics, and drugs to improve the health and yield of farmed fish and prevent disease [92]. The production and use of these factors have a relatively small impact on the carbon footprint of seafood products but a significant impact on other impact categories of the LCA analysis, like ecosystem toxicity and human toxicity, that were not considered in this study. Plant products have a lower carbon footprint than animal proteins, as expected. Meanwhile, agricultural practices, such as the use of machinery and equipment and the production and use of fertilizers and pesticides, contribute significantly to their environmental impact [93]. According to Nemecek et al. [94], the land use parameter for agricultural products in LCA studies includes not only land use but also all contributions indirectly associated with cultivation. Cultivating bivalve mollusks as feed for farmed finfish could counteract all of the negative aspects of current practices in aquaculture feed production [95].

Although the uncertainty ranges are still large, some early studies have reported that in vitro biomass cultivation may require fewer inputs and land use than livestock production [96]. The impacts of carbon footprint are due to biomass cultivation, bioreactor cleaning, and facilities, whereas the contribution to land use is due to the use of starch and soy hydrolysate, which are considered the main ingredients for the production of cultured meat [97]. Similarly, for insects, LCAs have only been released for mealworms, house crickets, black soldier flies, and houseflies [98]. Feed production is again the main driver of carbon footprints and land use. Total CO_2 eq. emissions are lower than conventional livestock's, but energy requirements are still high because insects require relatively high temperatures during rearing [99].

Specifically, shellfish have the lowest carbon footprint [75]. The fact that they are filter feeders contributes to their relatively low environmental impact. The main contributors to CO_2 eq emissions are plastic materials for farming and harvesting, fuel consumption, boat use and maintenance, and, in the case of mussels and oysters, farming facilities, such as off-shore long-line farms [100]. Bivalve mollusks also have a negligible land use contribution, because they occupy land only for land-based building and ancillary facilities and do not need soil to produce external feed [19]. Moreover, their contribution to the occupation of a marine area is also negligible, since it has been calculated in the range 1–2 m^2/a crop equivalent per kilograms of farmed bivalves [74]. Thus, bivalve aquaculture is generally characterized by a low energy input and established technologies, which lead to a low carbon footprint related to production [101,102]. A more detailed analysis of the carbon footprint methods and calculations related to the protein source production mentioned in this study can be found in the cited literature.

3.3. Carbon Capture Capacity

Agriculture is not only a major source of carbon but could also be a sink due to the functions of soils and crops [103]. Green plants absorb CO_2 and fix it in organic substances that can be retained in the soil or released back into the atmosphere through photosynthesis, which brings carbon into terrestrial ecosystems, including agriculture. The factors that influence carbon sources and sinks in agroecosystems are highly heterogeneous and complex, depending on climate patterns, soil types and properties, and management practices [104]. The issue of carbon sequestration and reduction through agriculture has received considerable attention, and it is beyond the scope of this study to look at it. Here, we limit ourselves to pointing out that crops can contribute to carbon storage in terrestrial ecosystems. Peas have a significant carbon sequestration capacity, similar to the amount of CO_2 eq. emitted for their production. The wide range of wheat depends on whether it is considered a winter cover crop [105,106]. Furthermore, photosynthetic organisms are only net carbon absorbers when photosynthesizing. At night, plants emit respiratory CO_2 like every other aerobic organism, and when the plant dies, carbon sequestered in its biomass is digested by a multitude of bacteria, fungi, and detritivores that release all the biomass carbon back to the atmosphere as respiratory CO_2 [107].

Terrestrial animals and seafood, as well as insects, are emitting organisms, both in terms of physiological metabolism and production process; hence, their contribution to this parameter is more doubtful. Similarly, there is no point in defining the parameter for cultured meat. On the contrary, due to the natural biocalcification process that leads to calcium carbonate formation, bivalves are the only animal protein source that have the potential to act as a carbon sink during life, leaving a legacy of deposited limestone when they die. In fact, their shell is principally composed of mineral calcium carbonate ($CaCO_3$) formed during a biological process called biogenic calcification or biomineralization. The stoichiometry of the chemical reaction that occurs during biocalcification provides that for every two CO_2 molecules captured from seawater, one is returned to the marine carbon cycle, and one is permanently stored in the shell as insoluble $CaCO_3$ [108]. Bivalve mollusks are farmed for harvest and human consumption; similarly, shells are permanently removed from the water at the end of the production cycle, resulting in net sequestration. Clams and oysters have a high carbon sequestration capacity due to their higher shell-to-flesh ratio than mussels (about 60% for clams and 75% for oysters vs. 35% of mussels).

4. Discussion

Based on the environmental Kuznets curve [109], developed countries show greater environmental awareness, while people in developing countries still aspire to Western consumption patterns. Bennett's law demonstrates that rising incomes lead to shifts from starchy staples toward refined grains, fruits, vegetables, meat, and dairy products [110]. These models explain the projected increase in livestock product demand by mid-century. Managing this requires modeling sustainable protein sources for human consumption to avoid exceeding planetary boundaries. Adequate dietary protein is essential for meeting the body's indispensable amino acid requirements, including phenylalanine, valine, threonine, tryptophan, isoleucine, methionine, histidine, leucine, and lysine [111]. Animal-derived proteins like meat, seafood, eggs, and dairy are considered "high-quality" due to their essential amino acid content [112], while plant-based proteins often contain some essential amino acids but are limited in others [113]. The issue of global protein supply extends beyond substituting animal proteins with plant sources. While this might reduce immediate impact, increased plant protein production requires new agricultural land, which is scarce and would impact forests. Intensive agriculture has become unsustainable due to fertilizer misuse. In 2018, synthetic nitrogen fertilizers generated 1.13 Gt CO_2 eq. emissions, comprising 10.6% of agricultural and 2.1% of global greenhouse emissions [114]. Expanding arable land increases CO_2 emissions and reduces biodiversity. While pulses and cereals can supplement animal proteins, they cannot fully replace them, as plant proteins alone are incomplete in amino acid composition for human health despite their vitamins and fiber content [115].

Insect production is touted as the future, offering low environmental impacts, land requirements, and feed/water inputs [116]. However, problems exist like high energy needs for heating, pest control, and maintaining hygienic conditions. Cultural barriers in Western societies make large-scale adoption difficult, except as hidden ingredients. Cultured meat allows for controlling the nutritional content by adjusting the production medium, such as increasing polyunsaturated fatty acids or micronutrients [117]. While requiring less land than livestock, its environmental benefits remain controversial. Beyond high costs, consumer acceptance and ethical–religious concerns need consideration when comparing current meat production. The production of farmed finfish and crustaceans has grown exponentially, severely impacting marine ecosystems through feed and chemical use [118]. Plant-derived feed affects terrestrial land use, while seafood fed with wild-caught fish worsens marine resource overexploitation [119]. Factory ships can destroy indigenous fishing industries. Aquaculture effluents containing fertilizers, drugs, and additives contribute to pollution and ecosystem toxicity in coastal regions. The high nutritional value of seafood and crustaceans makes them vital for human diets. Improving FCR and selecting appropriate feeds could reduce environmental impacts [120].

The data show the inadequacy of protein supply solutions ensuring food security and environmental protection. Bivalves offer a promising solution, though cultural and economic factors hinder widespread adoption. Limited knowledge, funding, infrastructure, and consumer acceptance remain barriers to expanding bivalve aquaculture. However, over 1,500,000 km^2 of undeveloped coastline is suitable for bivalve farming [121]. China's success since the 1950s, producing 85% of the global supply, demonstrates bivalve aquaculture's viability [122]. Expanding this model across Asia, Africa, the Americas, and Oceania could benefit health and sustainability.

According to Morales-Nin et al. [123], bivalve production's carbon footprint is 20 times lower than that of beef and 10 times lower than that of chicken and pork per serving. Substituting 10% of beef or chicken consumption with oysters could avoid 360 million and 120 million tons of CO_2-equivalent emissions annually. Carbon sequestration in shells could reduce atmospheric CO_2 by 2–3 mtons yearly, requiring 1.4–2.2 million hectares using terrestrial carbon capture systems [47]. While bivalves' carbon seizure is small compared to global emissions, it provides permanent sequestration, unlike reforestation or conservative agriculture [124]. Even advanced engineering solutions have not proven long-term storage capability, yet fossiliferous limestone strata were produced by calcifying organisms millions of years ago [125].

The key point worth highlighting is that the carbon sequestration capacity of shellfish can be considered a side-benefit at zero cost because it occurs while the mollusk is being cultivated to provide human food. Bivalve aquaculture could have an effective and significant role in fighting against climate change, especially in coastal countries, and help reach Goal #13 without creating environmental and economic issues. As global food demand continues to rise in the coming decades, aquaculture is likely to expand, driven by its human health benefits and increasing consumer preference for seafood [126]. Bivalve aquaculture has a minimal contribution to food production-related emissions and can be further enhanced through responsible practices. Ultimately, the advancement of aquaculture is a crucial strategy for meeting growing food demands while ensuring food security within planetary boundaries [127]. Recent estimates indicate that emissions from bivalve production account for only 7.6% of the average emissions associated with terrestrial protein sources [19]. As a result, bivalve aquaculture is now widely recognized as a sustainable, climate-friendly means of producing nutrient-rich protein for human consumption. Additionally, beyond its role in food production, bivalve farming contributes as a net carbon sink, with current

Manila clam production alone sequestering approximately 1 million tons of carbon per year, in support of global efforts toward UN Sustainable Development Goal #13.

A significant issue with bivalve consumption, particularly oysters, is the waste generated relative to the meat yield. Several options for waste shell valorization have been reported, showing increasing interest [113]. One possibility for using bivalve shells, combining carbon sequestration with ecosystem benefits, is creating coastal defense structures and submerged reefs [128]. The use of shells would promote bivalve larvae settlement, with additional carbon storage potential [129].

5. Conclusions

The growing need to ensure food security for an expanding global population, coupled with promoting the sustainable use of natural resources, has heightened interest in bivalve aquaculture. Unlike intensive fish farming, bivalve aquaculture is a zero-emission system that relies on natural phytoplankton, requiring minimal intervention with no additives or antibiotics that could affect its carbon footprint. The production of bivalves plays a significant economic role in coastal regions worldwide, particularly in developing countries. Notably, bivalve farming is the only food production method maintaining a positive net carbon balance in terms of sequestration capacity. Given the global demand for nutrient-rich food sources that do not strain environmental resources, coastal areas remain available for sustainable development. However, challenges must be addressed to unlock bivalve aquaculture's full potential. Harnessing innovation and technologies can help overcome these challenges, expanding bivalve aquaculture as a viable economic activity that provides sustainable income while feeding nearly a billion people in developing regions. By linking bivalve farming to environmental goals, this study supports climate-friendly aquaculture practices that generate ecological and economic benefits. Findings show that bivalve farming offers advantages in carbon sequestration and plays a crucial role in coastal marine ecosystems' carbon cycles.

Bivalves aquaculture could help achieve 5 of the 17 UN's SDGs of Agenda 2030: (1) SDG Goal #1 (no poverty), as sustainable bivalves aquaculture creates income opportunities for coastal communities; (2) SDG Goal #2 (zero hunger), as sustainable aquaculture ensures food security; (3) SDG Goal #3 (promoting health and wellbeing), due to nutritional benefits; (4) SDG Goal #12 (responsible consumption, and production), as it follows natural growing cycles without antibiotics and has lower environmental impacts than other foods; and (5) SDG Goal #14 (life below water), as bivalves aquaculture can reduce pressure on marine ecosystems while providing environmental benefits.

Author Contributions: Conceptualization and writing—review and editing, E.T.; validation and editing, D.M.; supervision, G.C. All authors have read and agreed to the published version of the manuscript.

Funding: This research received no external funding.

Institutional Review Board Statement: Not applicable.

Informed Consent Statement: Not applicable.

Data Availability Statement: The original contributions presented in the study are included in the article. Further inquiries can be directed to the corresponding author.

Conflicts of Interest: The authors declare no conflicts of interest.

References

1. Kc, S.; Dhakad, M.; Potančoková, M.; Adhikari, S.; Yildiz, D.; Mamolo, M.; Sobotka, T.; Zeman, K.; Abel, G.; Lutz, W.; et al. Updating the Shared Socioeconomic Pathways (SSPs) Global Population and Human Capital Projections. Available online: https://pure.iiasa.ac.at/id/eprint/19487/ (accessed on 13 June 2025).
2. Wei, L.; Li, C.; Wang, J.; Wang, X.; Wang, Z.; Cui, C.; Peng, S.; Liu, Y.; Yu, S.; Wang, L.; et al. Rising Middle and Rich Classes Drove China's Carbon Emissions. *Resour. Conserv. Recycl.* **2020**, *159*, 104839. [CrossRef]
3. Colgrave, M.L.; Dominik, S.; Tobin, A.B.; Stockmann, R.; Simon, C.; Howitt, C.A.; Belobrajdic, D.P.; Paull, C.; Vanhercke, T. Perspectives on Future Protein Production. *J. Agric. Food Chem.* **2021**, *69*, 15076–15083. [CrossRef] [PubMed]
4. Becker, S.; Fanzo, J. Population and Food Systems: What Does the Future Hold? *Popul. Environ.* **2023**, *45*, 20. [CrossRef]
5. Steenson, S.; Buttriss, J.L. The Challenges of Defining a Healthy and 'Sustainable' Diet. *Nutr. Bull.* **2020**, *45*, 206–222. [CrossRef]
6. Henchion, M.; Moloney, A.P.; Hyland, J.; Zimmermann, J.; McCarthy, S. Review: Trends for Meat, Milk and Egg Consumption for the next Decades and the Role Played by Livestock Systems in the Global Production of Proteins. *Animal* **2021**, *15*, 100287. [CrossRef]
7. Frezal, C.; Nenert, C.; Gay, H. *Meat Protein Alternatives: Opportunities and Challenges for Food Systems' Transformation*; OECD Publishing: Paris, France, 2022.
8. Malila, Y.; Owolabi, I.O.; Chotanaphuti, T.; Sakdibhornssup, N.; Elliott, C.T.; Visessanguan, W.; Karoonuthaisiri, N.; Petchkongkaew, A. Current Challenges of Alternative Proteins as Future Foods. *npj Sci. Food* **2024**, *8*, 53. [CrossRef]
9. Ziska, L.H. Rising Carbon Dioxide and Global Nutrition: Evidence and Action Needed. *Plants* **2022**, *11*, 1000. [CrossRef]
10. Mylan, J.; Andrews, J.; Maye, D. The Big Business of Sustainable Food Production and Consumption: Exploring the Transition to Alternative Proteins. *Proc. Natl. Acad. Sci. USA* **2023**, *120*, e2207782120. [CrossRef]
11. Hemathilake, D.M.K.S.; Gunathilake, D.M.C.C. Chapter 31—Agricultural Productivity and Food Supply to Meet Increased Demands. In *Future Foods*; Bhat, R., Ed.; Academic Press: Cambridge, MA, USA, 2022; pp. 539–553. ISBN 978-0-323-91001-9.
12. Su, C.-W.; Khan, K.; Umar, M.; Zhang, W. Does Renewable Energy Redefine Geopolitical Risks? *Energy Policy* **2021**, *158*, 112566. [CrossRef]
13. Iacobuţă, G.I.; Brandi, C.; Dzebo, A.; Elizalde Duron, S.D. Aligning Climate and Sustainable Development Finance through an SDG Lens. The Role of Development Assistance in Implementing the Paris Agreement. *Glob. Environ. Change* **2022**, *74*, 102509. [CrossRef]
14. Jeger, M.; Beresford, R.; Bock, C.; Brown, N.; Fox, A.; Newton, A.; Vicent, A.; Xu, X.; Yuen, J. Global Challenges Facing Plant Pathology: Multidisciplinary Approaches to Meet the Food Security and Environmental Challenges in the Mid-Twenty-First Century. *CABI Agric. Biosci.* **2021**, *2*, 20. [CrossRef]
15. Gil, M.; Rudy, M.; Duma-Kocan, P.; Stanisławczyk, R.; Krajewska, A.; Dziki, D.; Hassoon, W.H. Sustainability of Alternatives to Animal Protein Sources, a Comprehensive Review. *Sustainability* **2024**, *16*, 7701. [CrossRef]
16. Jain, S.; Srivastava, A.; Khadke, L.; Chatterjee, U.; Elbeltagi, A. Global-Scale Water Security and Desertification Management amidst Climate Change. *Environ. Sci. Pollut. Res.* **2024**, *31*, 58720–58744. [CrossRef]
17. Arshad, N.; Samat, N.; Lee, L.K. Insight Into the Relation Between Nutritional Benefits of Aquaculture Products and Its Consumption Hazards: A Global Viewpoint. *Front. Mar. Sci.* **2022**, *9*, 925463. [CrossRef]
18. Mesquita, A.F.; Gonçalves, F.J.M.; Gonçalves, A.M.M.; Mesquita, A.F.; Gonçalves, F.J.M.; Gonçalves, A.M.M. Marine Bivalves' Ecological Roles and Humans-Environmental Interactions to Achieve Sustainable Aquatic Ecosystems. In *Marine Ecosystems—Biodiversity, Ecosystem Services and Human Impacts*; IntechOpen: London, UK, 2023; ISBN 978-1-83968-460-9.
19. Zhang, H.; Cheong, K.-L.; Tan, K. Bivalves as Climate-Friendly High Quality Animal Protein: A Comprehensive Review. *Food Secur.* **2025**, *17*, 739–748. [CrossRef]
20. Cellura, M.; Cusenza, M.A.; Longo, S.; Luu, L.Q.; Skurk, T. Life Cycle Environmental Impacts and Health Effects of Protein-Rich Food as Meat Alternatives: A Review. *Sustainability* **2022**, *14*, 979. [CrossRef]
21. Detzel, A.; Krüger, M.; Busch, M.; Blanco-Gutiérrez, I.; Varela, C.; Manners, R.; Bez, J.; Zannini, E. Life Cycle Assessment of Animal-Based Foods and Plant-Based Protein-Rich Alternatives: An Environmental Perspective. *J. Sci. Food Agric.* **2022**, *102*, 5098–5110. [CrossRef]
22. Guetterman, T.C.; Fetters, M.D. Two Methodological Approaches to the Integration of Mixed Methods and Case Study Designs: A Systematic Review. *Am. Behav. Sci.* **2018**, *62*, 900–918. [CrossRef]
23. Food Balance Sheets 2010–2022. Global, Regional and Country Trends. Available online: https://www.fao.org/statistics/highlights-archive/highlights-detail/food-balance-sheets-2010-2022-global-regional-and-country-trends/en (accessed on 13 June 2025).
24. Drewnowski, A. The Nutrient Rich Foods Index Helps to Identify Healthy, Affordable Foods. *Am. J. Clin. Nutr.* **2010**, *91*, 1095S–1101S. [CrossRef]
25. Drewnowski, A.; Burton-Freeman, B. A New Category-Specific Nutrient Rich Food (NRF9f.3) Score Adds Flavonoids to Assess Nutrient Density of Fruit. *Food Funct.* **2020**, *11*, 123–130. [CrossRef]

26. Fry, J.P.; Mailloux, N.A.; Love, D.C.; Milli, M.C.; Cao, L. Feed Conversion Efficiency in Aquaculture: Do We Measure It Correctly? *Environ. Res. Lett.* **2018**, *13*, 024017. [CrossRef]
27. Jayathilakan, K.; Sultana, K.; Radhakrishna, K.; Bawa, A.S. Utilization of Byproducts and Waste Materials from Meat, Poultry and Fish Processing Industries: A Review. *J. Food Sci. Technol.* **2012**, *49*, 278–293. [CrossRef] [PubMed]
28. Lange, K.W.; Nakamura, Y. Edible Insects as Future Food: Chances and Challenges. *J. Future Foods* **2021**, *1*, 38–46. [CrossRef]
29. Wold, J.P.; Måge, I.; Løvland, A.; Sanden, K.W.; Ofstad, R. Near-Infrared Spectroscopy Detects Woody Breast Syndrome in Chicken Fillets by the Markers Protein Content and Degree of Water Binding. *Poult. Sci.* **2019**, *98*, 480–490. [CrossRef]
30. Bianchi, M.; Hallström, E.; Parker, R.W.R.; Mifflin, K.; Tyedmers, P.; Ziegler, F. Assessing Seafood Nutritional Diversity Together with Climate Impacts Informs More Comprehensive Dietary Advice. *Commun. Earth Environ.* **2022**, *3*, 188. [CrossRef]
31. Mir, N.A.; Rafiq, A.; Kumar, F.; Singh, V.; Shukla, V. Determinants of Broiler Chicken Meat Quality and Factors Affecting Them: A Review. *J. Food Sci. Technol.* **2017**, *54*, 2997–3009. [CrossRef]
32. Bohrer, B.M. Review: Nutrient Density and Nutritional Value of Meat Products and Non-Meat Foods High in Protein. *Trends Food Sci. Technol.* **2017**, *65*, 103–112. [CrossRef]
33. Cesevičienė, J.; Gorash, A.; Liatukas, Ž.; Armonienė, R.; Ruzgas, V.; Statkevičiūtė, G.; Jaškūnė, K.; Brazauskas, G. Grain Yield Performance and Quality Characteristics of Waxy and Non-Waxy Winter Wheat Cultivars under High and Low-Input Farming Systems. *Plants* **2022**, *11*, 882. [CrossRef]
34. Erenstein, O.; Jaleta, M.; Mottaleb, K.A.; Sonder, K.; Donovan, J.; Braun, H.-J. Global Trends in Wheat Production, Consumption and Trade. In *Wheat Improvement: Food Security in a Changing Climate*; Reynolds, M.P., Braun, H.-J., Eds.; Springer International Publishing: Cham, Switzerland, 2022; pp. 47–66. ISBN 978-3-030-90673-3.
35. Ritchie, H.; Roser, M.; Rosado, P. CO_2 and Greenhouse Gas Emissions. Our World Data 2020. Available online: https://ourworldindata.org/co2-and-greenhouse-gas-emissions (accessed on 12 December 2024).
36. Nile, S.H.; Venkidasamy, B.; Samynathan, R.; Nile, A.; Shao, K.; Chen, T.; Sun, M.; Khan, M.U.; Dutta, N.; Thiruvengadam, M.; et al. Soybean Processing Wastes: Novel Insights on Their Production, Extraction of Isoflavones, and Their Therapeutic Properties. *J. Agric. Food Chem.* **2022**, *70*, 6849–6863. [CrossRef]
37. Holliday, M.C.; Parsons, D.R.; Zein, S.H. Agricultural Pea Waste as a Low-Cost Pollutant Biosorbent for Methylene Blue Removal: Adsorption Kinetics, Isotherm And Thermodynamic Studies. *Biomass Convers. Biorefinery* **2022**, *14*, 6671–6685. [CrossRef]
38. Rocker, M.; Mock, T.; Turchini, G.; Francis, D. The Judicious Use of Finite Marine Resources Can Sustain Atlantic Salmon (*Salmo salar*) Aquaculture to 2100 and Beyond. *Nat. Food* **2022**, *3*, 644–649. [CrossRef] [PubMed]
39. Fouda, T. Using Green Cold Pressing to Produce High Quality Fish Oil From Industrial Salmon Waste. *J. Zool. Res.* **2020**, *2*, 14–19. [CrossRef]
40. Naylor, R.L.; Hardy, R.W.; Buschmann, A.H.; Bush, S.R.; Cao, L.; Klinger, D.H.; Little, D.C.; Lubchenco, J.; Shumway, S.E.; Troell, M. A 20-Year Retrospective Review of Global Aquaculture. *Nature* **2021**, *591*, 551–563. [CrossRef]
41. Kristiansen, T.S.; Madaro, A.; Stien, L.H.; Bracke, M.B.M.; Noble, C. Chapter 6—Theoretical Basis and Principles for Welfare Assessment of Farmed Fish. In *Fish Physiology*; Benfey, T.J., Farrell, A.P., Brauner, C.J., Eds.; Aquaculture; Academic Press: Cambridge, MA, USA, 2020; Volume 38, pp. 193–236.
42. Skałecki, P.; Florek, M.; Litwińczuk, A.; Staszowska, A.; Kaliniak, A. The Nutritional Value and Chemical Composition of Muscle Tissue of Carp (*Cyprinus carpio* L.) and Rainbow Trout (*Oncorhynchus mykiss* Walb.) Obtained from Fish Farms in the Lublin Region. *Sci. Ann. Pol. Soc. Anim. Prod.* **2013**, *9*, 57–62.
43. Arenna; Goddard, E.; Muringai, V. Consumer Purchase Intentions for Pork with Enhanced Carnosine–A Functional Food. *Can. J. Agric. Econ./Rev. Can. D'agroeconomie* **2019**, *67*, 151–169. [CrossRef]
44. Weru, J.; Chege, P.; Wanjoya, A.; Kinyuru, J. Comparison of Healthfulness of Conventional Meats and Edible Insects in Sub-Saharan Africa Using Three Nutrient Profiling Models. *Bull. Natl. Res. Cent.* **2022**, *46*, 43. [CrossRef]
45. Payne, C.L.R.; Scarborough, P.; Rayner, M.; Nonaka, K. A Systematic Review of Nutrient Composition Data Available for Twelve Commercially Available Edible Insects, and Comparison with Reference Values. *Trends Food Sci. Technol.* **2016**, *47*, 69–77. [CrossRef]
46. Chriki, S.; Hocquette, J.-F. The Myth of Cultured Meat: A Review. *Front. Nutr.* **2020**, *7*, 7. [CrossRef]
47. Tamburini, E.; Turolla, E.; Lanzoni, M.; Moore, D.; Castaldelli, G. Manila Clam and Mediterranean Mussel Aquaculture Is Sustainable and a Net Carbon Sink. *Sci. Total Environ.* **2022**, *848*, 157508. [CrossRef]
48. Filgueira, R.; Guyondet, T.; Comeau, L.A.; Tremblay, R. Bivalve Aquaculture-Environment Interactions in the Context of Climate Change. *Glob. Change Biol.* **2016**, *22*, 3901–3913. [CrossRef]
49. Jansen, H.; van den Bogaart, L. *Blue Carbon by Marine Bivalves: Perspective of Carbon Sequestration by Cultured and Wild Bivalve Stocks in the Dutch Coastal Areas*; Wageningen Marine Research: Yerseke, The Netherlands, 2020. [CrossRef]
50. Ahmed, N.; Thompson, S.; Glaser, M. Global Aquaculture Productivity, Environmental Sustainability, and Climate Change Adaptability. *Environ. Manag.* **2019**, *63*, 159–172. [CrossRef] [PubMed]

51. Moore, D.; Heilweck, M.; Petros, P. *Aquaculture: Ocean Blue Carbon Meets UN-SDGS*; Sustainable Development Goals Series; Springer International Publishing: Cham, Switzerland, 2022; ISBN 978-3-030-94845-0.
52. Nijdam, D.; Rood, T.; Westhoek, H. The Price of Protein: Review of Land Use and Carbon Footprints from Life Cycle Assessments of Animal Food Products and Their Substitutes. *Food Policy* **2012**, *37*, 760–770. [CrossRef]
53. Tetteh, H.; Bala, A.; Fullana-i-Palmer, P.; Balcells, M.; Margallo, M.; Aldaco, R.; Puig, R. Carbon Footprint: The Case of Four Chicken Meat Products Sold on the Spanish Market. *Foods* **2022**, *11*, 3712. [CrossRef] [PubMed]
54. Vitali, A.; Grossi, G.; Martino, G.; Bernabucci, U.; Nardone, A.; Lacetera, N. Carbon Footprint of Organic Beef Meat from Farm to Fork: A Case Study of Short Supply Chain. *J. Sci. Food Agric.* **2018**, *98*, 5518–5524. [CrossRef]
55. Buratti, C.; Fantozzi, F.; Barbanera, M.; Lascaro, E.; Chiorri, M.; Cecchini, L. Carbon Footprint of Conventional and Organic Beef Production Systems: An Italian Case Study. *Sci. Total Environ.* **2017**, *576*, 129–137. [CrossRef]
56. Nguyen, T.L.T.; Hermansen, J.E.; Mogensen, L. Environmental Consequences of Different Beef Production Systems in the EU. *J. Clean. Prod.* **2010**, *18*, 756–766. [CrossRef]
57. Chiriacò, M.V.; Grossi, G.; Castaldi, S.; Valentini, R. The Contribution to Climate Change of the Organic versus Conventional Wheat Farming: A Case Study on the Carbon Footprint of Wholemeal Bread Production in Italy. *J. Clean. Prod.* **2017**, *153*, 309–319. [CrossRef]
58. Brentrup, F.; Küsters, J.; Lammel, J.; Barraclough, P.; Kuhlmann, H. Environmental Impact Assessment of Agricultural Production Systems Using the Life Cycle Assessment (LCA) Methodology II. The Application to N Fertilizer Use in Winter Wheat Production Systems. *Eur. J. Agron.* **2004**, *20*, 265–279. [CrossRef]
59. Gan, Y.; Liang, C.; Chai, Q.; Lemke, R.L.; Campbell, C.A.; Zentner, R.P. Improving Farming Practices Reduces the Carbon Footprint of Spring Wheat Production. *Nat. Commun.* **2014**, *5*, 5012. [CrossRef]
60. Tidåker, P.; Karlsson Potter, H.; Carlsson, G.; Röös, E. Towards Sustainable Consumption of Legumes: How Origin, Processing and Transport Affect the Environmental Impact of Pulses. *Sustain. Prod. Consum.* **2021**, *27*, 496–508. [CrossRef]
61. Bernacchi, C.J.; Hollinger, S.E.; Meyers, T. The Conversion of the Corn/Soybean Ecosystem to No-till Agriculture May Result in a Carbon Sink. *Glob. Change Biol.* **2005**, *11*, 1867–1872. [CrossRef]
62. Wang, J.-K. Conceptual Design of a Microalgae-Based Recirculating Oyster and Shrimp System. *Aquac. Eng.* **2003**, *28*, 37–46. [CrossRef]
63. Liu, Y.; Rosten, T.W.; Henriksen, K.; Hognes, E.S.; Summerfelt, S.; Vinci, B. Comparative Economic Performance and Carbon Footprint of Two Farming Models for Producing Atlantic Salmon (*Salmo salar*): Land-Based Closed Containment System in Freshwater and Open Net Pen in Seawater. *Aquac. Eng.* **2016**, *71*, 1–12. [CrossRef]
64. Capuno, R.B.A. Environmental Performance of Milkfish Fry Production in Dagupan City, Pangasinan, Philippines. Master's Thesis, University of the Philippines Los Baños, Los Baños, Philippines, 2015.
65. Susanto, N.; Prastawa, H.; Sembiring, N.V.; Ulkhaq, M.M. Life Cycle Assessment of High Pressure-Cooked Smoked Milkfish Production: A Case Study in Semarang, Indonesia. *EnvironmentAsia* **2021**, *14*, 124–135.
66. Boyd, C.E.; D'Abramo, L.R.; Glencross, B.D.; Huyben, D.C.; Juarez, L.M.; Lockwood, G.S.; McNevin, A.A.; Tacon, A.G.; Teletchea, F.; Tomasso, J.R., Jr. Achieving Sustainable Aquaculture: Historical and Current Perspectives and Future Needs and Challenges. *J. World Aquac. Soc.* **2020**, *51*, 578–633. [CrossRef]
67. Samuel-Fitwi, B.; Nagel, F.; Meyer, S.; Schroeder, J.P.; Schulz, C. Comparative Life Cycle Assessment (LCA) of Raising Rainbow Trout (*Oncorhynchus mykiss*) in Different Production Systems. *Aquac. Eng.* **2013**, *54*, 85–92. [CrossRef]
68. Haslawati, B.; Saadiah, I.; Siti-Dina, R.P.; Othman, M.; Latif, M.T. Environmental Assessment of Giant Freshwater Prawn, Macrobrachium Rosenbergii Farming through Life Cycle Assessment. *Sustainability* **2022**, *14*, 14776. [CrossRef]
69. Cao, L.; Diana, J.S.; Keoleian, G.A.; Lai, Q. Life Cycle Assessment of Chinese Shrimp Farming Systems Targeted for Export and Domestic Sales. *Environ. Sci. Technol.* **2011**, *45*, 6531–6538. [CrossRef]
70. Oonincx, D.G.A.B.; Boer, I.J.M. de Environmental Impact of the Production of Mealworms as a Protein Source for Humans—A Life Cycle Assessment. *PLoS ONE* **2012**, *7*, e51145. [CrossRef]
71. Nikkhah, A.; Van Haute, S.; Jovanovic, V.; Jung, H.; Dewulf, J.; Cirkovic Velickovic, T.; Ghnimi, S. Life Cycle Assessment of Edible Insects (*Protaetia brevitarsis seulensis* larvae) as a Future Protein and Fat Source. *Sci. Rep.* **2021**, *11*, 14030. [CrossRef]
72. Mattick, C.S.; Landis, A.E.; Allenby, B.R.; Genovese, N.J. Anticipatory Life Cycle Analysis of In Vitro Biomass Cultivation for Cultured Meat Production in the United States. *Environ. Sci. Technol.* **2015**, *49*, 11941–11949. [CrossRef] [PubMed]
73. de Alvarenga, R.A.F.; Galindro, B.M.; de Fátima Helpa, C.; Soares, S.R. The Recycling of Oyster Shells: An Environmental Analysis Using Life Cycle Assessment. *J. Environ. Manag.* **2012**, *106*, 102–109. [CrossRef] [PubMed]
74. Martini, A.; Aguiari, L.; Capoccioni, F.; Martinoli, M.; Napolitano, R.; Pirlo, G.; Tonachella, N.; Pulcini, D. Is Manila Clam Farming Environmentally Sustainable? A Life Cycle Assessment (LCA) Approach Applied to an Italian *Ruditapes philippinarum* Hatchery. *Sustainability* **2023**, *15*, 3237. [CrossRef]
75. Vélez-Henao, J.A.; Weinland, F.; Reintjes, N. Life Cycle Assessment of Aquaculture Bivalve Shellfish Production—A Critical Review of Methodological Trends. *Int. J. Life Cycle Assess.* **2021**, *26*, 1943–1958. [CrossRef]

76. New World Bank Country Classifications by Income Level: 2021–2022. Available online: https://blogs.worldbank.org/opendata/new-world-bank-country-classifications-income-level-2021-2022 (accessed on 3 November 2022).
77. Tay, W.; Quek, R.; Lim, J.; Kaur, B.; Ponnalagu, S.; Henry, C.J. Plant-Based Alternative Proteins—Are They Nutritionally More Advantageous? *Eur. J. Clin. Nutr.* **2023**, *77*, 1051–1060. [CrossRef]
78. Sobczak, P.; Grochowicz, J.; Lusiak, P.; Żukiewicz-Sobczak, W. Development of Alternative Protein Sources in Terms of a Sustainable System. *Sustainability* **2023**, *15*, 12111. [CrossRef]
79. Daas, M.C.; van 't Veer, P.; Temme, E.H.M.; Kuijsten, A.; Gurinović, M.; Biesbroek, S. Diversity of Dietary Protein Patterns across Europe—Impact on Nutritional Quality and Environmental Sustainability. *Curr. Res. Food Sci.* **2025**, *10*, 101019. [CrossRef]
80. Sanchez-Sabate, R.; Sabaté, J. Consumer Attitudes Towards Environmental Concerns of Meat Consumption: A Systematic Review. *Int. J. Environ. Res. Public Health* **2019**, *16*, 1220. [CrossRef]
81. Guillen, J.; Natale, F.; Carvalho, N.; Casey, J.; Hofherr, J.; Druon, J.-N.; Fiore, G.; Gibin, M.; Zanzi, A.; Martinsohn, J.T. Global Seafood Consumption Footprint. *AMBIO* **2019**, *48*, 111–122. [CrossRef]
82. de Carvalho, N.M.; Madureira, A.R.; Pintado, M.E. The Potential of Insects as Food Sources—A Review. *Crit. Rev. Food Sci. Nutr.* **2020**, *60*, 3642–3652. [CrossRef]
83. van Huis, A. Edible Insects: Challenges and Prospects. *Entomol. Res.* **2022**, *52*, 161–177. [CrossRef]
84. Lee, H.J.; Yong, H.I.; Kim, M.; Choi, Y.-S.; Jo, C. Status of Meat Alternatives and Their Potential Role in the Future Meat Market—A Review. *Asian-Australas. J. Anim. Sci.* **2020**, *33*, 1533–1543. [CrossRef] [PubMed]
85. Kurek, M.A.; Onopiuk, A.; Pogorzelska-Nowicka, E.; Szpicer, A.; Zalewska, M.; Półtorak, A. Novel Protein Sources for Applications in Meat-Alternative Products—Insight and Challenges. *Foods* **2022**, *11*, 957. [CrossRef] [PubMed]
86. Tufail, T.; Bader Ul Ain, H.; Ashraf, J.; Mahmood, S.; Noreen, S.; Ijaz, A.; Ikram, A.; Arshad, M.T.; Abdullahi, M.A. Bioactive Compounds in Seafood: Implications for Health and Nutrition. *Food Sci. Nutr.* **2025**, *13*, e70181. [CrossRef]
87. Karakoltsidis, P.A.; Zotos, A.; Constantinides, S.M. Composition of the Commercially Important Mediterranean Finfish, Crustaceans, and Molluscs. *J. Food Compos. Anal.* **1995**, *8*, 258–273. [CrossRef]
88. Béné, C.; Barange, M.; Subasinghe, R.; Pinstrup-Andersen, P.; Merino, G.; Hemre, G.-I.; Williams, M. Feeding 9 Billion by 2050—Putting Fish Back on the Menu. *Food Secur.* **2015**, *7*, 261–274. [CrossRef]
89. Schön, A.-M.; Böhringer, M. Land Consumption for Current Diets Compared with That for the Planetary Health Diet—How Many People Can Our Land Feed? *Sustainability* **2023**, *15*, 8675. [CrossRef]
90. Rauw, W.M.; Gómez Izquierdo, E.; Torres, O.; García Gil, M.; de Miguel Beascoechea, E.; Rey Benayas, J.M.; Gomez-Raya, L. Future Farming: Protein Production for Livestock Feed in the EU. *Sustain. Earth Rev.* **2023**, *6*, 3. [CrossRef]
91. Cherubini, E.; Zanghelini, G.M.; Tavares, J.M.R.; Belettini, F.; Soares, S.R. The Finishing Stage in Swine Production: Influences of Feed Composition on Carbon Footprint. *Environ. Dev. Sustain.* **2015**, *17*, 1313–1328. [CrossRef]
92. Can, E.; Austin, B.; Steinberg, C.; Carboni, C.; Sağlam, N.; Thompson, K.; Yiğit, M.; Seyhaneyildiz Can, S.; Ergün, S. Best Practices for Fish Biosecurity, Well-Being and Sustainable Aquaculture. *Sustain. Aquat. Res.* **2023**, *2*, 221–267. [CrossRef]
93. Ma, L.; Zhang, Y.; Chen, S.; Yu, L.; Zhu, Y. Environmental Effects and Their Causes of Agricultural Production: Evidence from the Farming Regions of China. *Ecol. Indic.* **2022**, *144*, 109549. [CrossRef]
94. Nemecek, T.; Roesch, A.; Bystricky, M.; Jeanneret, P.; Lansche, J.; Stüssi, M.; Gaillard, G. Swiss Agricultural Life Cycle Assessment: A Method to Assess the Emissions and Environmental Impacts of Agricultural Systems and Products. *Int. J. Life Cycle Assess.* **2023**, *29*, 433–455. [CrossRef]
95. Heilweck, M. The High Seas Solution. In *Aquaculture: Ocean Blue Carbon Meets UN-SDGS*; Moore, D., Heilweck, M., Petros, P., Eds.; Springer International Publishing: Cham, Switzerland, 2022; pp. 97–130. ISBN 978-3-030-94846-7.
96. Kossmann, H.; Moess, T.; Breunig, P. The Climate Impact and Land Use of Cultivated Meat: Evaluating Agricultural Feedstock Production. *PLoS ONE* **2025**, *20*, e0316480. [CrossRef]
97. Trinidad, K.R.; Ashizawa, R.; Nikkhah, A.; Semper, C.; Casolaro, C.; Kaplan, D.L.; Savchenko, A.; Blackstone, N.T. Environmental Life Cycle Assessment of Recombinant Growth Factor Production for Cultivated Meat Applications. *J. Clean. Prod.* **2023**, *419*, 138153. [CrossRef]
98. Biteau, C.; Bry-Chevalier, T.; Crummett, D.; Loewy, K.; Ryba, R.; Jules, M.S. Have the Environmental Benefits of Insect Farming Been Overstated? A Critical Review. *EcoEvoRxiv* **2024**. [CrossRef]
99. Mantoam, E.J.; Angnes, G.; Mekonnen, M.M.; Romanelli, T.L. Energy, Carbon and Water Footprints on Agricultural Machinery. *Biosyst. Eng.* **2020**, *198*, 304–322. [CrossRef]
100. Moore, D.; Heilweck, M.; Fears, W.B.; Petros, P.; Squires, S.J.; Tamburini, E.; Waldron, R.P., Jr. An Assessment of the Potential Value for Climate Remediation of Ocean Calcifiers in Sequestration of Atmospheric Carbon. *ScienceOpen Preprints* **2022**. [CrossRef]
101. Alonso, A.A.; Álvarez-Salgado, X.A.; Antelo, L.T. Assessing the Impact of Bivalve Aquaculture on the Carbon Circular Economy. *J. Clean. Prod.* **2021**, *279*, 123873. [CrossRef]
102. Tamburini, E.; Fano, E.A.; Castaldelli, G.; Turolla, E. Life Cycle Assessment of Oyster Farming in the Po Delta, Northern Italy. *Resources* **2019**, *8*, 170. [CrossRef]

103. Lal, R.; Monger, C.; Nave, L.; Smith, P. The Role of Soil in Regulation of Climate. *Philos. Trans. R. Soc. B Biol. Sci.* **2021**, *376*, 20210084. [CrossRef]
104. Liu, X.; Wang, S.; Zhuang, Q.; Jin, X.; Bian, Z.; Zhou, M.; Meng, Z.; Han, C.; Guo, X.; Jin, W.; et al. A Review on Carbon Source and Sink in Arable Land Ecosystems. *Land* **2022**, *11*, 580. [CrossRef]
105. Shrestha, P.; Karim, R.A.; Sieverding, H.L.; Archer, D.W.; Kumar, S.; Nleya, T.; Graham, C.J.; Stone, J.J. Life Cycle Assessment of Wheat Production and Wheat-Based Crop Rotations. *J. Environ. Qual.* **2020**, *49*, 1515–1529. [CrossRef] [PubMed]
106. Du, X.; Xi, M.; Kong, L.; Chen, X.; Zhang, L.; Zhang, H.; Dai, Q.; Wu, W. Energy Budgeting and Carbon Footprint of Different Wheat–Rice Cropping Systems in China. *Sci. Total Environ.* **2023**, *879*, 163102. [CrossRef] [PubMed]
107. Fatichi, S.; Pappas, C.; Zscheischler, J.; Leuzinger, S. Modelling Carbon Sources and Sinks in Terrestrial Vegetation. *New Phytol.* **2019**, *221*, 652–668. [CrossRef]
108. Johnson, E.H. Experimental Tests of Bivalve Shell Shape Reveal Potential Tradeoffs between Mechanical and Behavioral Defenses. *Sci. Rep.* **2020**, *10*, 19425. [CrossRef]
109. Leal, P.H.; Marques, A.C. The Evolution of the Environmental Kuznets Curve Hypothesis Assessment: A Literature Review under a Critical Analysis Perspective. *Heliyon* **2022**, *8*, e11521. [CrossRef]
110. Drewnowski, A. Alternative Proteins in Low- and Middle-Income Countries Face a Questionable Future: Will Technology Negate Bennett's Law? *Curr. Dev. Nutr.* **2024**, *8*, 101994. [CrossRef]
111. Aggarwal, R.; Bains, K. Protein, Lysine and Vitamin D: Critical Role in Muscle and Bone Health. *Crit. Rev. Food Sci. Nutr.* **2022**, *62*, 2548–2559. [CrossRef]
112. Zeng, Y.; Chen, E.; Zhang, X.; Li, D.; Wang, Q.; Sun, Y. Nutritional Value and Physicochemical Characteristics of Alternative Protein for Meat and Dairy—A Review. *Foods* **2022**, *11*, 3326. [CrossRef]
113. Langyan, S.; Yadava, P.; Khan, F.N.; Dar, Z.A.; Singh, R.; Kumar, A. Sustaining Protein Nutrition Through Plant-Based Foods. *Front. Nutr.* **2022**, *8*, 772573. [CrossRef]
114. Litskas, V.D. Environmental Impact Assessment for Animal Waste, Organic and Synthetic Fertilizers. *Nitrogen* **2023**, *4*, 16–25. [CrossRef]
115. Chandran, A.S.; Suri, S.; Choudhary, P. Sustainable Plant Protein: An up-to-Date Overview of Sources, Extraction Techniques and Utilization. *Sustain. Food Technol.* **2023**, *1*, 466–483. [CrossRef]
116. Ardoin, R.; Prinyawiwatkul, W. Consumer Perceptions of Insect Consumption: A Review of Western Research since 2015. *Int. J. Food Sci. Technol.* **2021**, *56*, 4942–4958. [CrossRef]
117. Roy, N.K.; Panda, S.; Dey, G. Engineering a Sustainable Protein Revolution: Recent Advances in Cultured Meat Production. *Food Bioeng.* **2023**, *2*, 301–316. [CrossRef]
118. Bondad-Reantaso, M.G.; MacKinnon, B.; Karunasagar, I.; Fridman, S.; Alday-Sanz, V.; Brun, E.; Le Groumellec, M.; Li, A.; Surachetpong, W.; Karunasagar, I.; et al. Review of Alternatives to Antibiotic Use in Aquaculture. *Rev. Aquac.* **2023**, *15*, 1421–1451. [CrossRef]
119. Cojocaru, A.L.; Liu, Y.; Smith, M.D.; Akpalu, W.; Chávez, C.; Dey, M.M.; Dresdner, J.; Kahui, V.; Pincinato, R.B.M.; Tran, N. The "Seafood" System: Aquatic Foods, Food Security, and the Global South. *Rev. Environ. Econ. Policy* **2022**, *16*, 306–326. [CrossRef]
120. Nanda, P.K.; Das, A.K.; Dandapat, P.; Dhar, P.; Bandyopadhyay, S.; Dib, A.L.; Lorenzo, J.M.; Gagaoua, M. Nutritional Aspects, Flavour Profile and Health Benefits of Crab Meat Based Novel Food Products and Valorisation of Processing Waste to Wealth: A Review. *Trends Food Sci. Technol.* **2021**, *112*, 252–267. [CrossRef]
121. Willer, D.F.; Nicholls, R.J.; Aldridge, D.C. Opportunities and Challenges for Upscaled Global Bivalve Seafood Production. *Nat. Food* **2021**, *2*, 935–943. [CrossRef]
122. Dong, S.-L.; Dong, Y.-W. Sustainability of Aquaculture Production Systems. In *Aquaculture Ecology*; Dong, S.-L., Tian, X.-L., Gao, Q.-F., Dong, Y.-W., Eds.; Springer Nature: Singapore, 2023; pp. 491–530. ISBN 978-981-19-5486-3.
123. Morales-Nin, B.; Zimmermann, F.; Merino, G.; Antelo, L.T.; Murillas-Maza, A.; Moksness, E. Challenges on Blue Food Provision. *Front. Ocean Sustain.* **2024**, *2*, 1271783. [CrossRef]
124. Jones, A.R.; Alleway, H.K.; McAfee, D.; Reis-Santos, P.; Theuerkauf, S.J.; Jones, R.C. Climate-Friendly Seafood: The Potential for Emissions Reduction and Carbon Capture in Marine Aquaculture. *BioScience* **2022**, *72*, 123–143. [CrossRef]
125. O'Connell, B.; Wallace, M.W. The Palaeoenvironmental and Biological Significance of Marine Carbonate Depositional Surfaces. *Geol. Soc. Lond. Spec. Publ.* **2025**, *556*, 1–24. [CrossRef]
126. Farmery, A.K.; Alexander, K.; Anderson, K.; Blanchard, J.L.; Carter, C.G.; Evans, K.; Fischer, M.; Fleming, A.; Frusher, S.; Fulton, E.A.; et al. Food for All: Designing Sustainable and Secure Future Seafood Systems. *Rev. Fish Biol. Fish.* **2022**, *32*, 101–121. [CrossRef] [PubMed]
127. Troell, M.; Costa-Pierce, B.; Stead, S.; Cottrell, R.S.; Brugere, C.; Farmery, A.K.; Little, D.C.; Strand, Å.; Pullin, R.; Soto, D.; et al. Perspectives on Aquaculture's Contribution to the Sustainable Development Goals for Improved Human and Planetary Health. *J. World Aquac. Soc.* **2023**, *54*, 251–342. [CrossRef]

128. Brice, J.V. Wave Mechanics in Constructed Oyster Reefs and the Design of Nature-Based Coastal Adaptation. Bachelor's Thesis, Massachusetts Institute of Technology, Cambridge, MA, USA, 2024.
129. Song, J.; Wang, Y.; Huang, L.; Peng, Y.; Tan, K.; Tan, K. The Effects of Bivalve Aquaculture on Carbon Storage in the Water Column and Sediment of Aquaculture Areas. *Sci. Total Environ.* **2024**, *937*, 173538. [CrossRef]

foods

MDPI

Article

Moving Beyond Indices: A Systematic Approach Integrating Food System Performance and Characteristics for Comprehensive Food Security Assessment

Muna A. Al-Ansari, Hamad Nabeel *, Galal M. Abdella and Tarek El Mekkawy

Department of Mechanical and Industrial Engineering, College of Engineering, Qatar University, Doha P.O. Box 2713, Qatar; 200356160@qu.edu.qa (M.A.A.-A.); gmg5005@qu.edu.qa (G.M.A.); tmekkawy@qu.edu.qa (T.E.M.)
* Correspondence: hn2110478@qu.edu.qa

Abstract: Food security indices are widely used to support decision making and provide a structured assessment of countries' capacities to withstand global environmental and economic crises. However, these indices have inherent limitations, including potential biases in ranking and a lack of structural insights into food system dynamics. This study presents a systematic approach that combines elastic-net regression-based feature selection and two-step clustering to address some of these limitations and equip decision makers with structured procedures for making informed decisions and supporting food system management. The mathematical and operational procedures of the proposed approach were demonstrated through an illustrative example using the EIU dataset of 94 countries. The study investigated the sensitivity of composite indicators to extreme data points, relative weights, and dimensionality reduction. After applying elastic-net regression, 15 indicators were selected for Model 1 (M1) and 9 for Model 2 (M2) from an initial set of 25 indicators. Subsequently, two-step clustering grouped the countries into four distinct clusters, reflecting combinations of food system characteristics and income levels. The results demonstrate that countries with industrialized, consolidated food systems and high per capita income tend to exhibit greater food security. Conversely, countries with rural or traditional food systems and low-income levels are more vulnerable to food insecurity. By incorporating statistical rigor and empirical structure discovery, this methodology addresses key limitations of existing indices. It provides an adaptive, transparent framework that informs targeted policy by linking the structural characteristics of food systems to tangible food security outcomes.

Keywords: food security assessment; composite indicators; multiple regression analysis; two-step clustering

1. Introduction

Food security refers to a state in which all individuals, at all times, have reliable physical, social, and economic access to sufficient, safe, and nutritious food that satisfies their dietary requirements and food preferences, supporting an active and healthy lifestyle [1]. It is a multidimensional concept that plays a vital role in the well-being and survival of humanity [2–4]. The four primary pillars of food security are availability, accessibility, stability, and utilization. Food availability refers to the physical presence of sufficient food, whether through domestic production or a reliable external supply. This availability depends on factors such as storage capacity, transport infrastructure, and the integration of local or national markets [5]. Food access pertains to a household's ability to obtain sufficient and nutritious food through self-sufficiency, purchases, or food aid [6]. It is important to note that food may be available but not necessarily accessible to all households. Food stability is crucial for food security, as it reflects the system's resilience and capacity to maintain a consistent food supply amidst shocks or disruptions to the local or global economy [7,8]. Lastly, food utilization examines individuals' ability to use accessible food effectively. This can be achieved by ensuring an appropriate diet, safe drinking water, sanitation, and health care that meet nutritional and physiological needs. Thus, assessing food security comprehensively requires capturing these four dimensions, yet many existing indices struggle to reflect all pillars adequately.

Despite ongoing efforts by international institutions, global food systems face significant threats. Climate change poses a major challenge, disrupting agricultural productivity, water availability, and the overall stability of food systems. Rising temperatures, changing precipitation patterns, and more frequent extreme weather events reduce crop yields, alter growing seasons, and increase pest and disease pressures, compromising food availability and access [9–11]. Water scarcity is another critical concern, intensified by increasing global demand due to population growth, shifting consumption patterns, and energy production needs [12,13]. Recent data from the UN World Water Development Report (2025) indicate that agriculture still accounts for approximately 72% of global freshwater withdrawals, highlighting water's pivotal role in ensuring food security. Insufficient access to water constrains crop and livestock productivity and impairs food availability, stability, and utilization.

According to the 2023 State of Food Security and Nutrition in the World (SOFI) report, 735 million people experienced chronic hunger in 2022, a significant increase from 613 million in 2019. Projections indicate that nearly 600 million people may remain chronically undernourished by 2030, threatening the achievement of Sustainable Development Goal 2, i.e., zero hunger. Additional environmental stressors, such as air pollution and soil degradation, further jeopardize food security. These factors affect agriculture, livestock, and aquaculture by altering pest dynamics, diminishing crop yields, and degrading soil fertility through erosion and overgrazing [14,15]. These pressures are compounded by the socioeconomic aftermath of the COVID-19 pandemic, which disrupted food supply chains, drove up food prices, and disproportionately impacted low- and middle-income families [16–18]. Addressing these challenges requires effective policy and governance frameworks; however, many countries face institutional and regulatory barriers that limit the implementation of sustainable and resilient food systems [19,20].

Several initiatives have been developed and implemented worldwide to eliminate hunger and malnutrition at local and global levels. The growing focus on these efforts highlighted the need for appropriate tools to help provide a condensed view of progress and make proactive adjustments to ensure the fulfillment of the strategic goals of these initiatives. Food security assessment is increasingly recognized as crucial for making informed decisions to mitigate risks and provide adequate food security amid various threats [21–24]. This assessment helps understand the impacts of deteriorating living conditions on food security and evaluates food safety and nutritional quality, especially for vulnerable populations [25]. Various methods are employed for food security assessment, including expert-based surveys, technical reports, and indicators like anthropometry, which evaluates nutritional status through physical measures [26], and the Food Insecurity Experience Scale (FIES), which measures food security based on personal experiences [27].

Notable reports such as the Global Nutrition Report (GNR) and the State of Food Security and Nutrition (SOFI) report provide comprehensive insights into global food and nutrition issues [22]. Recently, there has been a notable trend in using composite indicators to assess the outcomes of several food initiatives. These indicators are widely known for making food system performance more understandable and interpretable by combining different indicators (or factors) into a single index or score [21–24]. The Global Hunger Index (GHI) assesses hunger through indicators like malnutrition and child mortality [28], while the Global Food Security Index (GFSI) evaluates food security across over 100 countries using affordability, availability, and quality metrics, with recent updates including natural resources and resilience as indicators [29].

Despite the widespread use of composite indices like the GFSI, GHI, and FIES, several methodological limitations undermine their effectiveness. Chief among these are subjective weighting schemes, limited incorporation of outcome-based indicators, and lack of contextual benchmarking. For instance, while the GFSI provides broad insights into food system performance, it often fails to reflect populations' actual consumption or nutritional outcomes [29,30]. Similarly, subjective weighting schemes usually rely on expert judgment, which can introduce bias, lack transparency, and reduce reproducibility in index construction [31]. These shortcomings reduce the reliability and policy relevance of food security assessments. To address these gaps, this study introduces a structured, data-driven framework that combines an input–output model, elastic-net regression, and a two-stage clustering method. The proposed approach enhances the interpretability of composite indices by selecting the most relevant indicators, incorporates outcome-based measures to link resources with real-world impact, and facilitates context-aware benchmarking through cluster analysis. Together, these methodological innovations aim to improve food security assessments' robustness, clarity, and policy relevance.

2. Literature Review

Food security assessment utilizes various methods such as expert-based surveys, annual technical reports, and food security indicators. Recently, there has been a growing trend towards employing composite indicators to evaluate the outcomes of food initiatives. These composite indicators are appreciated for their ability to simplify and enhance the interpretability of food system performance by integrating multiple indicators into a single index or score [21–24]. Three widely recognized food security indicators are discussed below.

The Global Food Security Index (GFSI), developed by the Economist Intelligence Unit (EIU) and sponsored by Corteva Agriscience in 2012, is a widely known measure of food security for over 100 countries. The GFSI initially involved three categories namely affordability, availability, and food quality and safety [29,32]. In 2020, the GFSI included a new category, natural resources and resilience, which was later renamed "Sustainability and Adaptation" in the 2022 GFSI Report. This category assesses the likelihood of natural disruptions and evaluates the ability of the system to adapt to their potential impacts and consequences. Several studies have assessed the procedures for computing the GFSI over the years (see [29,32,33]). These studies have revealed two main limitations. These are the use of subjective weighting methods and the lack of connecting the GFSI with indicators associated with food system outcomes of the associated countries, such as food consumption, food production, food supply, poverty or nutritional policies, and the population's nutritional status [29,30].

The Food Insecurity Experience Scale (FIES), developed by the FAO, is another measure of food security. The FIES assesses food security at the household or individual level by capturing personal experiences [27]. The FIES utilizes an eight-question-based survey module, FIES-SM, to assess the difficulty individuals or households usually encounter in reaching adequate amounts of food and nutrition. While the FIES offers valuable insights into the lived experiences of food insecurity, it has also faced criticism. Short reference periods, particularly the 30-day version, can yield severity assessments that differ substantially from the 12-month reference frame, complicating global comparisons [34]. Additionally, the FIES's focus on individual experiences may oversimplify the complexity of food insecurity by neglecting broader systemic factors such as sustainability, structural inequality, or democratic agency within food systems [35].

The Global Hunger Index (GHI), developed by the International Food Policy Research Institute (IFPRI), measures global hunger status. The GHI utilizes a combination of four indicators to determine caloric deficiency and malnutrition, namely malnutrition, child undernutrition, stunting and wasting in children, and child mortality. The GHI scores are calculated through a three-step process. The four indicators are first given standardized scores from 0 to 100, where 0 indicates "no hunger" and 100 indicates "severe hunger". Following this step, to calculate the Global Hunger Index (GHI) for each country, the standardized scores of the four key indicators are aggregated, with equal weights assigned to each indicator [28,32]. The data on the GHI are collected from the U.N. database and other multilateral agencies (for further details, visit https://www.globalhungerindex.org). Despite its widespread use, the GHI has been criticized for its methodology and conceptual framing. One notable concern is its narrow focus on caloric intake and child health outcomes, which may overlook broader dimensions of food insecurity, such as dietary quality, food access, and socioeconomic determinants [35].

While indices like the GFSI, FIES, and GHI offer useful snapshots of food security conditions, their structural and methodological limitations motivate the need for enhanced, data-driven approaches that better capture food system complexity.

Existing Weighting Methods

The weights are crucial to ensure the accuracy and credibility of the food security index. These weights refer to the importance of each indicator (or factor) to the overall score of the composite index. Several methods for estimating relative weights under various contexts have been conducted over the last years (for further reading, see [36–39]). Equal weighting is very popular in business and finance applications [40]. The equal weighting assigns equal importance to all indicators. However, several studies have criticized equal weighting for its shortcomings in accommodating double counting, which arises when two or more indicators measure the same aspect within the context of the GFSI calculations [41,42]. Common examples of methods used to address the issue of double counting in situations like these are the Analytical Hierarchy Process (AHP) and the Technique for Order of Preference by Similarity to the Ideal Solution (TOPSIS). These methods help mitigate the duplication of indicators measuring the same aspect within the GFSI framework. Over the years, some researchers have found that using expert judgments to assign weights, which is frequent in practice, may affect the credibility of the composite index [42,43]. While expert judgment remains a common approach, it is susceptible to cognitive and group-level biases that can distort weight assignments. For instance, cognitive biases such as overconfidence, anchoring, and availability can influence experts' decisions, leading to inconsistent or suboptimal outcomes [44–46]. Additionally, when experts work in groups, social dynamics like groupthink or social loafing may further undermine the objectivity of the process [45]. Beyond biases, challenges in calibrating expert input and the inherent uncertainty of subjective weights add further complexity, making cross-study comparisons difficult and often unreliable [47,48]. Moreover, over-reliance on experts assumes complete domain knowledge, which may not be realistic in multifaceted contexts such as sustainability assessment [49]. Other methods, such as Principal Component Analysis (PCA), Data Envelope Analysis (DEA), and Factor Analysis (FA), have also been utilized for estimating relative weights in various fields (see [50,51]). PCA is mainly known for its efficiency in finding a new set of components (features), called principal components (PCs), that capture the most significant variability in the original dataset. Mainly, it reduces the dimension of the original dataset (see [50,52]). However, despite its popularity, the new component may not have a clear real-world interpretation [53,54].

In practice, especially under high-dimensional settings, food security indicators are more likely to exhibit high correlation levels, a condition known as multicollinearity. This statistical phenomenon occurs when some indicators are interrelated, potentially influencing the generation of weights and leading to instability in regression model estimates. As a result, the robustness of the constructed composite indicators may be compromised. While multicollinearity is more pronounced in high-dimensional datasets, it can also arise in lower-dimensional contexts [55,56]. Classical regression methods have shown limited effectiveness in addressing the multicollinearity phenomenon. On the other hand, penalization regression methods, such as ridge, least absolute shrinkage and selection operator (lasso), and elastic net, are known for effectively managing multicollinearity and sparsity [38,57,58]. These methods stabilize model estimates by introducing regularization penalties, with elastic net offering the dual advantage of variable selection and multicollinearity management.

In parallel, several statistical and machine learning methods have also been explored in the context of food security. While widely adopted, econometric models often rely on historical assumptions and fail to adapt to the dynamic nature of food systems [59]. Machine learning models and neural networks offer strong predictive capabilities but typically demand extensive, high-quality datasets and often lack transparency [60,61]. Bayesian spatio-temporal models effectively capture spatial and temporal dependencies but require significant computational resources and careful specification [62]. Panel quantile regression captures heterogeneous effects but needs extensive panel data [63], while small-area estimation improves subnational insights but is sensitive to model and data assumptions [64,65]. Nonparametric methods like random forests can model nonlinear relationships but are less interpretable [66]. Similarly, although useful in benchmarking efficiency, data envelopment analysis (DEA) is susceptible to input selection and lacks mechanisms for handling noise or measurement error [67].

Despite the increasing popularity of these methods, the application of penalized regression models in food security, especially for developing composite indices, remains limited and is still an emerging area of research [38,68–71]. In this context, elastic-net regression offers a particularly suitable approach due to its ability to integrate the benefits of both ridge and lasso regularization, thereby enabling robust estimation while performing automatic feature selection. This is especially beneficial in identifying the most relevant food security indicators and mitigating the risk of overfitting, which is common in high-dimensional data scenarios [72].

Furthermore, existing food security indices predominantly rely on input-based models that focus on resource availability, failing to connect these inputs with the actual outputs of food systems. Consequently, these models do not adequately capture how efficiently resources are utilized, thus providing an incomplete picture of food security. Moreover, interpreting composite indices becomes challenging for stakeholders and decision makers, especially when multiple indicators are involved, leading to difficulties in identifying specific areas of strength and weakness within food systems. Moreover, the prevalent use of single-score-based rankings in food security assessments often overlooks crucial contextual information, such as geographical location, food system typology, and economic state, resulting in the loss of valuable insights necessary for realistic benchmarking and effective solution development.

To address these limitations, this study introduces a structured approach to promote the current practices of utilizing composite indices to assess food security locally and globally. This approach uses two well-established methods, feature selection-based regression and clustering methods, which provide deep insights into food security issues and lead to efficient resource allocation. The list below explains the significance of this approach's structured design compared to its counterparts.

1. Food security indices using input-based models, focusing on resource availability, do not relate the composite index to the food system outputs. In other words, these models do not clearly show how efficiently these resources are utilized. To overcome this limitation, this study introduces an input–output model that incorporates nutritional status as an output indicator. This output indicator is important to measure people's accessibility to enough nutrition and the availability of their daily dietary needs.
2. Interpreting composite indices is critical to highlighting specific areas of strength and weakness in food systems. However, stakeholders and decision-makers may encounter serious difficulties interpreting the results when many indicators are involved. To mitigate this, elastic-net regression is employed to reduce the dimension of the impact matrix and include only the indicators relevant to the food system's performance. This step will help understand the contribution of each indicator (or factor) and effectively address the root causes of food insecurity.
3. Using single-score-based rankings in food security assessments often disregards critical contextual factors, such as geographic location, food system typology, economic development levels, etc. The involvement of such characteristics is essential for conducting realistic benchmarking and developing short- and long-term solutions. This study proposes a two-stage clustering method to enhance food security analysis. First, entities (countries) are grouped based on their overall food security score. Following this, each cluster is further analyzed through a second round of clustering, considering key food system characteristics like typology and economic status. By creating more nuanced groupings, this approach aims to provide stakeholders with actionable insights for targeted interventions instead of relying solely on a single-score list.

The rest of this paper is organized as follows: Section 3 highlights the elastic-net regression method as a key element of the study's methodology and outlines the proposed approach. Section 4 reports the illustrative example used to assess the applicability of the new approach. Section 5 reports the results of implementing the two-step clustering approach. Section 6 is dedicated to discussing the main findings and conclusions extracted from the case study. In addition, this section provides recommendations and suggests future research directions.

3. Methodology

Addressing methodological gaps identified in the literature, this study introduces a new approach combining two well-known methods: feature selection and two-step clustering (see Figure 1). The proposed methodology addresses several limitations of traditional input-based models that focus solely on resource availability without linking the composite index to food system outputs. By incorporating an input–output model, the population's nutritional status is added as a crucial output indicator, representing the accessibility and availability of daily dietary needs. Furthermore, to improve the interpretability of composite indices, elastic-net regression is employed to reduce the dimensionality of the impact matrix, retaining only the most relevant indicators for food system performance. This technique was chosen for its ability to handle multicollinearity, a common issue in datasets with correlated predictors, ensuring that the selected indicators remain statistically independent. Unlike alternative methods such as Principal Component Analysis (PCA) or ridge regression, elastic-net regression allows for the simultaneous selection of a subset of predictors while maintaining model interpretability, a key feature for understanding the relationships between indicators in the food security context. This step aids in identifying and addressing the root causes of food insecurity.

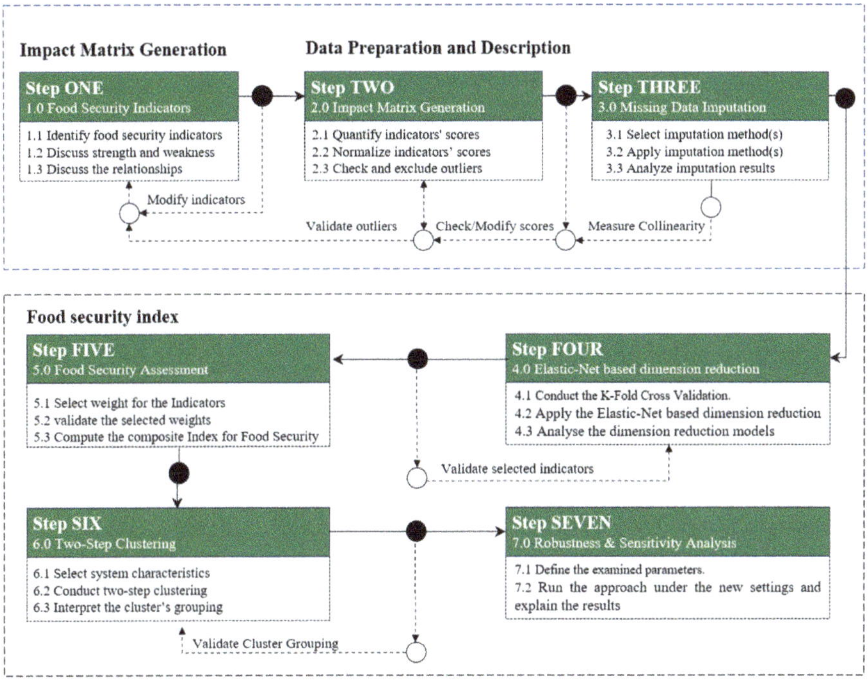

Figure 1. The outlines of the proposed approach.

Additionally, the study critiques the conventional single-score-based ranking system, which often overlooks valuable information related to specific characteristics of entities, such as geographical location, food system typology, and economic state. To overcome this, a two-stage clustering method is proposed: first, group entities based on their overall food security scores, and then, further, each cluster is analyzed through a second round of clustering that considers key food system characteristics. This nuanced clustering approach aims to provide stakeholders with actionable insights for targeted interventions, offering a more comprehensive understanding of food security dynamics than traditional methods.

The proposed approach consists of seven steps that were sequenced in a way to ensure a logical flow of computations and analysis as follows:

Steps 1–2: These initial steps involve identifying relevant food security indicators, assessing their impact within each category, and then integrating them to construct the initial impact matrix.

Step 3: Once the matrix is developed, its structure is examined for potential missing data. In consultation with domain experts, an appropriate data imputation method is selected to address any gaps and ensure data reliability.

Step 4: This step plays a pivotal role in the methodology by applying elastic-net regression to identify the most influential indicators contributing to food security performance. To enhance the accuracy and robustness of the results, the elastic-net regression is fine-tuned using a widely accepted optimization technique, i.e., *k*-fold cross-validation. A detailed explanation of this method is provided later in this section.

Step 5: In this step, various weighting methods are explored, and the most suitable approach is selected to compute relative weights for the indicators retained in the reduced impact matrix.

Step 6: Once each entity's food security score is calculated, two-step clustering is applied to examine the relationships between food security performance and selected food system characteristics.

Step 7: A sensitivity analysis is conducted to assess how variations in key methodological choices, such as data imputation and dimensionality reduction, affect the performance of the composite food security indicator. This step helps identify the composite indicator's robustness and highlights how methodological choices may lead to alternative rankings or interpretations, informing more transparent and evidence-based decision-making.

Elastic-net regression assumes linearity in the relationship between the predictors and the outcome variable, a common assumption in composite index modeling, where relationships are typically additive. However, while this method assumes linearity, non-linearities may be present but are not explicitly modeled in this framework. Additionally, elastic-net regression requires the careful handling of outliers, as extreme values can impact the regression results. To mitigate this, outlier detection and treatment are performed before the application of the model. A key advantage of the elastic net is its ability to handle multicollinearity by combining the strengths of both ridge and lasso, ensuring that correlated predictors do not overly influence the model.

To introduce the elastic-net regression, we begin with the generalized linear model, which describes the relationship between the dependent variable, y, and a set of independent variables, x.

$$y_i = \beta_0 + \sum_{j=1}^{p} \beta_j x_{ij} + \varepsilon_i \quad ; \qquad i = 1, 2, \ldots \tag{1}$$

where y_i is the response variable, and $\beta_j \in \mathbb{R}^p$ are the coefficients of the regression model associated with the jth independent variable. The term ε_i represents the error and is assumed to be normally distributed. The regression coefficients can be found using the following optimization problem [73–75]:

$$\hat{\beta^{net}} = (y - X\beta)^T (y - X\beta) + P_\lambda(\beta) \tag{2}$$

where $P_\lambda(\beta)$ is the elastic-net penalty function. Equation (2) can be also written as follows:

$$\hat{\beta^{net}} = (y - X\beta)^T (y - X\beta) + \lambda_1 \sum_{j=1}^{p} \left|\beta_j\right| + \lambda_2 \sum_{j=1}^{p} \beta_j^2 \tag{3}$$

In Equation (3), the term $\lambda_1 \sum_{j=1}^{p} \left|\beta_j\right|$ corresponds to the l_1(lasso) penalty, which promotes sparsity, and $\lambda_2 \sum_{j=1}^{p} \beta_j^2$ corresponds to the l_2(ridge) penalty, which addresses multicollinearity [57]. The tuning parameters $\lambda_1 \geq 0$ and $\lambda_2 \geq 0$ control the strength of the respective penalties. Notably, the elastic net has the advantage over other penalization methods because it does not require $n > p$. However, the parameter λ controls the penalization's strength. If λ is equal to zero, the regression problem in Equation (3) will become a general linear regression.

Numerous techniques for estimating the value of λ are described in the literature. The k-fold cross-validation (CV) technique is the most frequently used [75]. The initial dataset is divided randomly into k equal-sized subsets. Then, one subsample is used as a validation dataset, and the remaining datasets are used as training subsets. These procedures should be repeated k times. The k results can be averaged to obtain a single estimate. Each recommended value of λ_1 and λ_2 will be evaluated using the CV procedure. Finally, the most accurate value of λ is selected as the optimal for performing elastic-net regression.

4. Illustrative Example

This example details the implementation process for the new food security index. Guidelines and relevant remarks will be provided under each step.

4.1. Food Security Indicators

This study utilizes the GFSI framework developed by the Economist Intelligence Unit (EIU) and supported by Corteva Agriscience. The framework comprises four main categories: food affordability (six indicators), food availability (seven indicators), food quality and safety (five indicators), and natural resources and resilience (seven indicators). Table 1 presents the details on the specific indicators within each category. Detailed information regarding the indicators' specific weights and units can be found on the GFSI website [76].

Table 1. Structure of food security categories.

Category	No.	Indicators	Symbol
Affordability (AFF)	1	Change in the food costs	CFC
	2	Proportion of population under the global poverty line	PPUGP
	3	Inequality-adjusted income index	IAII
	4	Agricultural import tariffs	AIT
	5	Food safety net programs	FSNP
	6	Market access and agricultural financial services	MAAFS
Availability (AVA)	7	Sufficiency of supply	SOS
	8	Agricultural research and development	ARAD
	9	Agricultural infrastructure	AIS
	10	Volatility of agricultural production	VOAP
	11	Political and social barriers to access	PASBA
	12	Food loss	FL
	13	Food security and access policy commitments	FSAAPC
Quality and safety (QAS)	14	Dietary diversity	DD
	15	Nutritional standards	NS
	16	Micronutrient availability	MA
	17	Protein quality	PQ
	18	Food safety	FS
Natural resources and resilience (NR&R)	19	Exposure	E
	20	Water	W
	21	Land	L
	22	Oceans, rivers, and lakes	ORL
	23	Sensitivity	S
	24	Political commitment to adoption	PCTA
	25	Demographic stress	DS

This study focuses on 94 countries selected from the GFSI framework, based on the availability of data on the "prevalence of undernourishment as a percentage of the population". This indicator acts as the food system output (response variable) in the study.

4.2. Impact Matrix Generation

This study uses the 2019 GFSI dataset, which can be accessed on https://impact.economist.com/sustainability/project/food-security-index/ (accessed on 20 March 2022). To ensure meaningful comparability across all indicators, the dataset was normalized column-wise (i.e., per indicator) using the Min–Max (Feature Scaling) algorithm [77,78]. The Min–Max algorithm calculates the normalized scores using the following:

$$x_{ij}^{norm} = a_0 + \frac{\left(x_{ij} - \mathrm{M}_{min,j}\right)(a_1 - a_0)}{\left(\mathrm{M}_{max,j} - \mathrm{M}_{min,j}\right)}; i = 1, 2, \ldots, n, \quad j = 1, 2, \ldots, p \tag{4}$$

where x_{ij} represents the original score of the ith country under the jth indicator, $\mathrm{M}_{max,j}$ and $\mathrm{M}_{min,j}$ represent the maximum and minimum scores of the jth indicator, n represents the number of countries ($n = 94$), and p represents the number of indicators ($p = 25$). The constants a_0 and a_1 ($a_0 < a_1$) are predetermined min–max values for the range of x_{ij}^{norm}. The 2019 GFSI dataset is normalized using $a_0 = 0$ and $a_1 = 100$, such that the highest possible score is 100, while the smallest possible score is 0.

The correlation analysis is necessary for understanding the inter-relationship between the indicators of the composite index; [29,71] have analyzed the correlation structure of the GFSI, considering the first three categories. Consequently, we will dedicate our analysis primarily to the fourth category, "Natural resources and resilience". NR&R was first introduced into the GFSI in 2017 and was used as one of the main categories in 2020. This category consists of seven indicators, as shown in Table 1. The "Exposure" directly quantifies the impact of the crisis brought on by weather-related changes. This indicator may significantly impact countries in regions susceptible to frequent and severe climate changes and crises. The indicators "Water", "Land", and "Oceans, rivers, and lakes" describe the capacity of natural resources. It makes sense to include these three indicators, as they are essential for local and worldwide food producers and suppliers. Lastly, the "Political commitment to adoption" indicates the readiness for developing and executing effective food security solutions.

Pearson's correlation coefficient (r) is widely used in practice for measuring correlation [52]. The r values range between −1 and 1, such that the higher the r_{ij} value, the higher the correlation levels between the ith and jth indicators. When the ith and jth indicators are not significantly correlated, the r_{ij} value will be near or equal to 0. The correlation analysis among the NR&R indicators will be based on the normalized 2019 GFSI dataset. The correlation coefficients for all possible indicator pairs are presented in Table 2.

Table 2. Pearson correlation coefficients between the indicators under the Natural Resources and Resilience category and GFSI.

Variable	GFSI	NR&R	E	W	L	ORL	S	PCTA	DS
GFSI	1.000	0.684	−0.036	0.531	0.315	−0.352	0.243	0.535	0.803
NR&R	0.684	1.000	0.135	0.694	0.428	−0.158	0.399	0.796	0.571
E	−0.036	0.135	1.000	−0.061	0.064	−0.083	−0.016	−0.028	−0.199
W	0.531	0.694	−0.061	1.000	0.271	−0.098	0.038	0.390	0.385
L	0.315	0.428	0.064	0.271	1.000	−0.051	−0.075	0.094	0.227
ORL	−0.352	−0.158	−0.083	−0.098	−0.051	1.000	−0.386	−0.306	−0.401
S	0.243	0.399	−0.016	0.038	−0.075	−0.386	1.000	0.317	0.426
PCTA	0.535	0.796	−0.028	0.390	0.094	−0.306	0.317	1.000	0.406
DS	0.803	0.571	−0.199	0.385	0.227	−0.401	0.426	0.406	1.000

Note: For the abbreviation details of variables, please refer to Table 1.

The r values in Table 2 show that W, PCTA, and DS have the highest correlation levels (0.531, 0.535, and 0.803) with the 2019 GFSI scores. A two-sample t-test is applied to clarify the linearity between these indicators and the 2019 GFSI scores. The two-sample t-test is a statistical procedure used to determine if there is a significant difference between the means of two independent datasets. Two hypotheses will be examined against each other.

These are H_0: $r = 0$ versus H_1: $r \neq 0$, where r is the slope measure. The test is initiated by calculating the p-value. The decision in the t-test is made by comparing the p-value with the significance level (α), where α is a pre-defined threshold for rejecting the H_0. If the p-value < α, we should reject hypothesis H_0 and conclude that there is a linear relationship; otherwise, if the p-value > α; we should accept hypothesis H_0 and conclude that there is no linear relationship between the compared pairs of datasets.

The three pairs of indicators (W; GFSI), (PCTA; GFSI), and (DS; GFSI) were tested, and the t-test revealed p-values = 0.00001 when $\alpha = 0.05$. Since the three p-values < 0.05, we should reject H_0 and conclude that there is sufficient evidence of a linear relationship between the 2019 GFSI scores and the tested indicators (W, PCTA, and DS). It is important to remember that this decision is limited to the case when $\alpha = 0.05$. Other values of $\alpha < 0.05$ may reveal different decisions.

The pairwise correlations in Table 2 reveal that most indicators exhibit weak correlations with each other, with values ranging from −0.016 to 0.426. However, the same indicators exhibit slightly higher correlation values with the NR&R's score. The linearity between the three top indicators, W, PCTA, and DS, has been examined using the two-sample t-test. Similar to the previous t-test results, the three pairs of indicators (W; NR&R), (PCTA; NR&R), and (DS; NR&R) revealed p-values = 0.00001 when $\alpha = 0.05$. Hence, we can conclude that there is a significant linear relationship between these indicators and the NR&R's scores.

However, despite these findings highlighting the importance of including these indicators in food security assessment, other data quality and availability factors should be considered. The following section discusses one of the common data issues, mainly outliers (or extremes), and examines their impact on the GFSI ranking scores.

4.3. Outliers Detection and Imputation

Data points that significantly deviate from most data points in the dataset are known as outliers. Outliers can disrupt the proposed approach in two critical ways: (1) outliers may significantly influence the results of the elastic-net regression, leading to improper selection of the most relevant indicators, and (2) outliers may lead to an inflation of the composite index, leading to misrepresentation of the actual contribution of indicators on the food security scores of the corresponding country.

Several outlier detection methods exist in the literature, such as the Interquartile Range (IQR), Z-scores, and winsorized methods. Each method offers unique advantages and limitations, with the optimal choice depending on the data characteristics and analysis goals. The winsorized method detects outliers and replaces them with less extreme values. The authors of [29] examined this method to substitute the outliers in the 2016 GFSI dataset. The study reported that the change in countries ranking with or without winsorized-based replacement is not significant.

To expand the research context in this regard, this study uses the Z-score method to identify potential outliers in the 2019 GFSI datasets. The outliers using the Z-score method are defined as data points whose Z-scores fall outside the Kth standard deviation. First, we compute the Z-score for each x_{ij}^{norm} using $\left(x_{ij}^{norm} - \mu_i\right)/\sigma_j$, where μ_i, and σ_j represent the mean and standard deviation of the jth indicator. Once the data have been converted, the new dataset of the jth indicator is assumed to follow the standardized normal distribution $Z\left(\mu_j^Z, \sigma_j^Z\right) \cong Z(0,1)$; then, each Z-score is associated with each data point, x_{ij}^Z, represents the distance from the center in terms of σ_j^{norm}. According to the Z-score approach, the outliers are the countries whose absolute Z-score value is higher than the threshold $K\sigma_j$, where $K\sigma_j = 3$.

While the Z-score method with a threshold of ± 3 is widely adopted in the literature for detecting outliers in normally distributed data, we acknowledge that it assumes approximate normality and may be sensitive to extreme values in skewed distributions. Alternative approaches, such as the Median Absolute Deviation (MAD) method, are more robust in the presence of skewed data or outliers, as they rely on the median rather than the mean. However, preliminary data exploration showed that most indicators followed approximately symmetric distributions after normalization, making the Z-score method a suitable choice for this study. Additionally, using a ± 3 threshold ensures that only the most extreme values (less than 1% in a normal distribution) are flagged as outliers, aligning with related studies' practices [79].

The Z-score test was applied to all the indicators, and the results showed that the 2019 GFSI dataset contains twelve outliers distributed across six indicators, accounting for 24% of the indicators. The CFC and AIT each involve three outliers, accounting for 50% of the outliers. However, there are two possible scenarios in practice for dealing with outliers. These are outliers' exclusion or substitution. While excluding outliers is essential to have outlier-free datasets, it is important to carefully investigate how the number of excluded data points affects the credibility and applicability of the composite index.

Like [29], this study replaces the outliers with the nearest low/high non-outlier values. The calculations were performed, and the six indicators exhibited a pre- and post-replacement skewness as follows: CFC $\{-1.90, -1.74\}$, AIT$\{-1.61, -0.96\}$, FL$\{-1.61, -0.57\}$, MA$\{-1.00, -0.83\}$, E$\{-0.69, -0.47\}$, and W$\{1.45, 1.36\}$. The two-sample t-test was applied to statistically check the significance of differences between the Z-scores of the six indicators pre- and post-replacement. Let μ_{pre} and μ_{post} be the mean of the Z-scores of the pre- and post-replacements, respectively. Then, the two hypotheses of the t-test are set as H_0: $\mu_{pre} - \mu_{post} = 0$ and H_1: $\mu_{pre} - \mu_{post} \neq 0$. Table A1 in Appendix A reports the settings and outputs of the t-test conducted using $\alpha = 0.05$. However, since the p-value $= 0.517 > 0.05$, we fail to reject H_0 and conclude that the means of the pre- and post-replacement Z-scores are likely to have the same mean when $\alpha = 0.05$.

For further investigation of the influence of the replacement method on the food security rank, we compared the original GFSI ranking with its counterpart using the nearest low/high non-outliers. As shown in Table 3, while 25% of countries experienced a change in rank after outlier imputation, most of these shifts were limited to ± 1 position. This aligns with previous findings by [29], where the winsorization of GFSI indicators also led to minimal impact on rankings, with most countries shifting by only one or two positions. Based on these findings, we can roughly conclude that the imputation method utilized in this study has no significant effect on the 2019 GFSI ranking scores.

Table 3. Food security rank comparison based on the imputation method.

Rank Change	Number of Countries	Percentage, %
No change	70	75
+/− 1 rank	17	18
+/− 2 ranks	6	6
+/− 3 ranks	1	1
+/− ≥4 ranks	0	0

Like many others, the nearest low-/high-imputation method can be influenced by outlier characteristics such as the number, frequency, and distribution. Alternative approaches that account for these factors are often preferable. This is especially true in high-dimensional indicator settings, where traditional methods might encounter serious challenges, such as computational complexity and non-random patterns of missing values.

4.4. Elastic-Net-Based Dimension Reduction

Elastic-net regression is utilized to reduce the dimension of the impact matrix only to include indicators that significantly impact food security. To achieve an optimal outcome in the elastic-net regression, the shrinkage parameters (λ_1 and λ_2) were optimized using the *k-fold* cross-validation (CV) method. The CV was conducted using Statistical Package for the Social Sciences (IBM®-SPSS® v29) software, a widely recognized tool for advanced data management and statistical analysis.

Due to the moderate sample size (94 countries) and the structure of SPSS's elastic-net implementation, $k = 3$ was chosen to ensure sufficient data in each fold for coefficient stability. Preliminary tests with higher k values (e.g., 5 or 10) did not enhance model performance, supporting this choice. Similar practices have been recommended in the literature for modest datasets to manage variance and ensure convergence [80,81]. Two reduced models (M1 and M2) with dimensions ($p = 15$ and $p = 9$) were identified. However, given that the initial pool size of indicators is 25, we believe that 15 and 9 are still significant numbers. The ranges of λ_1 and λ_2 are set at (0.10 to 0.14) and (0.40 to 0.24), respectively. These narrow intervals were selected after broader preliminary tuning experiments showed that optimal performance was consistently obtained within these specific bounds. Constraining the range improved model precision and reduced unnecessary variability in parameter estimation.

Standard error (SE) is a widely used metric to assess the precision of regression models. In the context of this study, the SE reflects the average variation between the predicted prevalence of undernourishment, the % of the population (the dependent variable), and the actual observed prevalence in the population for a given combination of indicators. Based on the results presented in Table 3, one can say that M2 is a more optimized model for predicting malnutrition prevalence. The M2's lower SE implies that its predictions exhibit less deviation from the actual prevalence data compared to M1.

Table 4 reports the reduced dimensions of the four food security categories and the regression coefficients for the remaining indicators. The "Reduction/Category, %" and "Average Reduction, %" columns display the percentage and average reduction in each category's original number of indicators.

Table 4. Elastic-net regression coefficients and reduced dimension.

Category	Indicator	Coefficient Estimators		Reduction, %		Average Reduction, %
		M1	M2	M1	M2	
Affordability	CFC	-		50.00	66.67	58.34
	PPUGP	0.06	0.004			
	AIT	−0.02	-			
	FSNP	0.45	0.35			
Availability	SOS	−0.03	-	57.14	70.14	63.64
	ARAD	−0.05	-			
	PASBA	-	−0.003			
	FL	-	-			
	FSAAPC	0.09	0.15			
Quality and safety	DD	-	-	40.00	60.00	50.00
	NS	−0.02	-			
	MA	−0.13	−0.07			
	PQ	0.05	0.01			
	FS	-	-			

Table 4. Cont.

Category	Indicator	Coefficient Estimators		Reduction, %		Average Reduction, %
		M1	M2	M1	M2	
Natural resources and resilience	E	0.06	0.08	28.57	57.14	42.86
	L	−0.34	−0.27			
	ORL	0.02	-			
	S	−0.06	−0.01			
	PCTA	−0.05	-			
	DS	0.06	-			

The percentage of model reduction significantly plays a vital role in balancing the information retained and the resulting model complexity. To understand the trade-off between complexity and information content in the food security scores, we will compare the changes in the GFSI ranking scores of the two reduced models.

4.5. Food Security Score Calculation

This section illustrates the calculations of the proposed food security index. However, while reducing the number of indicators may appear initially advantageous, it can have two significant consequences. First, it can hide important information regarding a country's food security performance. For example, removing indicators such as political and social barriers to access (PASBA) or micronutrient availability (MA) could mask systemic access issues or hidden forms of malnutrition. Second, it can impact the ability of decision makers and stakeholders to understand the nature of the problem and develop effective, targeted solutions. For instance, without indicators like market access and agricultural financial services (MAAFS) or agricultural infrastructure (AIS), it becomes more difficult to distinguish whether food insecurity stems from logistical, financial, or policy-related barriers. Given that there is no specific metric for optimal dimensionality, this study will evaluate the new index utilizing the M1 model, which we believe has an acceptable dimension ($p = 15$) compared with M2 ($p = 9$) and is slightly larger than M2. However, the M2 model will be utilized later to examine the impact of the dimension reduction on the food security scores.

Also, while the regression coefficients in Table 3 provide clear insights into the direction and strength of the relationships between the indicators and malnutrition prevalence, it is essential to acknowledge that using these coefficients directly as weights remains an active area of research, especially when the relationships between indicators and response variable are not significantly linear. To avoid potential limitations associated with the regression coefficient-based weighting, this paper will utilize the two weighting approaches used by the existing GFSI: equal weighting and expert-based weighting.

The M1 model will be examined using two different weighting approaches: equal-weighting (W1) and expert-based weighting (W2). In the following, the examined models will be referred to as M1W1AA and M1W2AA, represented in Figure 2 using solid arrows. The percentages in Figure 2 (100%, 40%, and 64%) represent the percentage of remaining indicators.

4.5.1. Testing the Impact of the Weighting Method

This section will examine the impact of the weighting method on the GFSI scores and rank of the reduced models. The M1 model will be examined under two different weighting approaches: equal weighting (W1) and expert-based weighting (W2). In the following, the examined models will be referred to as M1W1AA and M1W2AA, where AA refers to accumulative aggregation.

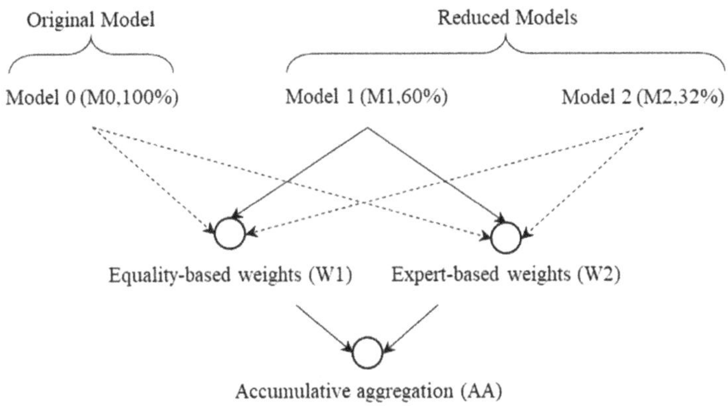

Figure 2. Models for calculating the elastic-net-based GFSI.

However, as the reduced model (M1) has fewer indicators (p = 15), the GFSI will redistribute the weight of each category to the new dimension of indicators using the same ratio. For instance, if a specific category has a weight equal to 0.3 and has three dimensions, then each indicator will be assigned a weight value equal to 0.1. If this category's dimension is reduced to two indicators, then the weight of each indicator will be reassigned as 0.15.

The food security scores using the two models (M1W1AA; M1W2AA) were calculated for 94 countries. The results indicated that these models' means and standard deviations are {61.58; 12.54} and {63.0; 15}, respectively. Another t-test was conducted to investigate the significance of the difference between these models' means of food security score. The t-test resulted in a p-value = 0.484. However, since the p-value > α = 0.05, we fail to reject H_0 and conclude that there is no statistically significant difference between the means of the food security scores of these models. Another rank change comparison examines how the food security scores are sensitive to the weighting method (see Table 5).

Table 5. Food security rank comparison based on the weighting method.

Rank Change	Number of Countries	Percentage, %
No change	18	19.15
+/− 1 rank	9	9.57
+/− 2 ranks	14	14.89
+/− 3 ranks	8	8.51
+/− ≥4 ranks	45	47.87

Despite the lack of statistical significance in the mean scores, the results of this comparison show that a substantial proportion of countries (80.83%) experienced changes in their food security rankings when the weighting scheme was altered from equal (W1) to expert-based (W2). Nearly half of the countries (47.87%) shifted by four or more ranks. These findings illustrate the dual implications of weighting choices. On the one hand, the expert-based method allows domain-specific priorities to shape the index, potentially giving more weight to underemphasized but strategically essential indicators. On the other hand, this approach can introduce subjectivity and increase the "judgment cost" for end-users who must interpret or justify the resulting scores. In particular, when expert weighting amplifies the influence of relatively minor indicators, rank shifts may reflect methodological decisions rather than actual performance differences, potentially complicating interpretation and reducing transparency. By contrast, equal weighting provides a neutral and easily interpretable baseline, although it may oversimplify the relative importance of indicators and ignore interdependencies. These trade-offs underscore the importance of adopting transparent, data-driven weighting strategies that balance interpretability, replicability, and policy relevance.

It is important to note that direct comparisons between these models and the original GFSI might be limited due to variations in indicator subsets and weights used by the GFSI. However, the importance of M1W1AA and M1W2AA lies in their focus on a smaller set of crucial indicators. This focus could benefit stakeholders and policymakers seeking to address food insecurity challenges and develop practical solutions.

4.5.2. Testing the Impact of the Dimension Reduction

This section investigates the influence of dimension reduction on the rankings derived from the food security scores. We will compare the performances of M1W1AA and M2W1AA using the same weighting technique. Two key factors are expected to affect model performance. These are the type and number of selected indicators. As shown in Table 3, M1 and M2 share only eight indicators (PPUGP, FSNP, FSAAPC, MA, PQ, E, L, and S). Notably, M1 utilizes eight additional indicators not included in M2. This difference in indicator selection might lead to significant variations in the resulting food security scores and rankings.

The food security scores were calculated for 94 countries using the two models. The analysis revealed that the means and standard deviations for the food security scores of these models were (61.58; 12.54) and (66.89; 14.25), respectively. A two-sample t-test was conducted to statistically assess the significance of the difference in the mean food security scores between the M1W1AA and M2W1AA models.

The t-test results showed that the p-value = 0.07 ($>\alpha = 0.05$). This leads to the conclusion that there is no significant difference between the means of the food security scores of M1W1AA and M2W1AA. In other words, there is no strong evidence of a substantial difference between the average food security scores produced by the two models. This finding can be attributed to two key factors. First, applying dimension reduction techniques can lead to different sets of indicators being included within the same food security category across various models. Second, even if the weight assigned to a specific category remains equal across all its indicators, variations in the indicator sets used by the compared models can still result in different weights being assigned, even to common indicators. In other words, the weight of a particular indicator on the food security scores and ranks can vary significantly between models that utilize different dimensions. However, unlike the PCA and FA, which aim to capture overall data variability through linear transformation reduction, elastic-net regression prioritizes selecting the most influential features. Consequently, this finding may not necessarily generalize to other dimensionality reduction techniques, which often prioritize capturing variance rather than feature selection.

5. Two-Step Cluster Analysis

The two-step cluster analysis represents a critical contribution to this paper. By applying this method, we can comprehensively understand the relationship between food security scores and food system characteristics. The two-step approach can help identify clusters of countries with comparable food security performance and identify key characteristics that most significantly contribute to a country's performance.

The two-step cluster is a data clustering technique that can handle large sets of data, and it uses both categorical and continuous variables to create data clusters. The objective is to group similar observations based on data attributes. Utilizing two-step clustering can help cluster countries not just on one basis but with different parameters, which allows us to understand the nature of clusters better and, hence, understand the countries' food security performance.

This two-step clustering will be applied to the M1W1AA model, using inputs of food security scores, food system type, and economic state of the countries involved. The last two inputs represent the characteristics of the food system. The food system type involves five categories: emerging and diversifying, industrial and consolidated, informal and expanding, modernizing and formalizing, and rural and traditional [82]. The economic state is classified into four categories based on the gross national income (GNI) [83]. These are low-income, low–middle–income, upper–middle–income, and high-income.

5.1. Step One: Food Security Score-Based Clustering

In the first step, countries are grouped based on their food security scores to identify clusters with similar performance levels. As shown in Figure 3, 14.9% of countries fall into the low-performing cluster (Cluster 1), with an average score of 43.05. Meanwhile, 23.4% belong to the high-performing cluster (Cluster 4), with a mean score of 73.73. The remaining 61.7% are distributed across Clusters 2 and 3, representing average-performing countries, with scores of 54.89 and 66.45, respectively. While this initial clustering helps categorize countries by performance, it does not reveal why such differences exist. Understanding the underlying food system characteristics is essential for developing targeted strategies and learning from better-performing peers. This motivation drives the second step of the analysis.

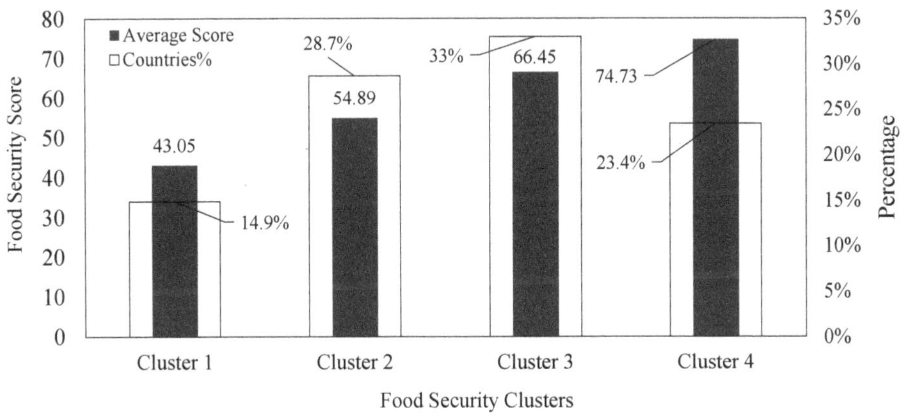

Figure 3. Food security clusters using the M1W1AA model.

5.2. Step Two: Characteristic-Based Clustering

The second step assesses the relationship between food system type, economic status, and food security outcomes. Based on these two characteristics, the clusters identified in the previous step will be analyzed to understand how the food system type and economic status may influence food security scores. This step explores how different combinations of food system characteristics and economic status contribute to variations in food security outcomes across the clusters.

Figure 4 shows that 92.9% of the countries in Cluster 1 adopt rural and traditional food systems. These systems often face challenges, inefficient transportation networks, financial shortages, and a lack of technological advancement, making them less secure than other systems. However, while adopting rural and traditional food systems in low-income countries may offer some environmental advantages, it might limit the capacity to maintain food security.

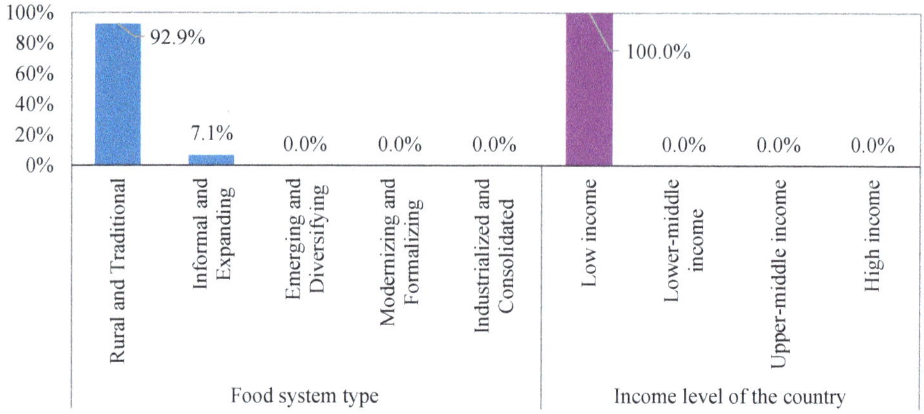

Figure 4. Food system characteristic-based grouping of Cluster 1.

Figure 5 reveals that 51.9% of the countries in Cluster 2 adopt informal and expanding food systems. The remaining 48.1% of the countries adopt either emerging and diversifying or rural and traditional food systems. Notably, although 33.3% of the countries in this cluster adopted rural and conventional food systems, such as Tanzania, Kenya, and Bangladesh, these countries demonstrated high security scores compared with Cluster 1. The economic strength of these countries likely explains this finding. All the countries in Cluster 2 have lower–middle income levels, which offer limited but essential resources to mitigate some food security challenges associated with rural and traditional food systems. This underscores the crucial role of economic status in shaping food security outcomes.

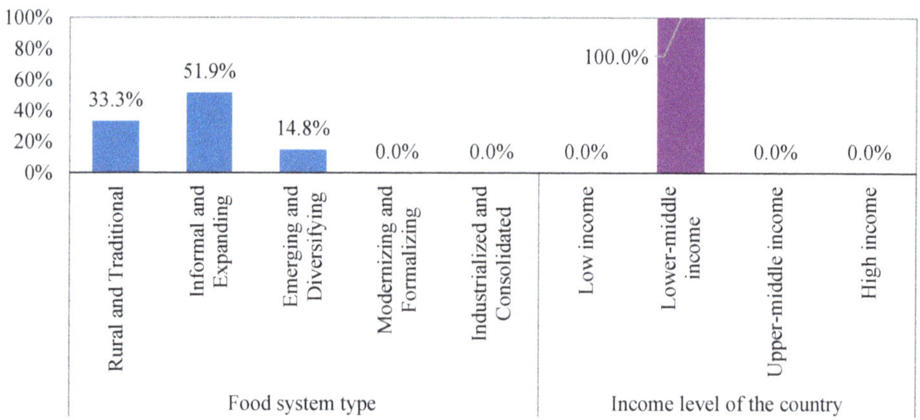

Figure 5. Food system characteristic-based grouping of Cluster 2.

Figure 6 shows that 45.2% of the countries within Cluster 3 adopt the modernizing and formalizing food system type, while 35.5% adopt the emerging and diversifying food system type. The economic situation regarding the countries' GDP in this cluster varies between upper–middle (77.4%) and high income (22.6%). The rise in the income level of the countries in Cluster 3 significantly contributes to improving the food systems of these countries, such as Indonesia, Azerbaijan, and Thailand.

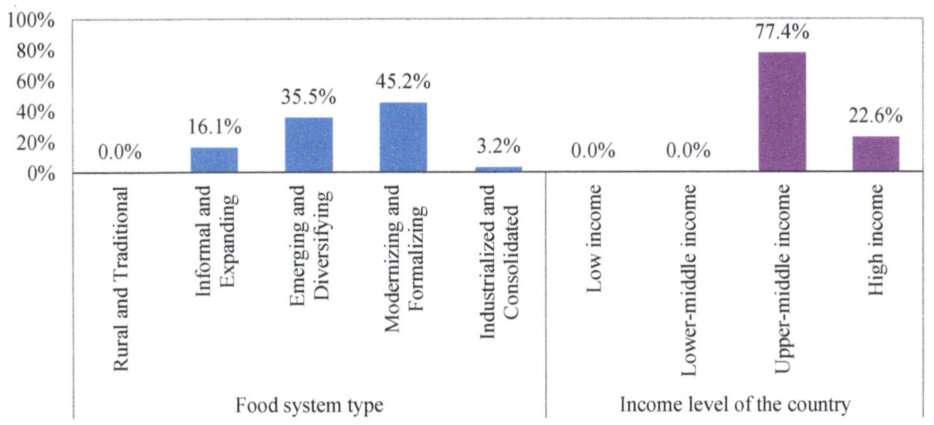

Figure 6. Food system characteristic-based grouping of Cluster 3.

Figure 7 shows that 100% of the countries in Cluster 4 adopt industrialized and consolidated food system types, including Switzerland, Germany, and Finland. These systems are well known for utilizing advanced technology and automation to maximize their productivity and efficiency, which are important for maintaining economic growth and ensuring self-sufficiency in food and nutrition. The results also show that 100% of these countries are classified as high-income countries. This finding is unsurprising, as many jobs across various sectors, from manufacturing and engineering to logistics and transportation, created by these systems likely contribute to their economic status.

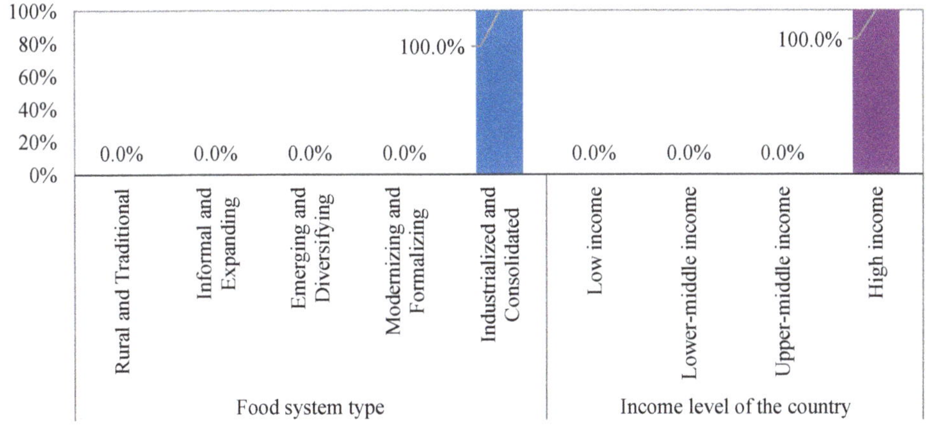

Figure 7. Food system characteristic-based grouping of Cluster 4.

The two-step cluster analysis unveiled a significant correlation between food system type, economic status, and food security level. This revelation is crucial for understanding the dynamics of food security. While industrialized and consolidated food systems have demonstrated adequate food security levels, rural and traditional food systems face serious challenges. This is particularly true for countries with low income levels, underscoring the importance of our research in shedding light on these critical issues.

Most of the existing food security analyses use traditional score-based ranking, which does not consider the characteristics of the food system. Scores provide a single numeric value, which can oversimplify the multifaceted nature of food security. They may not capture underlying issues and may even ignore context-specific factors like cultural food preferences, regional disparities, and the impact of local policies, which are crucial for a holistic understanding of food security. They also lack nuance in policymaking, wherein relying solely on scores may lead to generic policy recommendations that may be ineffective for all regions or populations. Different areas might have similar scores but require distinct interventions due to their unique challenges. For instance, despite similar food security scores, i.e., 76, Poland and China require different food security strategies due to distinct challenges, as they have different income levels and types of food systems, too. Poland, a high-income country with an aging population and EU policy constraints, should focus on agricultural innovation, rural development, and adapting EU policies for competitiveness. Meanwhile, China, an upper–middle-income nation facing rapid urbanization, environmental degradation, and food safety issues, must prioritize sustainable farming, stringent food safety regulations, and optimizing supply chains. These strategies reflect each country's unique socioeconomic and environmental contexts, highlighting the need for tailored approaches to achieve food security. While different areas having similar scores may need different interventions because of their unique challenges, vice versa also holds true. Kenya was awarded a food security score of 47.4, while Cambodia could only manage to secure a food security score 57.7. Both countries have rural and traditional food systems and belong to the same "lower–middle income" category. Both countries need investments in rural infrastructure and access to markets. Despite Cambodia's higher score, both face challenges related to rural poverty, underdeveloped infrastructure, and limited market access. Interventions in road building, market access improvements, and rural development programs would benefit both.

To assess the external relevance of the identified clusters, we compared our results with country classifications from the 2019 Global Hunger Index (GHI) [84], a widely recognized indicator of food insecurity severity. While our clustering framework was designed to capture structural characteristics, namely food system typologies and economic status, rather than outcome measures, this comparative analysis provides useful context.

The comparison revealed a notable alignment between our clusters and the GHI categories. Countries grouped into Cluster 1, characterized by rural and traditional food systems and low-income status, predominantly fall under the "Serious" or "Alarming" GHI categories (e.g., Yemen, Madagascar, and Chad). In contrast, Cluster 4, composed mainly of high-income countries with industrialized and consolidated food systems, corresponds closely with countries categorized as "Low" in the GHI (e.g., Germany, Canada, and Australia). Intermediate clusters (Clusters 2 and 3) include countries with lower–middle or upper–middle income levels and a broader diversity of food system structures, reflecting a wider spread across the "Moderate" to "Low" GHI categories.

This alignment suggests that while our typology does not directly mirror score-based indices, it captures meaningful structural dimensions correlating with food security outcomes. The results support the view that food insecurity is not only a matter of outcomes but is also deeply rooted in systemic configurations, reinforcing the potential of structural clustering approaches as complementary tools to traditional classification systems.

Relying solely on score-based rankings may result in misguided prioritization by policymakers, as these scores often lack the nuance required to target specific needs. Identifying the underlying drivers of food insecurity becomes challenging without incorporating structural characteristics such as the food system type, income level, infrastructure, market access, etc. Structural clustering, thus, presents a more comprehensive and actionable framework for designing effective, context-specific interventions.

6. Conclusions and Remarks

This study presented a new food security assessment approach that combines two well-established methods, feature selection and two-step clustering, to address some of the limitations of its existing counterparts and provide decision makers and stakeholders with systematic procedures for making informed decisions and supporting food system management. The new approach's mathematical and operational procedures were assessed using the EIU dataset of 94 food-producing countries. Below are valuable findings and remarks extracted from this study.

The analysis reveals the influence of the food system typology and economic state on the food security levels of the underlying countries. Countries with industrialized and consolidated food systems, typically high-income economies such as Switzerland and Germany, consistently demonstrate the highest levels of food security. In contrast, countries with rural and traditional food systems and low-income status exhibited significantly lower performance. For instance, 92.9% of the countries in Cluster 1 adopted rural and traditional food systems and faced numerous challenges, including limited infrastructure and technological capabilities.

The cluster-based output format of the proposed method enhances the interpretability and practical utility of food security assessments. It enables policymakers to link aggregated scores with underlying country characteristics and context-specific design interventions. For example, countries in clusters dominated by rural and traditional food systems could benefit from policies to improve transportation infrastructure, expand access to agricultural finance, and foster agro-technological innovation. Development partners might prioritize investments in areas such as farmer training, mobile extension services, and climate-resilient crop systems. Conversely, in countries with industrialized food systems but persistent food insecurity, policy efforts could focus on improving food distribution equity, supporting smallholder inclusion, or reducing food waste. Tailoring interventions based on cluster identity facilitates more efficient resource allocation and ensures that policy responses are context-sensitive and impactful. Importantly, the study highlights that similar food security scores can mask different underlying realities. For instance, although Kenya and Cambodia received different food security scores (47.4 and 57.7, respectively), both share the same income level and food system type, indicating similar policy needs. Conversely, despite having comparable scores, countries like Poland and China require distinct strategies due to differences in their economic profiles and food system typologies. These examples underscore the added value of the clustering approach in uncovering policy-relevant patterns that a single score cannot.

While clustering enhances interpretability and supports tailored policy design, it is essential to acknowledge that typologies may not capture the full complexity of country-specific contexts. Over-reliance on aggregated cluster identities risks overlooking unique socio-political, cultural, or institutional dynamics influencing food security. Therefore, while the clustering framework provides a helpful starting point for targeted intervention design, policymakers are encouraged to complement it with country-level diagnostics and stakeholder engagement to ensure relevance and effectiveness.

The new approach has shown appropriate applicability in assessing food security. By adjusting the two shrinkage parameters of the elastic-net regression, the practitioner can examine and analyze food security performance using different subsets of indicators and dimensions. This feature can help to investigate how the inclusion of specific single or multiple groups of indicators influences the overall performance of food system security. Although the proposed approach effectively captures food security performance, some limitations should be acknowledged. First, the analysis is based solely on cross-sectional data from GFSI 2019, which limits the ability to track changes or trends over time. Expanding to multi-year datasets would allow for a more dynamic understanding. Second, relying on a single data source may constrain indicator diversity and coverage; incorporating other datasets could offer a more comprehensive view. Third, elastic-net regression assumes linear relationships between predictors and outcomes, which may not fully capture the complex nature of food systems. This limitation could be addressed using non-linear models such as random forests or neural networks. Although missing data were handled through imputation, some sensitivity to outlier handling and imputation choices persists, and future work should systematically compare alternative methods. Finally, the clustering relied on only two country-level features; involving experts to guide the inclusion of more relevant characteristics could enhance the contextual relevance of the groupings. While this study used country-level data for cross-sectional analysis, future work could test the framework at the subnational level to better capture local disparities. Moreover, incorporating additional dimensions such as food system resilience or environmental stressors could enhance the model's capacity to inform targeted, region-specific interventions. Additionally, while the method may not be immediately transferable to regions with limited data, it provides a flexible framework that can be adapted over time. In low-data contexts, the methodology can prioritize available key indicators and gradually improve its robustness as more data becomes accessible, making it applicable to a broader range of settings.

Author Contributions: Conceptualization, M.A.A.-A. and G.M.A.; Formal analysis, M.A.A.-A. and H.N.; Methodology, M.A.A.-A. and G.M.A.; Software, H.N.; Supervision, G.M.A. and T.E.M.; Validation, H.N.; Writing—original draft, M.A.A.-A. and H.N.; Writing—review and editing, H.N. and G.M.A. All authors have read and agreed to the published version of the manuscript.

Funding: The publication of this article was funded by the Qatar National Library. This research was funded by Qatar National Research Fund (GSRA6-1-0614-19088) and Qatar University (I-GA-1030).

Institutional Review Board Statement: Not Applicable.

Informed Consent Statement: Not Applicable.

Data Availability Statement: The original contributions presented in this study are included in the article. Further inquiries can be directed to the corresponding author.

Conflicts of Interest: The authors declare that they have no known competing financial interests or personal relationships that could have appeared to influence the work reported in this paper.

Abbreviations

Acronym	Definition
IQR	Interquartile Range
AHP	Analytical Hierarchy Process
CV	Cross-Validation
DEA	Data Envelope Analysis
EIU	Economist Intelligence Unit
FA	Factor Analysis
FAO	Food and Agriculture Organization
FIES	Food Insecurity Experience Scale
GFSI	Global Food Security Index
GHI	Global Hunger Index
GNI	Gross national income
IFPRI	International Food Policy Research Institute
LASSO	Least Absolute Squared Shrinkage Operator
NR&R	Natural resources and resilience

PCA	Principal Component Analysis
SE	Standard error
SPSS	Statistical Package for the Social Sciences
TOPSIS	Technique for Order of Preference by Similarity to the Ideal Solution

List of Mathematical Notations/Abbreviations

Symbol	Description
y	The dependent variable for the elastic-net regression
x	The independent variables for the elastic-net regression
β_j	The coefficients of the regression model
ε	Normally distributed random error term
$P_\lambda(\beta)$	Penalty function
r	Pearson's correlation coefficient
l_1 and l_2	Penalty functions 1 and 2
λ_1 and λ_2	Shrinkage parameters
x_{ij}	The original score of the ith country under the jth indicator
$M_{max,j}$ and $M_{min,j}$	Represent the maximum and minimum scores of the jth indicator

Appendix A

Table A1. Two-sided t-test: two-sample Z-score.

Statistic	Value
Difference ($\mu_{pre} - \mu_{post}$)	−0.630
Observed t	−0.658
Critical $\lvert t \rvert$	2.074
Degrees of freedom	22
p-value (two-tailed)	0.517
Significance level (α)	0.05

References

1. Food and Agriculture Organization of the United Nations. Food Security: Concepts and Measurement. Available online: https://www.fao.org/4/y4671e/y4671e06.htm (accessed on 22 April 2025).
2. Stringer, R. Food Security Global Overview. In *Food Poverty and Insecurity: International Food Inequalities*; Springer: Berlin/Heidelberg, Germany, 2016; pp. 11–18. [CrossRef]
3. Upton, J.B.; Denno Cissé, J.; Barrett, C.B. Food Security as Resilience: Reconciling definition and measurement. *Agric. Econ.* **2016**, *47*, 135–147. [CrossRef]
4. Henry, B.; Murphy, B.; Cowie, A. *Sustainable Land Management for Environmental Benefits and Food Security*; A Synthesis Report for the GEF: Washington, DC, USA, 2018.
5. Swaminathan, M.S.; Bhavani, R.V. Food production & availability—Essential prerequisites for sustainable food security. *Indian J. Med. Res.* **2013**, *138*, 383. [CrossRef]
6. Chamdimba, O.Y.; Ortmann, G.F.; Wale, E. Resilience to shocks and food insecurity: Determinants, and the impact of smallholder *Jatropha curcas* cultivation in Southern Malawi. *Afr. J. Sci. Technol. Innov. Dev.* **2021**, *13*, 619–633. [CrossRef]
7. Hwalla, N.; El Labban, S.; Bahn, R.A. Nutrition security is an integral component of food security. *Front. Life Sci.* **2016**, *9*, 167–172. [CrossRef]
8. Nahid, N.; Lashgarara, F.; Hosseini, S.J.F.; Mirdamadi, S.M.; Rezaei-Moghaddam, K. Determining the Resilience of Rural Households to Food Insecurity during Drought Conditions in Fars Province, Iran. *Sustainability* **2021**, *13*, 8384. [CrossRef]
9. Wiebe, K.; Robinson, S.; Cattaneo, A. Climate Change, Agriculture and Food Security: Impacts and the Potential for Adaptation and Mitigation. In *Sustainable Food and Agriculture—An Integrated Approach*; Campanhola, C., Pandey, S., Eds.; The Food and Agricultural Organisation of the United Nations (FAO): Rome, Italy, 2019; pp. 55–74. [CrossRef]
10. Lee, C.C.; Zeng, M.; Luo, K. How does climate change affect food security? Evidence from China. *Environ. Impact Assess. Rev.* **2024**, *104*, 107324. [CrossRef]
11. Eswaran, S.; Anand, A.; Lairenjam, G.; Mohan, G.; Sharma, N.; Khare, A.; Bhargavi, A. Climate Change Impacts on Agricultural Systems Mitigation and Adaptation Strategies: A Review. *J. Exp. Agric. Int.* **2024**, *46*, 1–12. [CrossRef]

12. Dinar, A.; Tieu, A.; Huynh, H. Water scarcity impacts on global food production. *Glob. Food Secur.* **2019**, *23*, 212–226. [CrossRef]
13. Kwasek, M. Threats To Food Security And Common Agricultural Policy. *Econ. Agric.* **2013**, *59*, 701–713. [CrossRef]
14. Bullock, J.M.; Dhanjal-Adams, K.L.; Milne, A.; Oliver, T.H.; Todman, L.C.; Whitmore, A.P.; Pywell, R.F. Resilience and food security: Rethinking an ecological concept. *J. Ecol.* **2017**, *105*, 880–884. [CrossRef]
15. Campbell, B.M.; Vermeulen, S.J.; Aggarwal, P.K.; Corner-Dolloff, C.; Girvetz, E.; Loboguerrero, A.M.; Ramirez-Villegas, J.; Rosenstock, T.; Sebastian, L.; Thornton, P.K.; et al. Reducing risks to food security from climate change. *Glob. Food Secur.* **2016**, *11*, 34–43. [CrossRef]
16. Korneiko, O.V.; Rogaleva, N.L. The threats to food security. *Amazon. Investig.* **2018**, *7*, 232–242. Available online: https://amazoniainvestiga.info/index.php/amazonia/article/view/398 (accessed on 22 April 2025).
17. Nechifor, V.; Ramos, M.P.; Ferrari, E.; Laichena, J.; Kihiu, E.; Omanyo, D.; Musamali, R.; Kiriga, B. Food security and welfare changes under COVID-19 in Sub-Saharan Africa: Impacts and responses in Kenya. *Glob. Food Secur.* **2021**, *28*, 100514. [CrossRef]
18. Mandal, S.C.; Boidya, P.; Haque, I.-M.; Hossain, A.; Shams, Z.; Mamun, A.-A. The impact of the COVID-19 pandemic on fish consumption and household food security in Dhaka city, Bangladesh. *Glob. Food Secur.* **2021**, *29*, 100526. [CrossRef]
19. Erezi, E.; Ehi, O.E.; Ayodeji, O.T. Promoting Sustainable Agriculture and Climate Resilience in African Nations. *Int. J. Agric. Earth Sci.* **2023**, *9*, 27–45. [CrossRef]
20. Al-Muqdadi, S.W.; Abdalwahab, N.S.; Almallah, I.A.; Hussein, S.F.; Yassin, A.; Salim, S.O.; Alfarraji, M.J.; Mishbak, H.H.; Altuma, L.R.; Ab-Dulrahman, B. The complex system of climate change security and the ripple effect of water–food–socioeconomic nexus. *J. Infrastruct. Policy Dev.* **2024**, *8*, 3928. [CrossRef]
21. Kuei, C.-H.; Lu, M.H. Integrating quality management principles into sustainability management. *Total Qual. Manag. Bus. Excell.* **2013**, *24*, 62–78. [CrossRef]
22. Meade, B.; Thome, K. International Food Security Assessment, 2017–2027. In *Amber Waves: The Economics of Food, Farming, Natural Resources, and Rural America*; U.S. Department of Agriculture, Economic Research Service: Washington, DC, USA, 2017. [CrossRef]
23. Cole, M.B.; Augustin, M.A.; Robertson, M.J.; Manners, J.M. The science of food security. *NPJ Sci. Food* **2018**, *2*, 14. [CrossRef]
24. Irtyshcheva, I.; Ponomarova, M.; Dolzhykova, I. Conceptual Fundamentals of Development of the Food Security System. *Balt. J. Econ. Stud.* **2019**, *5*, 57–64. [CrossRef]
25. Rosen, S.; Meade, B.; Fuglie, K.; Rada, N.E. *International Food Security Assessment, 2014–2024*; United States Department of Agriculture: Washington, DC, USA, 2014.
26. Pérez-Escamilla, R.; Gubert, M.B.; Rogers, B.; Hromi-Fiedler, A. Food security measurement and governance: Assessment of the usefulness of diverse food insecurity indicators for policy makers. *Glob. Food Secur.* **2017**, *14*, 96–104. [CrossRef]
27. Rohr, V.; Blakley, J.; Loring, P. A framework to assess food security in regional strategic environmental assessment. *Environ. Impact Assess. Rev.* **2021**, *91*, 106674. [CrossRef]
28. von Grebmer, K.; Bernstein, J.; Hossain, N.; Brown, T.; Prasai, N.; Yohannes, Y.; Patterson, F.; Sonntag, A.; Zimmerman, S.-M.; Towey, O. *2017 Global Hunger Index: The Inequalities of Hunger*; International Food Policy Research Institute: Washington, DC, USA, 2017. Available online: https://ideas.repec.org/b/fpr/ifprib/9780896292710.html (accessed on 22 April 2025).
29. Thomas, A.-C.; Dhombres, B.; Casubolo, C.; Saisana, M.; Kayitakire, F. *The Use of the Global Food Security Index to Inform the Situation in Food Insecure Countries*; EUR 28885 EN; The Joint Research Centre: Ispra, Italy, 2017. [CrossRef]
30. Maricic, M.; Bulajic, M.; Dobrota, M.; Jeremic, V. Redesigning The Global Food Security Index: A Multivariate Composite I-Distance Indicator Approach. *Int. J. Food Agric. Econ.* **2016**, *4*, 69–86. [CrossRef]
31. Booysen, F. An Overview and Evaluation of Composite Indices of Development. *Soc. Indic. Res.* **2002**, *59*, 115–151. [CrossRef]
32. Poudel, D.; Gopinath, M. Exploring the disparity in global food security indicators. *Glob. Food Secur.* **2021**, *29*, 100549. [CrossRef]
33. Pangaribowo, E.H.; Gerber, N.; Torero, M.A. Food and Nutrition Security Indicators: A Review. *SSRN* **2013**. [CrossRef]
34. Cafiero, C.; Gheri, F.; Viviani, S. Validating the Food Insecurity Experience Scale for Use In Analyses of Recent Food Insecurity. *Glob. Food Secur.* **2024**, *42*, 100783. [CrossRef]
35. Iversen, T.O.; Westengen, O.; Jerven, M. Measuring the end of hunger: Knowledge politics in the selection of SDG food security indicators. *Agric. Hum. Values* **2023**, *40*, 1273–1286. [CrossRef]
36. Abdella, G.M.; Kucukvar, M.; Kutty, A.A.; Abdelsalam, A.G.; Sen, B.; Bulak, M.E.; Onat, N.C. A novel approach for developing composite eco-efficiency indicators: The case for US food consumption. *J. Clean. Prod.* **2021**, *299*, 126931. [CrossRef]
37. Mori, K.; Christodoulou, A. Review of sustainability indices and indicators: Towards a new City Sustainability Index (CSI). *Environ. Impact Assess. Rev.* **2012**, *32*, 94–106. [CrossRef]
38. Gan, X.; Fernandez, I.C.; Guo, J.; Wilson, M.; Zhao, Y.; Zhou, B.; Wu, J. When to use what: Methods for weighting and aggregating sustainability indicators. *Ecol. Indic.* **2017**, *81*, 491–502. [CrossRef]
39. Al-Ansari, M.A.; Nabeel, H.; Abdella, G.M.; El Mekkawy, T.; Kutty, A.A. A Non-Parametric Approach-Based Trade-Off between Food System Efficiency and Robustness. *Sustainability* **2024**, *16*, 6480. [CrossRef]
40. Lee, Y.J.; Huang, C.M. Sustainability index for Taipei. *Environ. Impact Assess. Rev.* **2007**, *27*, 505–521. [CrossRef]

41. Freudenberg, M. *Composite Indicators of Country Performance: A Critical Assessment*; OECD Science, Technology and Industry Working Papers: Paris, France, 2003. [CrossRef]
42. Reisi, M.; Aye, L.; Rajabifard, A.; Ngo, T. Transport sustainability index: Melbourne case study. *Ecol. Indic.* **2014**, *43*, 288–296. [CrossRef]
43. Kutty, A.A.; Kucukvar, M.; Abdella, G.M.; Bulak, M.E.; Onat, N.C. Sustainability Performance of European Smart Cities: A Novel DEA Approach with Double Frontiers. *Sustain. Cities Soc.* **2022**, *81*, 103777. [CrossRef]
44. Tversky, A.; Kahneman, D. Judgment under uncertainty: Heuristics and biases. *Science* **1974**, *185*, 1124–1131. [CrossRef] [PubMed]
45. Montibeller, G.; von Winterfeldt, D. Individual and Group Biases in Value and Uncertainty Judgments. In *International Series in Operations Research and Management Science*; Dias, L., Morton, A., Quigley, J., Eds.; Springer: Cham, Switzerland, 2018; pp. 377–392. [CrossRef]
46. Meyer, M.; Grisar, C.; Kuhnert, F. The impact of biases on simulation-based risk aggregation: Modeling cognitive influences on risk assessment. *J. Manag. Control* **2011**, *22*, 79–105. [CrossRef]
47. Hemming, V.; Hanea, A.M.; Burgman, M.A. What is a Good Calibration Question? *Risk Anal.* **2021**, *42*, 264–278. [CrossRef]
48. Dharmarathne, G.; Nane, G.F.; Robinson, A.; Hanea, A.M. Shrinking the Variance in Experts' "Classical" Weights Used in Expert Judgment Aggregation. *Forecasting* **2023**, *5*, 522–535. [CrossRef]
49. Bolger, F.; Rowe, G. There is Data, and then there is Data: Only Experimental Evidence will Determine the Utility of Differential Weighting of Expert Judgment. *Risk Anal.* **2015**, *35*, 21–26. [CrossRef]
50. Zhang, X.; Austin, S.; Glass, J.; Mills, G. Toward collective organizational values: A case study in UK construction. *Constr. Manag. Econ.* **2008**, *26*, 1009–1028. [CrossRef]
51. Abdella, G.M.; Kucukvar, M.; Ismail, R.; Abdelsalam, A.G.; Onat, N.C.; Dawoud, O. A mixed model-based Johnson's relative weights for eco-efficiency assessment: The case for global food consumption. *Environ. Impact Assess. Rev.* **2021**, *89*, 106588. [CrossRef]
52. Abdella, G.M.; Kim, J.; Al-Khalifa, K.N.; Hamouda, A.M. Penalized Conway-Maxwell-Poisson regression for modelling dispersed discrete data: The case study of motor vehicle crash frequency. *Saf. Sci.* **2019**, *120*, 157–163. [CrossRef]
53. Park, Y.S.; Egilmez, G.; Kucukvar, M. Emergy and end-point impact assessment of agricultural and food production in the United States: A supply chain-linked Ecologically-based Life Cycle Assessment. *Ecol. Indic.* **2016**, *62*, 117–137. [CrossRef]
54. Jolliffe, I.T.; Cadima, J. Principal component analysis: A review and recent developments. *Philos. Trans. R. Soc. A Math. Phys. Eng. Sci.* **2016**, *374*, 20150202. [CrossRef]
55. Kutty, A.A.; Wakjira, T.G.; Kucukvar, M.; Abdella, G.M.; Onat, N.C. Urban resilience and livability performance of European smart cities: A novel machine learning approach. *J. Clean. Prod.* **2022**, *378*, 134203. [CrossRef]
56. Kutty, A.A.; Kucukvar, M.; Onat, N.C.; Ayvaz, B.; Abdella, G.M. Measuring sustainability, resilience and livability performance of European smart cities: A novel fuzzy expert-based multi-criteria decision support model. *Cities* **2023**, *137*, 104293. [CrossRef]
57. Zou, H.; Hastie, T. Regularization and variable selection via the elastic net. *J. R. Stat. Soc. Ser. B Stat. Methodol.* **2005**, *67*, 301–320. [CrossRef]
58. Ajana, S.; Acar, N.; Bretillon, L.; Hejblum, B.P.; Jacqmin-Gadda, H.; Delcourt, C.; BLISAR Study Group. Benefits of dimension reduction in penalized regression methods for high-dimensional grouped data: A case study in low sample size. *Bioinformatics* **2019**, *35*, 3628–3634. [CrossRef]
59. Riepina, I. Comparative analysis of modern concepts in modeling global food security. *Sci. Notes* **2024**, *37*, 249–259. [CrossRef]
60. Ammar, K.A.; Kheir, A.M.S.; Manikas, I. Agricultural big data and methods and models for food security analysis—A mini-review. *PeerJ* **2022**, *10*, e13674. [CrossRef]
61. Herteux, J.; Räth, C.; Martini, G.; Baha, A.; Koupparis, K.; Lauzana, I.; Piovani, D. Forecasting Trends in Food Security: A Reservoir Computing Approach. *arXiv* **2023**. [CrossRef]
62. Bofa, A.; Zewotir, T. A bayesian spatio-temporal dynamic analysis of food security in Africa. *Sci. Rep.* **2024**, *14*, 15132. [CrossRef] [PubMed]
63. Asim, H.; Rehman, H.U. Panel Quantile Approach towards Exigent Determinants of Food Security in the Selected Developing Countries. *Pak. J. Humanit. Soc. Sci.* **2024**, *12*, 612–626. [CrossRef]
64. Guha, S.; Chandra, H. Measuring disaggregate level food insecurity via multivariate small area modelling: Evidence from rural districts of Uttar Pradesh, India. *Food Secur.* **2021**, *13*, 597–615. [CrossRef]
65. Yilema, S.A.; Shiferaw, Y.A.; Zewotir, T.; Muluneh, E.K. Multivariate small area estimation of undernutrition for children under five using official statistics. *Stat. J. IAOS* **2022**, *38*, 625–636. [CrossRef]
66. Bakhtsiyarava, M.; Williams, T.G.; Verdin, A.; Guikema, S.D. A nonparametric analysis of household-level food insecurity and its determinant factors: Exploratory study in Ethiopia and Nigeria. *Food Secur.* **2021**, *13*, 55–70. [CrossRef]
67. Nabeel, H.; Alansari, M.; Abdella, G.M.; Kutty, A.A.; Al Mohannadi, I.M. An Input-Output-Based Assessment of the Global Food Security Index. In Proceedings of the 6th European International Conference on Industrial Engineering and Operations Management, Lisbon, Portugal, 18–20 July 2023. [CrossRef]

68. Paruolo, P.; Saisana, M.; Saltelli, A. Ratings and rankings: Voodoo or science? *J. R. Stat. Soc. Ser. A Stat. Soc.* **2013**, *176*, 609–634. [CrossRef]
69. Azzurra, A.; Massimiliano, A.; Angela, M. Measuring sustainable food consumption: A case study on organic food. *Sustain. Prod. Consum.* **2019**, *17*, 95–107. [CrossRef]
70. Kucukvar, M.; Alawi, K.A.; Abdella, G.M.; Bulak, M.E.; Onat, N.C.; Bulu, M.; Yalçıntaş, M. A frontier-based managerial approach for relative sustainability performance assessment of the world's airports. *Sustain. Dev.* **2020**, *29*, 89–107. [CrossRef]
71. Abdella, G.M.; Kucukvar, M.; Onat, N.C.; Al-Yafay, H.M.; Bulak, M.E. Sustainability assessment and modeling based on supervised machine learning techniques: The case for food consumption. *J. Clean. Prod.* **2020**, *251*, 119661. [CrossRef]
72. Forero-Cantor, G.; Ribal, J.; Sanjuán, N. Measuring regional differences in food security from access and stability dimensions: A methodological proposal based on elasticities. *Agric. Econ.* **2020**, *66*, 112–119. [CrossRef]
73. Tibshirani, R. Regression Shrinkage and Selection via the Lasso. *J. R. Stat. Soc. Ser. B Stat. Methodol.* **1996**, *58*, 267–288. [CrossRef]
74. Hastie, T.; Taylor, J.; Tibshirani, R.; Walther, G. Stagewise algorithms and lasso-type problems. *arXiv* **2007**, arXiv:0705.0269.
75. Oyeyemi, G.M.; Ogunjobi, E.O.; Folorunsho, A.I. On Performance of Shrinkage Methods—A Monte Carlo Study. *Int. J. Stat. Appl.* **2015**, *5*, 72–76. Available online: http://article.sapub.org/10.5923.j.statistics.20150502.04.html (accessed on 22 April 2025).
76. Global Food Security Index (GFSI). Available online: https://impact.economist.com/sustainability/project/food-security-index/ (accessed on 22 April 2025).
77. Kutty, A.A.; Abdella, G.M.; Kucukvar, M.; Onat, N.C.; Bulu, M. A system thinking approach for harmonizing smart and sustainable city initiatives with United Nations sustainable development goals. *Sustain. Dev.* **2020**, *28*, 1347–1365. [CrossRef]
78. Kutty, A.A.; Yetiskin, Z.; Abraham, M.M.; Nooh, M.A.; Kucukvar, M.; Abdella, G.M. An Empirical Assessment on the Transportation Sustainability Indicators and their Impact on Economic Productivity, 2020. Available online: http://qspace.qu.edu.qa/handle/10576/31868 (accessed on 22 April 2025).
79. Majeed, W.A.; Hamad, M.M. Exploring High-Dimensional Outlier Detection: A Comprehensive Study on Methods and Applications Using PCA and k-NN Algorithm. In Proceedings of the 2024 21st International Multi-Conference on Systems, Signals and Devices (SSD), Erbil, Iraq, 22–25 April 2024; pp. 699–704. [CrossRef]
80. Kohavi, R. A Study of Cross-Validation and Bootstrap for Accuracy Estimation and Model Selection. In Proceedings of the 14th International Joint Conference on Artificial Intelligence, Montreal, QB, Canada, 20–25 August 1995. Available online: https://www.ijcai.org/Proceedings/95-2/Papers/016.pdf (accessed on 23 April 2025).
81. Wong, T.T. Performance evaluation of classification algorithms by k-fold and leave-one-out cross validation. *Pattern Recognit.* **2015**, *48*, 2839–2846. [CrossRef]
82. Food Systems Dashboard. Available online: https://www.foodsystemsdashboard.org/ (accessed on 22 April 2025).
83. World Bank, WDI—The World by Income and Region. Available online: https://datatopics.worldbank.org/world-development-indicators/the-world-by-income-and-region.html (accessed on 22 April 2025).
84. von Grebmer, K.; Bernstein, J.; Patterson, F.; Wiemers, M.; Chéilleachair, R.N.; Foley, C.; Ekstrom, K.; Fritschel, H. *2019 Global Hunger Index: The Challenge of Hunger and Climate Change*; Concern Worldwide: Dublin, Ireland, 2019.

foods

MDPI

Article

HRMS Characterization and Antioxidant Evaluation of Costa Rican Spent Coffee Grounds as a Source of Bioactive Polyphenolic Extracts

Mirtha Navarro-Hoyos [1,2,3,*], Luis Felipe Vargas-Huertas [1,2], Juan Diego Chacón-Vargas [1,2], Valeria Leandro-Aguilar [1,2], Diego Alvarado-Corella [1], Jose Roberto Vega-Baudrit [4], Luis Guillermo Romero-Esquivel [5], Andrés Sánchez-Kopper [6], Andrea Monge-Navarro [1] and Andrea Mariela Araya-Sibaja [2,4,7]

[1] BIODESS, Department of Chemistry, University of Costa Rica, San Pedro de Montes de Oca 11501, Costa Rica; luis.vargashuertas@ucr.ac.cr (L.F.V.-H.); juan.chaconvargas@ucr.ac.cr (J.D.C.-V.); valeria.leandroaguilar@ucr.ac.cr (V.L.-A.); luis.alvaradocorella@ucr.ac.cr (D.A.-C.); tpc@codeti.org (A.M.-N.)
[2] I&D+i Department, INNOBIOTIQ, Pavas 10107, Costa Rica; aaraya@cenat.ac.cr
[3] Chemistry Department, Georgetown University, Washington, DC 20057, USA
[4] National Laboratory of Nanotechnology, CENAT, Pavas 10109, Costa Rica; jvegab@gmail.com
[5] Costa Rica Institute of Technology, Department of Chemistry, Environmental Protection Research Center, Cartago 30109, Costa Rica; lromero@itcr.ac.cr
[6] Department of Chemistry, Chemistry and Microbiology Research and Services Center, Costa Rica Institute of Technology, Cartago 30109, Costa Rica; ansanchez@itcr.ac.cr
[7] National Technical University, Central Campus, Alajuela 20101, Costa Rica
* Correspondence: mnavarro@codeti.org; Tel.: +1-202-415-8548

Abstract: Spent coffee grounds constitute a waste product that has attracted potential interest as a rich source of secondary metabolites such as polyphenolic compounds with antioxidant properties. In this work, aqueous extracts from samples of different spent coffee grounds from Costa Rica were prepared and analyzed using ultra-performance liquid chromatography coupled with high-resolution mass spectrometry using a quadrupole time-of-flight analyzer (UPLC-QTOF-ESI MS). This allowed for the identification of twenty-one compounds, including fourteen phenolic acids, three caffeoylquinic lactones, and four atractyligenin diterpenes. In addition, using UPLC coupled with a diode array detector (UPLC-DAD), we quantified the levels of caffeine (0.55–3.42 mg/g dry weight [DW]) and six caffeoylquinic and feruloylquinic acids (0.47–5.34 mg/g DW). The highest value was found for the fine-grind sample (EXP), both for phenolic acids and for total polyphenols (9.59 mg gallic acid equivalents [GAE]/g DW), compared to 2.13 and 1.70 mg GAE/g DW for the medium-grind (GR) and coarse-grind samples (PCR), respectively. The results obtained from the antioxidant evaluations using the 2,2-diphenyl-1-picrylhydrazyl assay (IC_{50} 0.0964–6.005 g DW/L), the ferric-reducing antioxidant power (PFRAP) analysis (0.0215–0.1385 mmol $FeSO_4$/g DW), the oxygen radical absorbance capacity (ORAC) assessment (45.7–309.7 μmol Trolox/g DW), and the Trolox equivalent antioxidant capacity (TEAC) assay (3.94–23.47 mg Trolox/g DW) also showed the best values for the fine-grind sample, with results similar to or higher than those reported in the literature. Statistical Pearson correlation analysis ($p < 0.05$) indicated a high correlation ($R \geq 0.842$) between all antioxidant analyses, the total polyphenols, and the phenolic acid quantification using UPLC-DAD. These results show the potential for further studies aiming to exploit this waste product's bioactive properties, constituting the first detailed study of spent coffee grounds from Costa Rica.

Keywords: spent coffee grounds; waste; polyphenols; caffeoylquinic acids; feruloylquinic acids; atractylegenins; antioxidant activity

1. Introduction

Coffee is one of the most consumed beverages in the world and is the second-most traded commodity after petroleum, which highlights its immense global market share [1]. It has been estimated that roughly 2.25 billion cups of coffee are served every day and that more than 80% of the world's population consumes at least one caffeine-containing beverage on a daily basis [2].

The growth of coffee consumption at the global level has in turn increased the volume of spent coffee grounds (SCGs) generated as waste, therefore creating a challenge for waste management [3]. A diversity of constituents including chlorogenic acids, pyrazines, furans, alkaloids, melanoidins, and others are responsible for coffee's organoleptic characteristics [4].

Despite being a waste material, SCGs' chemical composition reveals them to be a rich source of antioxidants and other high-value compounds [5]. SCGs have been identified as a source of organic compounds such as fatty acids, lignin, cellulose, hemicellulose, and other polysaccharides [6]. SCGs have also been explored for applications such as the production of bioethanol [7], biodiesel [8], bio-adsorbents for water treatment [9,10], and extracts of bioactive molecules such as polyphenols and caffeine [11,12].

In addition, the extraction of antioxidants from coffee waste has been applied in cosmetic products and several food products and formulations as either a preservation treatment or an enrichment step [13]. In the food industry in particular, considering the adverse effects of synthetic antioxidants at high concentrations and their low thermal stability in heat processing, there is a growing trend of substituting synthetic antioxidants with natural ones [14].

Chlorogenic acids (CGAs) are the main phenolic compounds in coffee, formed by the esterification of quinic acid with hydroxycinnamic acids, such as caffeic, ferulic, and p-coumaric acids [15,16]. Their metabolism follows four main pathways: direct absorption, absorption with or without hydrolysis followed by conjugation or further metabolism, microbial catabolite absorption without alteration, and microbial catabolite absorption with subsequent mammalian phase II metabolism [17,18].

Among the health benefits attributed to CGAs, their antioxidant capacity has been widely studied. After donating hydrogen atoms, CGAs are oxidized to their respective phenoxyl radicals, which are quickly stabilized through resonance. Evidence also indicates that CGAs exhibit anti-inflammatory activity by downregulating pro-inflammatory cytokines through the modulation of key transcription factors [19].

In vivo experimentation with animals has demonstrated that caffeoylquinic acid isomers (CQAs), one subgroup of CGAs, are capable of decreasing various plasma and liver lipids and improving glucose tolerance [20]. CQAs have also been evaluated for their capacity to decrease blood pressure, including the stimulation of NO production through the endothelial-dependent pathway, the reduction of free radicals through inhibiting NAD(P)H oxidase expression and activity, and the inhibition of angiotensin-converting enzyme [21].

Moreover, CQAs have been identified as a potential treatment for cancer and were approved by the China Food and Drug Administration for phase I and II clinical trials in glioma patients, which showed that CQAs function as a safe differentiation inducer for solid tumors [22].

Other important but less explored CGAs present in coffee are feruloylquinic acid isomers (FQAs), which have been reported to exhibit anti-inflammatory effects by inhibiting LPS-induced NO production and IL-1β, IL-6, iNOS, COX-2, and NF-κB expression in a dose-dependent manner in RAW 264.7 cells [23]. Another study on the isomers of both CQAs and FQAs also described them as potent antioxidants with a greater capacity to scavenge hydroperoxyl radicals in polar and lipidic media than other potential antioxidants [24].

In turn, coumaroylquinic acid isomers (CoQAs) have been identified as potential astringent compounds [25] and also as compounds with antioxidant activity when evaluating their hydroxyl radical-scavenging capacity [26]. A study dealing with the effect of in vitro gastrointestinal digestion on the bio-accessibility of polyphenols showed that 3-CoQA was easily released from food matrices after incubation with digestive fluids, contributing to the antiproliferative activity against two models of human colon adenocarcinoma cell lines [27].

On the other hand, atractyloside I, a diterpenoid glycoside that has been reported in coffee beans [28], has been found to block ANT2 expression, promoting the activation of adenosylate-activated protein kinase (AMPK), decreasing mTOR activity, and promoting autophagy activation, thus accelerating the degradation of accumulated lipids in the liver induced by high-fat-content diets [29].

In addition, treatment with atractyloside has been reported to reduce blood glucose, possibly as a consequence of restored genes and pathways, providing evidence for a new therapy for type 2 diabetes [30]. In turn, atractyligenin, the aglycone isolated from coffee silverskin, has been shown to exhibit anti-photoaging effects by recovering the altered fibroblast morphology induced by UV light and suppressing UVA-mediated matrix metalloprotease generation via the MAPK and AP-1 signaling pathways [31].

Hence, the objective of this work was to obtain extracts of the most common types of Costa Rican SCGs in order to characterize their main secondary metabolites through ultra-performance liquid chromatography coupled with high-resolution mass spectrometry using a quadrupole time-of-flight analyzer (UPLC-QTOF-ESI MS) and to determine their content using UPLC coupled with a diode array detector (UPLC-DAD). In addition, this study aimed to evaluate antioxidant activity using the 2,2-diphenyl-1-picrylhydrazyl (DPPH) assay, the potassium ferricyanide-reducing antioxidant power (PFRAP) assay, the oxygen radical absorbance capacity (ORAC) assay, and the Trolox equivalent antioxidant capacity (TEAC) assay methods, applying a correlation analysis to the data obtained. This assessment aimed to contribute to the potential application of this waste product based on its bioactive properties. To the best of our knowledge, this is the first detailed study of SCGs from Costa Rica.

2. Materials and Methodology

2.1. Reagents and Solvents

Solvents of ACS or HPLC grade, for instance, methanol and acetonitrile, were acquired from Baker (Center Valley, PA, USA). Reagents such as gallic acid, DPPH, Folin–Ciocalteu reagent, sodium carbonate, 2,2′-azinobis (3-ethylbenzothiazoline-6-sulphonate) diammonium salt (ABTS), potassium ferricyanide, iron (III) chloride, and 2,2′-azobis (2-amidinopropane) dihydrochloride (AAPH) were obtained from Sigma-Aldrich (St. Louis, MO, USA).

2.2. Sample Preparation

The spent coffee samples corresponded to the three main types of coffee grind from the most consumed brand in Costa Rica: fine-grind espresso (EXP) coffee, with a mean particle diameter (MPD) of 0.42 ± 0.01 mm; medium-grind (GR) coffee, with an MPD of 0.58 ± 0.01 mm; and coarse-grind percolator (PRC) coffee, with an MPD of 1.18 ± 0.02 mm. For each of the three samples, 25 g was extracted with 200 mL of boiling water using a coffee maker (Oster, Coahuila, Mexico) and a new cotton fabric filter (RCR Ruteo, Cartago, Costa Rica). The extraction was repeated once, and the 400 mL of liquid was concentrated to 25 mL in a Rotavapor® R-100 with a vacuum pump V-100 rotary evaporator (Buchi, DE, USA) at 40 °C and 35 mmHg. The humidity in the spent coffee samples was measured by drying 3 g of the wet material in an oven at 50 °C until a constant mass was achieved.

2.3. UPLC-QTOF-ESI MS Analysis

The UPLC high-resolution mass spectrometry (UPLC-HRMS) system used to analyze the composition of spent coffee extracts consisted of a Xevo G2-XS Q-TOF (Waters, Wilmslow, UK) coupled with an AQUITY H Class UPLC system with a quaternary pump. The ESI source parameters were set to a capillary voltage of 2 kV, sampling cone of 20 eV, source temperature of 150 °C, and source offset of 10 °C. The desolvation temperature was set at 450 °C, the cone gas flow at 10 L/h, and the desolvation gas flow at 900 L/h.

Measurements were performed in the MSe high-resolution negative mode using an acquisition mass range from 100 m/z to 2000 m/z and a scan rate of 0.5 s, while fragmentation was carried out using independent data acquisition with a collision energy ramp with 20 V to 30 V stored in the high-energy function. Instrument calibration was performed in the mass range of the measurements with sodium formate. Lock mass correction was applied directly to the measurements using a leucine enkephalin infusion measured every 30 s during the run. The data were analyzed using MassLynx V4.2 software from Waters (Wilmslow, UK) and mzmine 3 from mzio.

Separation was carried out on a Luna RP-C18 column (150 mm × 4.6 mm i.d. × 4 µm, Phenomenex, Torrance, CA, USA). The solvents used in the mobile phase were water with 0.1% formic acid (A) and acetonitrile (B). In this step, 5 µL of the sample was injected with a flow rate of 0.5 mL/min at 30 °C. The chromatographic gradient started at 5% B, changing to 10% B at 12 min and holding it for 22 min, then changing to 30% B at 40 min, and finally changing to 50% B at 45 min and holding it for 5 min.

2.4. Total Phenolics

The determination of the total phenolics was performed using a modified Singleton and Rossi method, employing the Folin–Ciocalteu (FC) reagent, which is composed of a mixture of phosphotungstic and phosphomolybdic acids. As previously reported [32], the assay comprised mixing 10 mL of Na_2CO_3 (7.5%) and 0.5 mL of the FC reagent with 0.5 mL of the respective extract. Subsequently, the volume was completed to 25 mL with water. A blank was prepared following the same procedure using 0.5 mL of H_2O in place of the extract. Both the extract mixtures and the blank were kept in the dark for 1 h, and afterwards the absorbance was measured at 750 nm. The absorbance measurements obtained were extrapolated in a gallic acid calibration curve to obtain the FC-reducing capacity results, further expressed as the mg of gallic acid equivalents (GAE)/g of the extract. Each determination was performed in triplicate.

2.5. UPLC-DAD Analysis

For quantification purposes, measurements were performed using a Thermo PDA photo DAD coupled with a Thermo UltiMate U3000 UPLC system and a simple quadrupole MSQ Plus instrument (Thermo Fisher Scientific, San Jose, CA, USA). A Luna RP-C18 column (150 mm × 4.6 mm i.d. × 4 µm, Phenomenex, Torrance, CA, USA) with a pre-column filter (Phenomenex, Torrance, CA, USA) was used. Mobile phases A and B consisted of a combination of 0.1% formic acid in water (v/v) and 0.1% formic acid in acetonitrile (v/v), respectively. The gradient started at 5% B and increased to 8% B at 8 min, then to 9% B at 25 min, 30% B at 40 min, and finally 50% B at 45 min, holding it for 5 min. The PDA acquisition was set at 275 nm for caffeine and 325 nm for phenol acids. Calibration curves of caffeine (10–1000 ppm) and CGA (5–150 ppm) were prepared. In this step, 20 µL of the standard or sample was injected. The results were expressed as the mg of caffeine equivalents or mg of CGA equivalents per gram of the extract.

2.6. Antioxidant Activity Evaluation

2.6.1. DPPH Analysis

A DPPH evaluation of the samples was conducted according to previously reported methods [33]. In brief, a solution of DPPH (0.25 mM) was set in ethanol. A volume of 0.5 mL of this solution was mixed with 1 mL of the sample under evaluation at different concentrations and incubated at 25 °C in the dark for 30 min. The DPPH absorbance was measured at 517 nm. Blanks were prepared for each concentration, and the DPPH absorbance was measured at 517 nm. The inhibition percentage was determined as follows:

$$\text{Inhibition percentage } (\%) = \frac{\left(Abs_{blank} - Abs_{sample}\right)}{Abs_{blank}} * 100 \quad (1)$$

The percentage of the radical-scavenging activity of the sample was plotted against its concentration to calculate IC_{50}, which corresponded to the amount of the sample necessary to reach 50% radical-scavenging activity. Each sample was analyzed in three independent assays.

2.6.2. PFRAP Analysis

The PFRAP assay was conducted following a modified version of the protocol described by Işil Berker et al. [34]. In brief, an aliquot of 1 mL of the sample at a correct dilution was mixed with 0.2 mL of 1 mol/L HCl, 1.5 mL of 1% potassium ferricyanide, 0.5 mL of sodium dodecyl sulfate, and 0.5 mL of 0.1% iron (III) chloride. The volume was completed to 10 mL with distilled water and incubated at room temperature for 30 min prior to measuring the absorbance at 750 nm against a blank. Calibration curves of Trolox (50–600 µmol/L) and $FeSO_4$ (35–1000 µmol/L) were made. The results were expressed as the µmol Trolox equivalent or µmol $FeSO_4$ equivalent per gram of the extract.

2.6.3. TEAC Analysis

For the TEAC analysis, Thaipong et al. [35]'s method was followed. In brief, an equal volume of 7.4 mmol/L ABTS (2,2′-azinobis (3-ethylbenzothiazoline-6-sulphonate) diammonium salt) and 2.6 mmol/L potassium persulfate were mixed 12 h prior to analysis to form a solution of the radical cation $ABTS^{*+}$. Before the analysis, an aliquot of this solution was diluted with distilled water to obtain a working solution with an absorbance of approximately 1.1 mAU at 734 nm. A volume of 250 µL of the sample was mixed with 4750 µL of the working solution of $ABTS^{*+}$ and incubated at room temperature in the dark for 2 h. The absorbance was measured at 734 nm against a blank. The inhibition percentage was calculated, and the result was interpolated in a calibration curve for Trolox (5–150 mg/L). The results were expressed as the mg Trolox equivalent (TE) per gram of the extract.

2.6.4. ORAC Analysis

The ORAC analysis used to determine the antioxidant activity followed a previously described method [32,36] using fluorescein as a fluorescence probe. The reaction was performed in a 75 mM phosphate buffer (pH 7.4) at 37 °C. The final assay mixture consisted of AAPH (12 mM), fluorescein (70 nM), and either Trolox (1–8 µM) or the extract at different concentrations. The fluorescence was recorded every minute for 98 min in black 96-well untreated microplates using a Fluoroskan Ascent plate reader (ThermoFisher Scientific, San Jose, CA, USA) with 485-P excitation and 520-P emission filters. Ascent Software version 2.6 (Thermo Labsystems, Helsinki, Finland) was used to measure the fluorescence. Fluorescein was diluted from a stock solution (1.17 mM) in a 75 mM phosphate buffer (pH 7.4), while the AAPH and Trolox solutions were freshly prepared. All the reaction mixtures were prepared in duplicate, and three independent runs were completed for each extract. The fluorescence measurements were normalized to the curve of the blank (no antioxidant). From the normalized curves, the area under the fluorescence decay curve (AUC) was calculated. The net AUC corresponding to a sample was calculated as the difference between the sample or Trolox and the blank. The regression equation between the net AUC and the antioxidant concentration was calculated. The ORAC value was estimated by dividing the slope of the latter equation by the slope of the Trolox line obtained for the same assay. The final ORAC values were expressed as the mmol of TE/g of the phenolic extract.

2.7. Statistical Analysis

A one-way analysis of variance (ANOVA), followed by Tukey's post hoc test, was applied to the polyphenol quantification values, and differences were considered significant at $p < 0.05$. Pearson correlation analyses were performed ($p < 0.05$) for all four antioxidant activities, the quantified metabolites, and the total polyphenols. R program version 4.4.2 (2024-10-31 ucrt) was used for statistical analysis.

3. Results and Discussion

3.1. UPLC-QTOF-ESI MS Analysis

The results of the UPLC-QTOF-ESI MS analysis, performed as described in Section 2.3, are summarized in Table 1, including the identified compounds, their molecular formulas, the m/z of the $[M-H]^-$ ions, the main ms^2 fragments, and the respective retention times (Rts) in the chromatogram. Figure 1 shows the ion extraction chromatograms (XICs) for each compound.

Table 1. Main compounds in spent coffee extracts identified through UPLC-Q-TOF analysis.

No.	Compound	Rt (min)	$[M-H]^-$	Molecular Formula	ms2 Fragments
1	Caffeolylquinic acid (isomer 1 of 4)	16.43	353.0876	$C_{16}H_{17}O_9$	201, 191, 179, 135
2	Caffeoylquinic acid (isomer 2 of 4)	20.96	353.0869	$C_{16}H_{17}O_9$	191
3	Coumaroylquinic acid (isomer 1 of 3)	22.69	337.0921	$C_{16}H_{17}O_8$	163
4	Caffeoylquinic acid (isomer 3 of 4)	24.86	353.0872	$C_{16}H_{17}O_9$	191
5	Feruloylquinic acid (isomer 1 of 3)	27.31	367.1026	$C_{17}H_{19}O_9$	193, 173, 134
6	Caffeoylquinic acid (isomer 4 of 4)	27.84	353.0869	$C_{16}H_{17}O_9$	201, 191, 179, 173, 135
7	Coumaroylquinic acid (isomer 2 of 3)	33.17	337.0923	$C_{16}H_{17}O_8$	191, 163
8	Feruloylquinic acid (isomer 2 of 3)	33.54	367.1025	$C_{17}H_{19}O_9$	191, 173, 134
9	Coumaroylquinic acid (isomer 2 of 3)	33.61	337.0921	$C_{16}H_{17}O_8$	191, 173, 163
10	Caffeoylshikimic acid	34.78	335.0767	$C_{16}H_{15}O_8$	173, 161, 135,133
11	Feruloylquinic acid (isomer 3 of 3)	35.08	367.1026	$C_{17}H_{19}O_9$	191, 173, 134
12	Caffeoylquinic lactone (isomer 1 of 3)	35.57	335.0766	$C_{16}H_{15}O_8$	161, 135, 133
13	Caffeoylquinic lactone (isomer 2 of 3)	36.30	335.0763	$C_{16}H_{15}O_8$	161, 135, 133
14	Caffeoylquinic lactone (isomer 3 of 3)	36.71	335.0768	$C_{16}H_{15}O_8$	161, 133
15	Atractyligenin-2-O-glucopyranoside	37.08	481.2448	$C_{25}H_{37}O_9$	301, 119
16	Dicaffeoylquinic acid (isomer 1 of 3)	41.15	515.1190	$C_{25}H_{23}O_{12}$	353, 335, 191, 179, 173, 161, 135
17	Dicaffeoylquinic acid (isomer 2 of 3)	41.89	515.1187	$C_{25}H_{23}O_{12}$	375, 353, 191, 179, 135
18	Dicaffeoylquinic acid (isomer 3 of 3)	42.91	515.1185	$C_{25}H_{23}O_{12}$	353, 191, 179, 173, 135
19	Atractyligenin	43.10	319.1903	$C_{19}H_{27}O_4$	275, 273
20	Atractyligenin-3′-O-glucopyranosyl-2′-isovaleryl-2-O-glucopiranoside	45.24	727.3541	$C_{36}H_{55}O_{15}$	643, 625
21	Atractyligenin-2′-isovaleryl-2-O-glucopiranoside	47.46	565.3007	$C_{30}H_{45}O_{10}$	473, 3

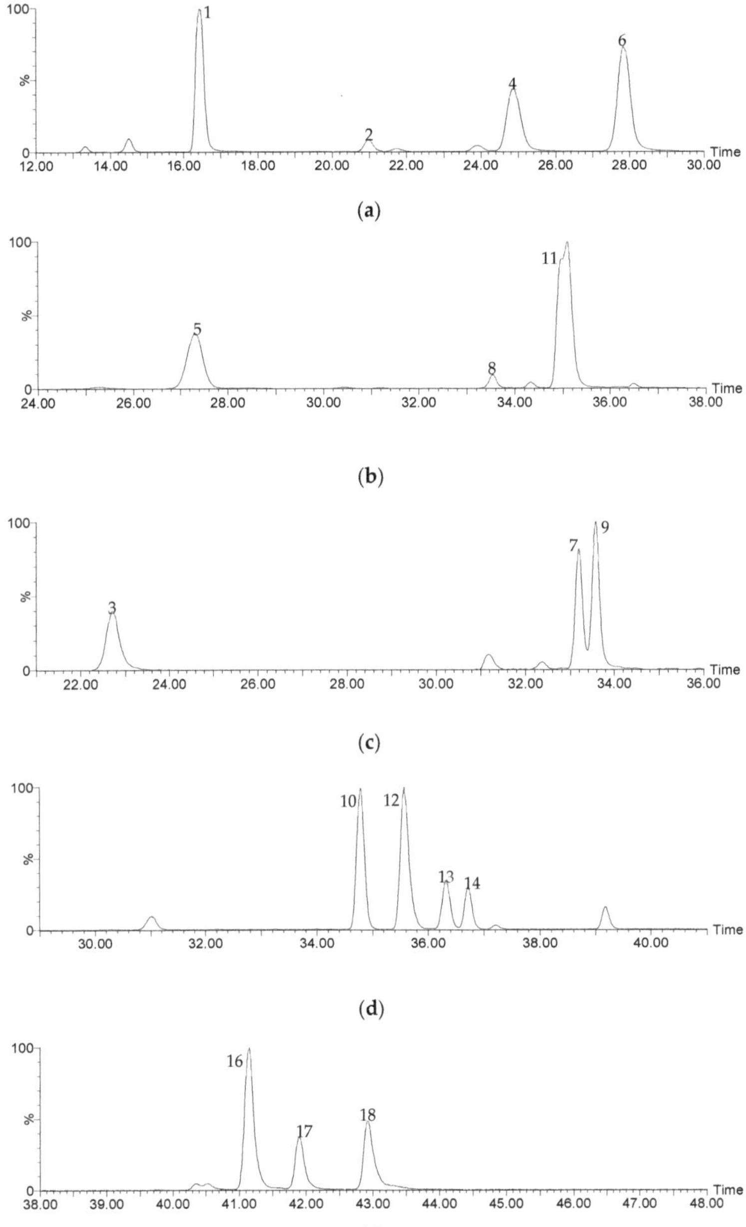

Figure 1. Extracted ion chromatograms (XICs) for the phenolic acids identified through the UPLC-QTOF-ESI MS analysis of spent coffee extracts. (**a**) Isomers of caffeoylquinic acid, (**b**) isomers of feruloylquinic acid, (**c**) isomers of coumaroylquinic acid, (**d**) caffeoylshikimic acid and caffeoylquinic lactone isomers, and (**e**) isomers of dicaffeoylquinic acid.

As described, 21 different compounds were identified, including CQAs, FQAs, CoQAs, and additional types of polyphenolic acids such as caffeoylshikimic acids and higher molecular mass acids such as dicaffeoylquinic acids. Likewise, three caffeoylquinic lactones and four diterpene atractyligenins were identified.

Figure 2 shows the main fragmentations for peaks 1, 2, 4, and 6, with Rts of 16.43 min, 20.96 min, 24.86 min, and 27.84 min, respectively, which were identified as isomers of caffeoylquinic acid which have previously been reported in the literature [37]. Indeed, these peaks had a pseudomolecular ion $[M\text{-}H]^-$ = 353.0873 Da, which corresponded to the molecular formula $C_{16}H_{17}O_9$. Also, a peak was observed at 707 Da corresponding to the dimer $[2M\text{-}H]^-$. The main fragments in these peaks were observed at 191 Da (quinic acid), 179 Da (caffeic acid), and 135 Da, derived from carbonyl α-cleavage [38].

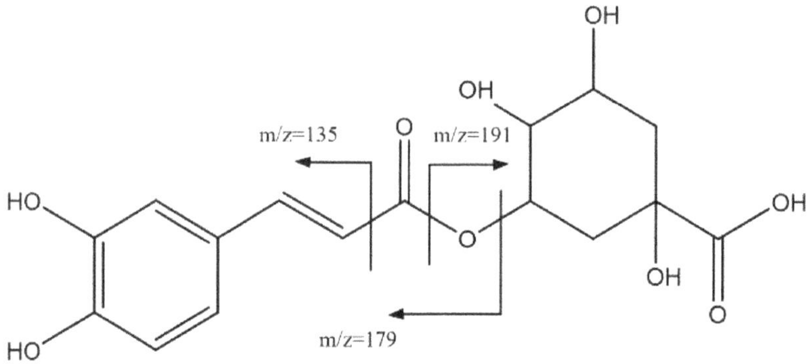

Figure 2. Main fragmentations of caffeoylquinic acid isomers, peaks 1, 2, 4 and 6.

Figure 3 shows the main fragmentations for peaks 5, 8, and 11, with Rts of 27.31 min, 33.54 min, and 35.08 min, respectively, with the ion $[M\text{-}H]^-$ = 367.1026 Da ($C_{17}H_{19}O_9$) corresponding to feruloylquinic acid isomers. This was confirmed by the presence of the adduct $[2M\text{-}H]^-$ at 735 Da, as well as by the fragmentation pattern, which included peaks at 193 Da (ferulic acid), 191 Da (quinic acid), 173 Da for the loss of H_2O from quinic acid, and 134 Da for the loss of CH_3 in conjunction with carbonyl α-cleavage [38].

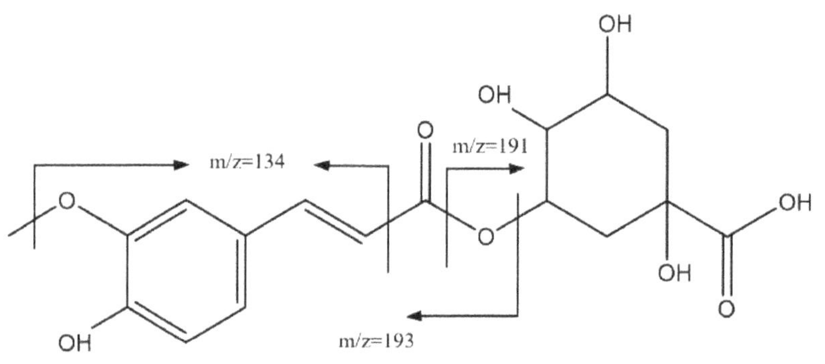

Figure 3. Main fragmentations of isomers of feruloylquinic acid 5, 8, and 11.

Peaks 3, 7, and 9 (Figure 4), with Rts of 22.69 min, 33.17 min, and 33.61 min, respectively, and $[M\text{-}H]^-$ = 337.0921 Da ($C_{16}H_{17}O_8$), were assigned as isomers of coumaroylquinic acid. Accordingly, fragments at 191 Da (quinic acid) and 163 Da (coumaric acid) were observed [38].

Figure 5 shows the main fragmentations of peak 10 (34.78 min) with a mass of $[M\text{-}H]^-$ = 335.0767 Da ($C_{16}H_{15}O_8$). This peak was identified as a caffeoylshikimic acid isomer, with fragments at 173 Da due to shikimic acid, 161 Da from caffeic acid with the loss of an additional H_2O molecule, and 135 Da from carbonyl α-cleavage [39,40].

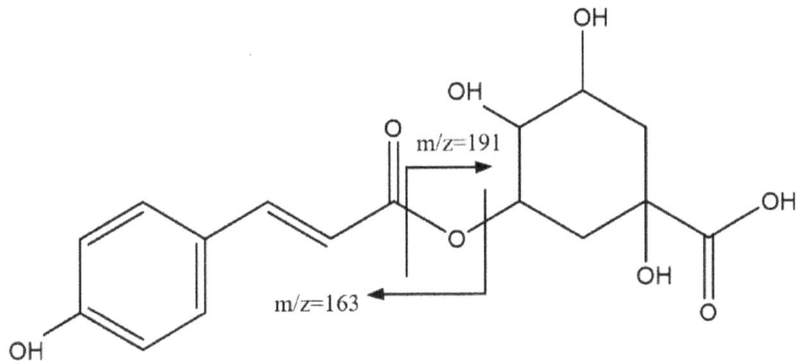

Figure 4. Main fragmentations of isomers of coumaroylquinic acid 3, 7, and 9.

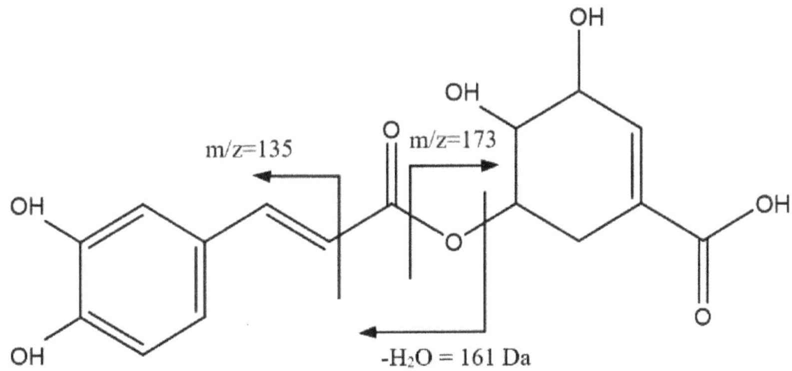

Figure 5. Main fragmentations of an isomer of caffeoylshikimic acid, 10.

Figure 6 shows the main fragmentations of peaks 12, 13, and 14, for which a mass of [M-H]$^-$ = 335.0765 Da ($C_{16}H_{15}O_8$) was observed, with Rts of 35.57, 36.30, and 36.71 min, respectively. These peaks were identified as isomers of caffeoylquinic lactones, with a fragment at 135 Da due to carbonyl α-cleavage and at 161 Da from caffeic acid with the loss of an additional H_2O molecule, which in turn generated the fragment at 133 Da due to the additional loss of CO, in line with the literature [39].

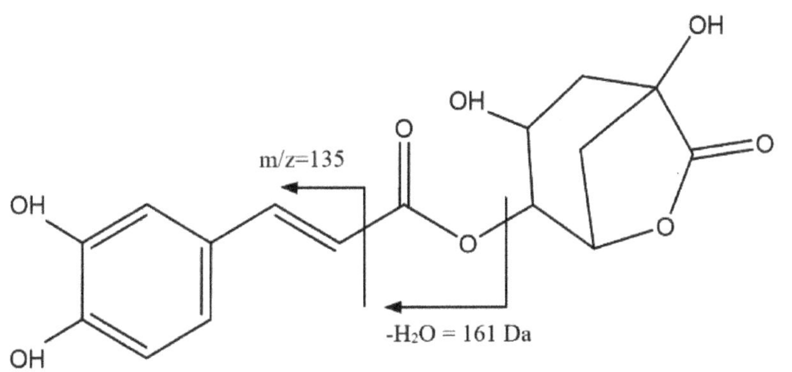

Figure 6. Main fragmentations of caffeoylquinic lactone isomers 12, 13, and 14.

For peaks 16, 17, and 18, with Rts of 41.15 min, 41.89 min, and 42.91 min, respectively, there was an observed ion [M-H]$^-$ = 515.1187 Da ($C_{25}H_{23}O_{12}$), corresponding to dicaffeoylquinic acid isomers (Figure 7). The main fragments were observed at 353 Da due to the loss of one of the caffeoyl moieties, at 191 Da corresponding to quinic acid, at 179 Da due to the caffeic acid fragment, and at 135 Da from carbonyl α-cleavage [38].

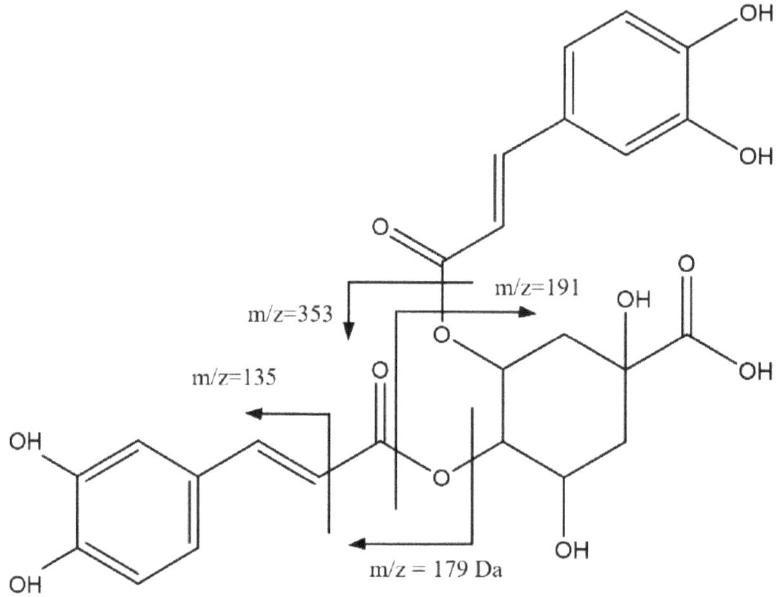

Figure 7. Main fragmentations of isomers of dicaffeoylquinic acid 15, 16, and 17.

Figure 8 below shows the different ion extraction chromatograms (XICs) for the atractyligenin-type diterpenes.

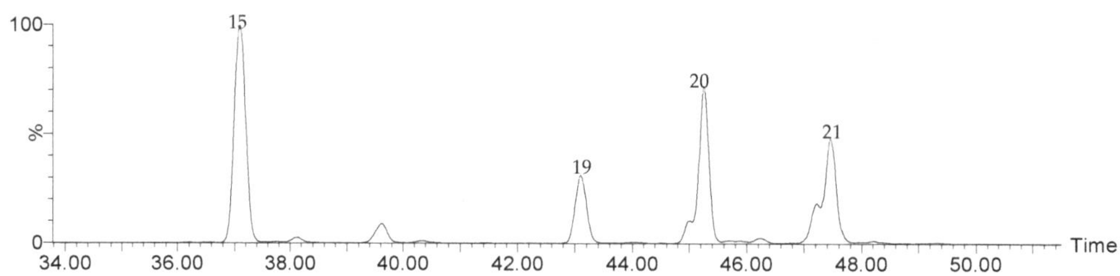

Figure 8. Ion extraction chromatograms (XICs) for the actratylogenins 15, 19, 20, and 21 identified through the UPLC-QTOF-ESI MS analysis of spent coffee extracts.

Peak 19 at 43.10 min had an ion $[M-H]^-$ = 319.1903 Da, which corresponded to the formula $C_{19}H_{27}O_4$ (Figure 9). This peak was assigned to the atractyligenin aglycone, with fragments at 275 Da and 273 Da due to the loss of the acid group in the form of CO_2 and HCO_2H, respectively [41,42].

Peak 15, at 37.08 min, had an $[M-H]^-$ ion at 481.2448 Da ($C_{25}H_{37}O_9$), corresponding to atractyligenin-2-*O*-glucoside (Figure 10), with the main fragment at 301 Da due to the loss of the glycoside along with an additional H_2O molecule [41,43]. Peak 20 (45.24 min, $[M-H]^-$ = 727.3541 Da, $C_{36}H_{55}O_{15}$) was assigned to atractyligenin-3′-*O*-glucopyranosyl-2′-isovaleryl-2-*O*-glucopyranoside, with fragments at 643 Da due to the loss of the isovaleryl group and at 625 Da from the additional loss of a molecule of H_2O [42,43].

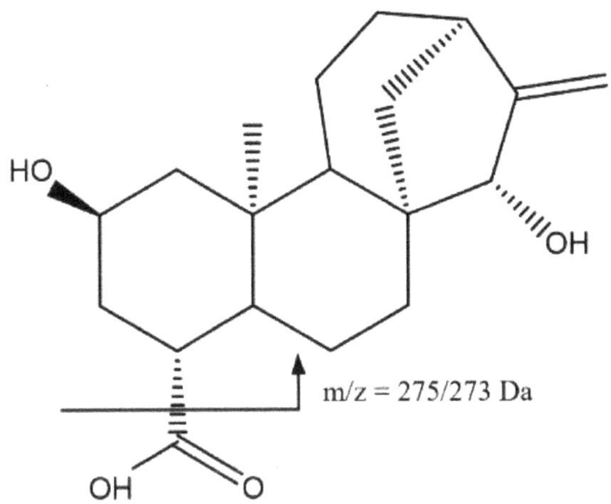

Figure 9. Main fragmentations for the aglycone atractyligenin, 19.

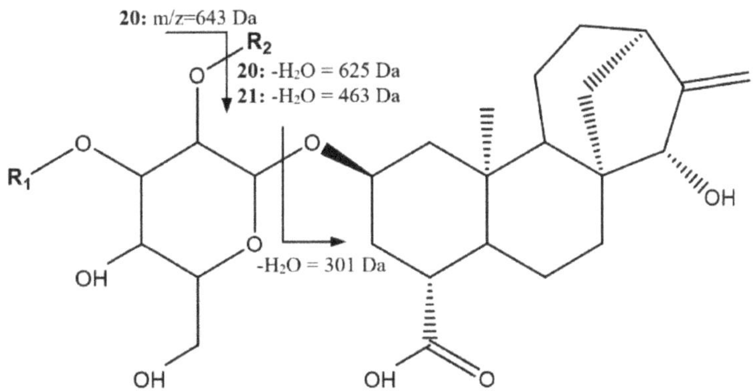

Compound	R_1	R_2
15	H	H
20	Glucose	$-COCH_2CH(CH_3)_2$
21	H	$-COCH_2CH(CH_3)_2$

Figure 10. Structure and main fragmentation for attractyligenin glycosides 15, 20, and 21.

Finally, peak 21 at 47.46 min was assigned to atractyligenin-2′-*O*-isovaleryl-2-*O*-glucopyranoside, with $[M\text{-}H]^-$ = 565.3007 Da ($C_{30}H_{45}O_{10}$) and characteristic fragments at 463 Da due to the loss of isovaleryl and an additional H_2O molecule and at 301 Da from the cleavage of the glucoside along with a molecule of H_2O [42].

The presence of these metabolites makes SCGs a valuable material for food applications with potential health advantages. For instance, lipid-lowering and glucose-modulating effects have been reported for the CGAs identified in this study [20,21], indicating their potential application in functional foods targeting metabolic health, such as weight management or diabetes-friendly products. Additionally, their antihypertensive and anticancer properties further broaden their potential in designing therapeutic foods or supplements [22]. In turn, the atractyligenins identified in this work have been reported to offer novel opportunities for creating functional ingredients that support liver health, glucose regulation, and skin protection [28–31], aligning with consumer trends for holistic wellness solutions. These bioactive compounds exemplify the potential of spent coffee-derived ingredients in addressing health concerns through functional and therapeutic food innovations.

3.2. Total Polyphenols

The Folin–Ciocalteu (FC) method is widely used to quantify the total polyphenols in a sample. The method consists of reducing a complex of phosphomolybdic acid and phosphotungstic acid through the donation of electrons made by phenolic groups. Due to the mechanism by which the reaction occurs, the FC method is a single electron transfer (SET) method, which is why it is also considered a means of analyzing the antioxidant capacity [44].

Section 2.4 describes the procedure used in this study, and Table 2 summarizes the values obtained.

Table 2. Total polyphenols of spent coffee samples EXP, GR, and PRC.

Sample	Total Polyphenols (mg GAE/g Extract) [1,2]	Total Polyphenols (mg GAE/g Dry Weight) [1,2]
EXP	$88.1^{a} \pm 3.8$	$9.59^{a} \pm 0.42$
GR	$70.0^{b} \pm 0.9$	$2.13^{b} \pm 0.03$
PRC	$53.1^{c} \pm 0.5$	$1.70^{c} \pm 0.01$

[1] Values are expressed as average ± standard deviation. [2] Different superscript letters in same column indicate significant differences ($p < 0.05$) using one-way analysis of variance (ANOVA) with post hoc Tukey test as statistical analysis test.

The above results showed that the polyphenol content of the GR and PRC samples was significantly lower than that of the EXP sample. The EXP sample had a similar value to that found by Wu et al. [45], who reported 9.44 mg GAE/g dry weight (DW), and a higher value than those reported by Mussatto et al. [46], which ranged from 6.0 to 7.4 mg GAE/g DW. Other references using spent coffee beans' hydroalcoholic extracts indicated slightly higher values of the polyphenolic content [37,47,48].

3.3. UPLC-DAD Analyses

A UPLD-DAD analysis was conducted as described in Section 2.5. The results for the quantification of caffeine in the spent coffee samples are presented in Table 3.

Table 3. Caffeine quantification using UPLC-DAD on samples of spent coffee.

Sample	Concentration (mg/g Extract) [1,2]	Concentration (mg/g DW) [1,2]
EXP	$31.435^{a} \pm 0.012$	$3.4201^{a} \pm 0.0014$
GR	$18.040^{b} \pm 0.008$	$0.5496^{b} \pm 0.0002$
PRC	$54.020^{c} \pm 0.065$	$1.7294^{c} \pm 0.0021$

[1] Values are expressed as average ± standard deviation. [2] Different superscript letters in same column indicate significant differences ($p < 0.05$) using one-way analysis of variance (ANOVA) with post hoc Tukey test as statistical analysis test.

The caffeine content of the EXP sample was higher than that of the GR and PRC samples. In fact, the EXP content showed a value within the range of 2.5–7.9 mg/g DW reported by Cruz et al. [49] for spent espresso coffee. Other previous results obtained for extracts in aqueous ethanol showed a wider range of 0.3–20 mg/g DW [48], which include the values obtained in the present work.

A UPLC-DAD analysis was also carried out to quantify the phenolic compounds, as described in Section 2.5. Table 4 shows the results obtained for the spent coffee samples EXP, GR, and PRC.

As shown in Table 4, the fine-grind sample (EXP) had the highest content of CQAs, (peaks 1, 4, and 6) corresponding to 87% of the total quantifiable phenolic acids, while FQAs (peaks 5, 8, and 11) accounted for 13% of the total. This result differs from that of the medium-grind sample (GR), which showed a similar content of both types of acids, with 56% CQAs and 44% FQAs. Meanwhile, CQAs and FQAs accounted for 36% and 64%, respectively, in the coarse-grind sample (PRC).

Table 4. Quantification of caffeoylquinic and feruloylquinic acids in spent coffee samples using UPLC-DAD.

Compound No.	Concentration (mg/g Extract) [1,2]			Concentration (mg/g DW) [1,2]		
	EXP	GR	PRC	EXP	GR	PRC
1	12.026 [a] ± 0.026	5.495 [a] ± 0.036	1.854 [a] ± 0.060	1.3085 [a] ± 0.0026	0.1674 [a] ± 0.0014	0.0593 [a] ± 0.0024
4	18.101 [b] ± 0.13	3.800 [b] ± 0.033	1.861 [a] ± 0.016	1.9695 [b] ± 0.0141	0.1158 [b] ± 0.0012	0.0596 [a] ± 0.0006
5	1.617 [c] ± 0.003	4.098 [c] ± 0.001	3.742 [b] ± 0.043	0.1760 [c] ± 0.0003	0.1249 [c] ± 0.0001	0.1198 [b] ± 0.0017
6	12.768 [d] ± 0.070	4.127 [c] ± 0.010	1.554 [c] ± 0.013	1.3893 [d] ± 0.0070	0.1257 [c] ± 0.0004	0.0497 [c] ± 0.0005
8	2.697 [e] ± 0.022	3.001 [d] ± 0.017	2.628 [d] ± 0.042	0.2934 [e] ± 0.0022	0.0914 [d] ± 0.0006	0.0842 [d] ± 0.0016
11	1.859 [f] ± 0.039	3.524 [e] ± 0.021	2.897 [e] ± 0.068	0.2023 [f] ± 0.0039	0.1074 [e] ± 0.0008	0.0927 [e] ± 0.0008
Total	49.068	24.045	14.536	5.3390	0.7326	0.4653

[1] Values are expressed as average ± standard deviation. [2] Different superscript letters in same column indicate significant differences ($p < 0.05$) using one-way analysis of variance (ANOVA) with post hoc Tukey test as statistical analysis test.

Compared with the literature, the total concentration of phenolics obtained in the GR and PRC samples was within the range of results previously reported for spent coffee, ranging between 0.21 and 0.77 mg/g DW [46,49], while the EXP sample had significantly higher values than these reports. Angeloni et al. [40] reported one sample of Costa Rican spent coffee with a phenolic content of 1.64 mg/g DW in a hydroalcoholic extract, which is lower than the value obtained in our study for the EXP sample. Another study reported a methanolic extract with a phenolic content of 0.58 mg/g DW [50], which is similar to the values we found for the GR and PRC samples.

3.4. Antioxidant Activity Determination

Table 5 summarizes the results obtained from analyzing the antioxidant activities of the different SCG extracts.

Table 5. Antioxidant activities measured for extracts of spent coffee samples.

Sample	DPPH	PFRAP		ORAC	TEAC
	IC_{50} (mg Extract/L) [1,2]	mg Trolox/g Extract [1,2]	mmol $FeSO_4$/g Extract [1,2]	µmol Trolox/mg Extract [1,2]	mg Trolox/g Extract [1,2]
EXP	10.5 [a] ± 0.4	158.1 [a] ± 1.7	1.273 [a] ± 0.020	2.847 [a] ± 0.087	215.7 [a] ± 1.2
GR	78.1 [b] ± 1.0	106.6 [b] ± 2.1	0.921 [b] ± 0.025	1.590 [b] ± 0.033	152.2 [b] ± 2.4
PRC	192.2 [c] ± 3.3	78.5 [c] ± 0.5	0.672 [c] ± 0.006	1.427 [b] ± 0.012	123.1 [c] ± 0.4

[1] Values are expressed as average ± standard deviation. [2] Different superscript letters in same column indicate significant differences ($p < 0.05$) using one-way analysis of variance (ANOVA) with post hoc Tukey test as statistical analysis test.

The above results are discussed for each antioxidant methodology in the following sections. Table 6 summarizes the data in relation to the dry mass of the spent coffee samples.

Table 6. Antioxidant activities measured for samples of spent coffee, expressed per gram of dry weight.

Sample	DPPH	PFRAP		ORAC	TEAC
	IC_{50} (g DW/L) [1,2]	mg Trolox/g DW [1,2]	mmol $FeSO_4$/g DW [1,2]	µmol Trolox/g DW [1,2]	mg Trolox/g DW [1,2]
EXP	0.096 ^{a} ± 0.003	17.20 ^{a} ± 0.19	0.1385 ^{a} ± 0.0022	309.7 ^{a} ± 9.4	23.47 ^{a} ± 0.14
GR	2.562 ^{b} ± 0.034	3.25 ^{b} ± 0.06	0.0281 ^{b} ± 0.0008	48.4 ^{b} ± 1.0	4.64 ^{b} ± 0.07
PRC	6.005 ^{c} ± 0.104	2.51 ^{c} ± 0.02	0.0215 ^{c} ± 0.0002	45.7 ^{b} ± 0.4	3.94 ^{c} ± 0.01

[1] Values are expressed as average ± standard deviation. [2] Different superscript letters in same column indicate significant differences ($p < 0.05$) using one-way analysis of variance (ANOVA) with post hoc Tukey test as statistical analysis test.

3.4.1. DPPH

DPPH analysis uses the free radical 2,2-diphenyl-1-picrylhydrazyl. In this test, the intensity of the purple color produced by the DPPH* radical, which decreases due to the action of antioxidant compounds, is quantified using spectrophotometry. The mechanism of the reaction is mixed, being carried out by both electron transfer (SET) and hydrogen atom transfer (HAT) [44]. As previously mentioned, the results are expressed as the IC_{50} value, which indicates the sample concentration that produces a 50% inhibition of the DPPH* free radical.

The results showed a lower IC_{50} value for the fine-grind EXP sample compared to the GR and PRC samples, again indicating an antioxidant value that surpasses those of previous reports in the literature for spent espresso coffee, which range between 0.4 and 1.4 g DW/L [37,45]. The present findings are consistent with the literature for both coffee and spent coffee extracts, with a higher polyphenol content aligning with a better result regarding the antioxidant activity [46,51].

3.4.2. PFRAP

The PFRAP method is an analysis that employs the SET mechanism and is based on the reduction of an iron (III) complex to iron (II) to form Prussian blue, which is observed through the formation of an intense blue color measured using spectrophotometry. Unlike the other antioxidant analyses carried out, PFRAP does not involve radicals, which is why it constitutes a complementary technique in the evaluation of antioxidant activity [44].

As with the previous tests, the PFRAP results revealed the EXP sample as having the highest antioxidant activity compared to the GR and PRC samples. The EXP sample also showed similar values to those reported in the literature, which range from 0.20 to 0.22 mmol $FeSO_4$/g DW [46] when using the similar FRAP (ferric-reducing antioxidant power) technique.

3.4.3. ORAC

The ORAC assay is performed by measuring the kinetics of fluorescence at a known concentration of sodium fluoresceinate. The radicals are peroxide groups emitted by the AAPH reagent, which react with fluorescein, and then the loss of fluorescence is recorded. This loss is inhibited in the presence of an antioxidant compound, which reacts through a HAT mechanism with the peroxyl radicals, allowing the fluorescence signal to last longer [52].

In this analysis, the GR and PRC samples' results were similar, with the GR being slightly higher in its antioxidant capacity, while the EXP sample gave a significantly higher value, following a similar trend to the antioxidant activity evaluations described above. In the literature, a value of 735–823 µmol Trolox/g DW was reported for coffee fruits [53], indicating that the EXP spent coffee sample retained nearly half of its antioxidant activity in the ORAC assay.

3.4.4. TEAC

In TEAC analysis, the intensity of the coloration of the blueish-green ABTS$^{\bullet+}$ radical is measured using spectrophotometry. This radical is formed by the reaction between an ABTS solution and an oxidant such as potassium persulfate. In the presence of antioxidants, these react with ABTS$^{\bullet+}$ radicals and cause the discoloration of the solution, which is then compared against a curve prepared using Trolox as the standard [54].

In this analysis, the GR and PRC samples showed similar results to those obtained by Panusa et al. [43] for coffee silverskin, with values ranging from 6.14 to 15.63 mg Trolox/g D. The fine-grind EXP sample in our study showed higher values, which aligned with the 58.62 mg Trolox/g DW reported by Panusa et al. for a green coffee bean sample. Again, these results show that the EXP spent coffee sample retained almost half of its antioxidant activity. Finally, a Pearson correlation study was performed to evaluate the relationship between the polyphenolic content and the antioxidant activity evaluations. The results are summarized in Table 7 and Figure 11.

Table 7. Pearson correlation values ($p < 0.05$) obtained for the quantification of polyphenolic acids and antioxidant activity evaluations.

	HPLC-DAD	TP	DPPH	PFRAP	ORAC
TP	0.962				
DPPH	−0.920	−0.975			
PFRAP	0.995	0.981	−0.950		
ORAC	0.984	0.920	−0.842	0.967	
TEAC	0.998	0.973	−0.936	0.999	0.977

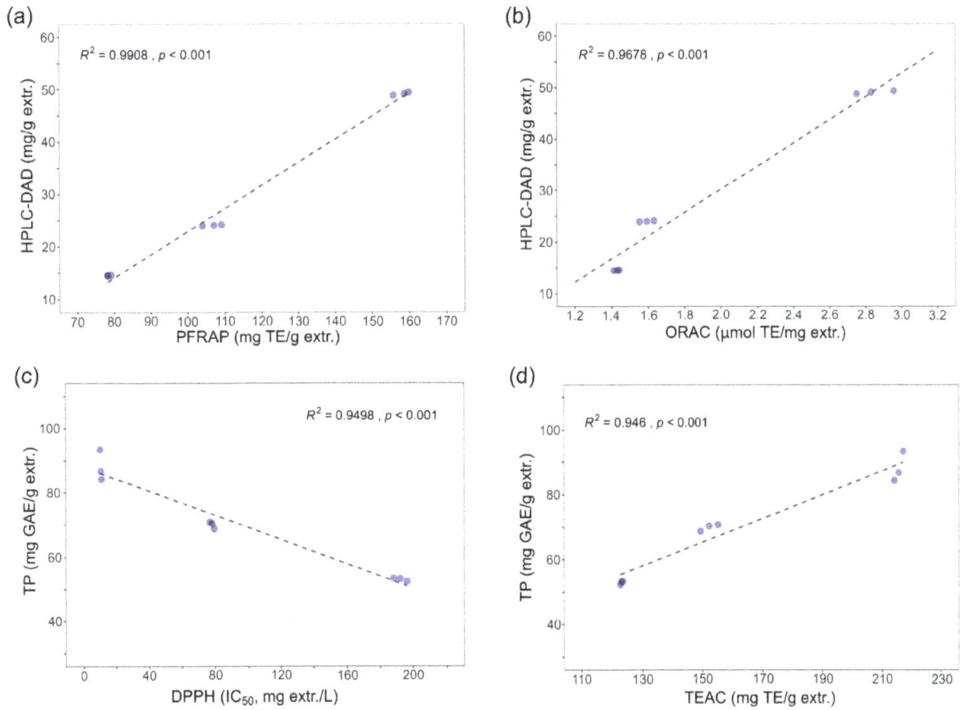

Figure 11. Correlations between phenolics content and antioxidant activity evaluations: (**a**) UPLC-DAD quantification and PFRAP antioxidant values, (**b**) UPLC-DAD quantification and ORAC antioxidant values, (**c**) total polyphenols (TP) and DPPH antioxidant values, and (**d**) TP and TEAC antioxidant values.

As can be observed in Table 7, significant positive correlation values ($p < 0.05$) were obtained for the relationship between the antioxidant activities as indicated by the ORAC, TEAC, and PFRAP and the total polyphenols evaluation (using the FC method) and the quantification of polyphenolic acids using HPLC-DAD. This finding aligns with previous studies showing that the phenolic compounds content correlates with the antioxidant activity [55,56].

Likewise, significant negative correlation values ($p < 0.05$) were found between the results of the DPPH assay and the other antioxidant activities, as well as between the TP values and the polyphenolic acid quantification results. This negative correlation is due to the DPPH method's measurement of an inhibition value, which means a lower numerical result implies a better antioxidant capacity, thus cohering with a higher amount of phenolics [57].

These promising results on the phenolic content and antioxidant potential align with previous reports on other natural Costan Rican products [58] that call for further phytochemical and bioactivity studies of these types of extracts. CGAs and related compounds in coffee hold significant potential for innovative food applications due to their diverse bioactive properties. As natural antioxidants, CGAs can enhance the shelf life and stability of food products by reducing oxidative damage, while their anti-inflammatory effects offer functional health benefits [15–19], appealing to the growing demand for nutraceuticals. FQAs and CoQAs, with their potent antioxidative and anti-inflammatory capacities [23–27], could enhance the nutritional profile of beverages, snacks, or dietary supplements.

However, while in vitro antioxidant assays provide valuable insights, they do not fully represent the bioavailability, metabolism, and physiological effects of the compounds in vivo. Therefore, other studies are needed to capture the full spectrum of bioactive compounds present in SCGs. Moreover, the nature of the spent coffee, for instance the variability in the polyphenol content based on the coffee type, grind size, and processing conditions, may impact the consistency in food formulations. In addition, large-scale extraction and processing methods for SCG utilization must address compliance with food safety regulations, the cost effectiveness, and infrastructure limitations [59]. Therefore, despite the importance of the present results, further work is crucial to fully assess the potential and viability of these SCG extracts for commercial applications.

4. Conclusions

This work constitutes the first comprehensive report on extracts of spent coffee grounds from Costa Rica, with a total of twenty-one compounds tentatively identified using UPLC-QTOF-ESI MS, including fourteen phenolic acids, three caffeoylquinic lactones, and four atractyligenin diterpenes. The sample with the highest value for the total polyphenols, the fine-grind spent coffee, also showed a higher UPLC-DAD phenolic acid content and better performance in the DPPH, PFRAP, TEAC, and ORAC antioxidant evaluations compared to the medium-grind sample, the coarse-grind sample, and even previous reports in the literature. Even though these are promising results, further studies are needed to determine the potential of this waste material and its antioxidant properties.

Author Contributions: Conceptualization, M.N.-H. and A.M.A.-S.; data curation, M.N.-H., A.M.-N. and A.M.A.-S.; formal analysis, L.F.V.-H., J.D.C.-V., V.L.-A. and A.S.-K.; funding acquisition, M.N.-H. and A.M.A.-S.; investigation, M.N.-H., D.A.-C., J.R.V.-B., L.G.R.-E. and A.M.A.-S.; methodology, M.N.-H., L.F.V.-H. and A.M.-N.; project administration, M.N.-H., L.G.R.-E. and A.M.A.-S.; resources, M.N.-H., L.G.R.-E. and A.M.A.-S.; writing—original draft, M.N.-H., L.F.V.-H. and D.A.-C.; writing—review and editing, M.N.-H., L.F.V.-H., J.D.C.-V., V.L.-A., D.A.-C., J.R.V.-B., L.G.R.-E., A.S.-K., A.M.-N. and A.M.A.-S. All authors have read and agreed to the published version of the manuscript.

Funding: This research was partially funded by the Costa Rican Innovation and Research Promoting Agency (grant FI-055B-19) and the University of Costa Rica (Refs 115C2501, 115C3005, 115C4132, and 115C0800).

Institutional Review Board Statement: Not applicable.

Informed Consent Statement: Not applicable.

Data Availability Statement: The original contributions presented in this study are included in the article. Further inquiries can be directed to the corresponding author.

Acknowledgments: The authors acknowledge the support of the University of Costa Rica UCR-Helice Program for InnoBiotiQ SRL, the National Laboratory of Nanotechnology (LANOTEC-CENAT), the Costa Rican Institute of Technology (TEC), and the National Technical University (UTN).

Conflicts of Interest: The authors declare no conflicts of interest.

References

1. Murthy, P.S.; Naidu, M.M. Sustainable management of coffee industry by-products and value addition—A review. *Resour. Conserv. Recycl.* **2012**, *66*, 45–58. [CrossRef]
2. Heckman, M.A.; Weil, J.; De Mejia, E.G. Caffeine (1,3,7-trimethylxanthine) in Foods: A Comprehensive Review on Consumption, Functionality, Safety, and Regulatory Matters. *J. Food Sci.* **2010**, *75*, R77–R87. [CrossRef]
3. Stylianou, M.; Agapiou, A.; Omirou, M.; Vyrides, I.; Ioannides, I.M.; Maratheftis, G.; Fasoula, D. Converting environmental risks to benefits by using spent coffee grounds (SCG) as a valuable resource. *Environ. Sci. Pollut. Res.* **2018**, *25*, 35776–35790. [CrossRef]
4. Andrade, C.; Perestrelo, R.; Câmara, J.S. Valorization of Spent Coffee Grounds as a Natural Source of Bioactive Compounds for Several Industrial Applications—A Volatilomic Approach. *Foods* **2022**, *11*, 1731. [CrossRef] [PubMed]
5. Barbosa, H.M.A.; De Melo, M.M.R.; Coimbra, M.A.; Passos, C.P.; Silva, C.M. Optimization of the supercritical fluid coextraction of oil and diterpenes from spent coffee grounds using experimental design and response surface methodology. *J. Supercrit. Fluids* **2014**, *85*, 165–172. [CrossRef]
6. Pujol, D.; Liu, C.; Gominho, J.; Olivella, M.À.; Fiol, N.; Villaescusa, I. The chemical composition of exhausted coffee waste. *Ind. Crops Prod.* **2013**, *50*, 423–429. [CrossRef]
7. Burniol-Figols, A.; Cenian, K.; Skiadas, I.V.; Gavala, H.N. Integration of chlorogenic acid recovery and bioethanol production from spent coffee grounds. *Biochem. Eng. J.* **2016**, *116*, 54–64. [CrossRef]
8. Lee, X.J.; Ong, H.C.; Gao, W.; Ok, Y.S.; Chen, W.-H.; Goh, B.H.H.; Chong, C.T. Solid biofuel production from spent coffee ground wastes: Process optimization, characterization and kinetic studies. *Fuel* **2021**, *292*, 120309. [CrossRef]
9. Giraldo, L.; Moreno-Piraján, J.C. Synthesis of Activated Carbon Mesoporous from Coffee Waste and Its Application in Adsorption Zinc and Mercury Ions from Aqueous Solution. *J. Chem.* **2012**, *9*, 938–948. [CrossRef]
10. Figueroa Campos, G.A.; Perez, J.P.H.; Block, I.; Sagu, S.T.; Saravia Celis, P.; Taubert, A.; Rawe, H.M. Preparation of Activated Carbons from Spent Coffee Grounds and Coffee Parchment and Assessment of Their Adsorbent Efficiency. *Processes* **2021**, *9*, 1396. [CrossRef]
11. García-Roldán, A.; Piriou, L.; Jauregi, P. Natural deep eutectic solvents as a green extraction of polyphenols from spent coffee ground with enhanced bioactivities. *Front. Plant Sci.* **2023**, *13*, 1072592. [CrossRef] [PubMed]
12. Vandeponseele, A.; Draye, M.; Piot, C.; Chatel, G. Study of Influential Parameters of the Caffeine Extraction from Spent Coffee Grounds: From Brewing Coffee Method to the Waste Treatment Conditions. *Clean Technol.* **2021**, *3*, 335–350. [CrossRef]
13. Mirón-Mérida, V.A.; Barragán-Huerta, B.E.; Gutiérrez-Macías, P. Coffee waste: A source of valuable technologies for sustainable development. In *Valorization of Agri-Food Wastes and By-Products*; Bhat, R., Ed.; Academic Press: London, UK, 2021; pp. 173–198. ISBN 978-0-12-824044-1.
14. Taghvaei, M.; Jafari, S.M. Application and stability of natural antioxidants in edible oils in order to substitute synthetic additives. *J. Food Sci. Technol.* **2015**, *52*, 1272–1282. [CrossRef]
15. De Rosso, M.; Colomban, S.; Flamini, R.; Navarini, L. UHPLC-ESI-QqTOF-MS/MS characterization of minor chlorogenic acids in roasted *Coffea arabica* from different geographical origin. *J. Mass Spectrom.* **2018**, *53*, 763–771. [CrossRef]
16. Clifford, M.N. Chlorogenic acids and other cinnamates-nature, occurrence, dietary burden, absorption and metabolism. *J. Sci. Food Agric.* **2000**, *80*, 1033–1043. [CrossRef]
17. Clifford, M.N.; Kerimi, A.; Williamson, G. Bioavailability and metabolism of chlorogenic acids (acyl-quinic acids) in humans. *Comp. Rev. Food Sci. Food Safe* **2020**, *19*, 1299–1352. [CrossRef] [PubMed]
18. Monteiro, M.; Farah, A.; Perrone, D.; Trugo, L.C.; Donangelo, C. Chlorogenic Acid Compounds from Coffee Are Differentially Absorbed and Metabolized in Humans. *J. Nut.* **2007**, *137*, 2196–2201. [CrossRef]

19. Liang, N.; Kitts, D. Role of Chlorogenic Acids in Controlling Oxidative and Inflammatory Stress Conditions. *Nutrients* **2015**, *8*, 16. [CrossRef]
20. Rodriguez De Sotillo, D.V.; Hadley, M. Chlorogenic acid modifies plasma and liver concentrations of: Cholesterol, triacylglycerol, and minerals in (fa/fa) Zucker rats. *J. Nutr. Biochem.* **2022**, *13*, 717–726. [CrossRef] [PubMed]
21. Suzuki, A.; Fujii, A.; Jokura, H.; Tokimitsu, I.; Hase, T.; Saito, I. Hydroxyhydroquinone Interferes with the Chlorogenic Acid-induced Restoration of Endothelial Function in Spontaneously Hypertensive Rats. *Am. J. Hypertens.* **2008**, *21*, 23–27. [CrossRef]
22. Huang, S.; Wang, L.L.; Xue, N.N.; Li, C.; Guo, H.H.; Ren, T.K.; Zhan, Y.; Li, W.B.; Zhang, J.; Chen, X.G.; et al. Chlorogenic acid effectively treats cancers through induction of cancer cell differentiation. *Theranostics* **2019**, *9*, 6745–6763. [CrossRef]
23. Ao, X.; Yan, J.; Liu, S.; Chen, S.; Zou, L.; Yang, Y.; He, L.; Li, S.; Liu, A.; Zhao, K. Extraction, isolation and identification of four phenolic compounds from Pleioblastus amarus shoots and their antioxidant and anti-inflammatory properties in vitro. *Food Chem.* **2022**, *374*, 131743. [CrossRef]
24. Boulebd, H.; Carmena-Bargueño, M.; Pérez-Sánchez, H. Exploring the Antioxidant Properties of Caffeoylquinic and Feruloylquinic Acids: A Computational Study on Hydroperoxyl Radical Scavenging and Xanthine Oxidase Inhibition. *Antioxidants* **2023**, *12*, 1669. [CrossRef]
25. Wen, M.; Han, Z.; Cui, Y.; Ho, C.-T.; Wan, X.; Zhang, L. Identification of 4-*O*-*p*-coumaroylquinic acid as astringent compound of Keemun black tea by efficient integrated approaches of mass spectrometry, turbidity analysis and sensory evaluation. *Food Chem.* **2022**, *368*, 130803. [CrossRef] [PubMed]
26. Zhao, Q.; Li, Y.; Li, S.; He, X.; Gu, R. Comparative bioactivity evaluation and metabolic profiling of different parts of Duhaldea nervosa based on GC-MS and LC-MS. *Front. Nutr.* **2023**, *10*, 1301715. [CrossRef]
27. Martini, S.; Conte, A.; Tagliazucchi, D. Bioactivity and cell metabolism of in vitro digested sweet cherry (*Prunus avium*) phenolic compounds. *Int. J. Food Sci. Nutr.* **2019**, *70*, 335–348. [CrossRef] [PubMed]
28. Obatomi, D.K.; Bach, P.H. Biochemistry and Toxicology of the Diterpenoid Glycoside Atractyloside. *Food Chem. Toxicol.* **1998**, *36*, 335–346. [CrossRef]
29. Zhang, P.; Cheng, X.; Sun, H.; Li, Y.; Mei, W.; Zeng, C. Atractyloside Protect Mice Against Liver Steatosis by Activation of Autophagy via ANT-AMPK-mTORC1 Signaling Pathway. *Front. Pharmacol.* **2021**, *12*, 736655. [CrossRef]
30. Li, H.; Shi, X.; Jiang, H.; Kang, J.; Yu, M.; Li, Q.; Yu, K.; Chen, Z.; Pan, H.; Chen, W. CMap analysis identifies Atractyloside as a potential drug candidate for type 2 diabetes based on integration of metabolomics and transcriptomics. *J. Cell. Mol. Med.* **2020**, *24*, 7417–7426. [CrossRef] [PubMed]
31. Xuan, S.H.; Lee, N.H.; Park, S.N. Atractyligenin, a terpenoid isolated from coffee silverskin, inhibits cutaneous photoaging. *J. Photochem. Photobiol. B* **2019**, *194*, 166–173. [CrossRef] [PubMed]
32. Navarro, M.; Arnaez, E.; Moreira, I.; Hurtado, A.; Monge, D.; Monagas, M. Polyphenolic Composition and Antioxidant Activity of *Uncaria tomentosa* Commercial Bark Products. *Antioxidants* **2019**, *8*, 339. [CrossRef] [PubMed]
33. Quirós-Fallas, M.I.; Wilhelm-Romero, K.; Quesada-Mora, S.; Azofeifa-Cordero, G.; Vargas-Huertas, L.F.; Alvarado-Corella, D.; Mora-Román, J.J.; Vega-Baudrit, J.R.; Navarro, M.; Araya-Sibaja, A.M. Curcumin Hybrid Lipid Polymeric Nanoparticles: Antioxidant Activity, Immune Cellular Response, and Cytotoxicity Evaluation. *Biomedicines* **2022**, *10*, 2431. [CrossRef] [PubMed]
34. Işil Berker, K.; Güçlü, K.; Tor, I.; Demirata, B.; Apak, R. Total antioxidant capacity assay using optimized ferricyanide/Prussian blue method. *Food Anal. Methods* **2010**, *3*, 154–168. [CrossRef]
35. Thaipong, K.; Boonprakob, U.; Crosby, K.; Cisneros-Zevallos, L.; Hawkins Byrne, D. Comparison of ABTS, DPPH, FRAP, and ORAC assays for estimating antioxidant activity from guava fruit extracts. *J. Food Compos. Anal.* **2006**, *19*, 669–675. [CrossRef]
36. Dávalos, A.; Gómez-Cordovés, C.; Bartolomé, B. Extending Applicability of the Oxygen Radical Absorbance Capacity (ORAC-Fluorescein) Assay. *J. Agric. Food Chem.* **2004**, *52*, 48–54. [CrossRef]
37. Panusa, A.; Zuorro, A.; Lavecchia, R.; Marrosu, G.; Petrucci, R. Recovery of natural antioxidants from spent coffee grounds. *J. Agric. Food Chem.* **2013**, *61*, 4162–4168. [CrossRef]
38. Fang, N.; Yu, S.; Prior, R.L. LC/MS/MS Characterization of Phenolic Constituents in Dried Plums. *J. Agric. Food Chem.* **2002**, *50*, 3579–3585. [CrossRef]
39. Jaiswal, R.; Matei, M.F.; Ullrich, F.; Kuhnert, N. How to distinguish between cinnamoylshikimate esters and chlorogenic acid lactones by liquid chromatography-tandem mass spectrometry. *J. Mass Spectrom.* **2011**, *46*, 933–942. [CrossRef] [PubMed]
40. Angeloni, S.; Nzekoue, F.K.; Navarini, L.; Sagratini, G.; Torregiani, E.; Vittori, S.; Caprioli, G. An analytical method for the simultaneous quantification of 30 bioactive compounds in spent coffee ground by HPLC-MS/MS. *J. Mass Spectrom.* **2020**, *55*, e4519. [CrossRef]
41. Lang, R.; Beusch, A.; Dirndorfer, S. Metabolites of dietary atractyligenin glucoside in coffee drinkers' urine. *Food Chem.* **2023**, *405 Pt B*, 135026. [CrossRef]
42. Tsukui, A.; Vendramini, P.H.; Garrett, R.; Scholz, M.B.S.; Eberlin, M.N.; Bizzo, H.R.; Rezende, C.M. Direct-infusion electrospray ionization-mass spectrometry analysis reveals atractyligenin derivatives as potential markers for green coffee postharvest discrimination. *LWT—Food Sci. Technol.* **2019**, *103*, 205–211. [CrossRef]

43. Panusa, A.; Petrucci, R.; Lavecchia, R.; Zuorro, A. UHPLC-PDA-ESI-TOF/MS metabolic profiling and antioxidant capacity of arabica and robusta coffee silverskin: Antioxidants vs. phytotoxins. *Food Res. Int.* **2017**, *99*, 155–165. [CrossRef] [PubMed]
44. Munteanu, I.G.; Apetrei, C. Analytical methods used in determining antioxidant activity: A review. *Int. J. Mol. Sci.* **2021**, *22*, 3380. [CrossRef]
45. Wu, C.T.; Agrawal, D.C.; Huang, W.Y.; Hsu, H.C.; Yang, S.J.; Huang, S.L.; Lin, Y.S. Functionality Analysis of Spent Coffee Ground Extracts Obtained by the Hydrothermal Method. *J. Chem.* **2019**, *10*, 4671438. [CrossRef]
46. Mussatto, S.I.; Ballesteros, L.F.; Martins, S.; Teixeira, J.A. Extraction of antioxidant phenolic compounds from spent coffee grounds. *Sep. Purif. Technol.* **2011**, *83*, 173–179. [CrossRef]
47. Choi, B.; Koh, E. Spent coffee as a rich source of antioxidative compounds. *Food Sci. Biotechnol.* **2017**, *26*, 921–927. [CrossRef]
48. Ramón-Gonçalves, M.; Gómez-Mejía, E.; Rosales-Conrado, N.; León-González, M.E.; Madrid, Y. Extraction, identification and quantification of polyphenols from spent coffee grounds by chromatographic methods and chemometric analyses. *Waste Manag.* **2019**, *96*, 15–24. [CrossRef] [PubMed]
49. Cruz, R.; Cardoso, M.M.; Fernandes, L.; Oliveira, M.; Mendes, E.; Baptista, P.; Morais, S.; Casal, S. Espresso coffee residues: A valuable source of unextracted compounds. *J. Agric. Food Chem.* **2012**, *60*, 7777–7784. [CrossRef] [PubMed]
50. Ho, K.V.; Schreiber, K.L.; Park, J.; Vo, P.H.; Lei, Z.; Sumner, L.W.; Brown, C.R.; Lin, C.H. Identification and Quantification of Bioactive Molecules Inhibiting Pro-Inflammatory Cytokine Production in Spent Coffee Grounds Using Metabolomics Analyses. *Front. Pharmacol.* **2020**, *11*, 229. [CrossRef] [PubMed]
51. Ludwig, I.A.; Sanchez, L.; Caemmerer, B.; Kroh, L.W.; De Peña, M.P.; Cid, C. Extraction of coffee antioxidants: Impact of brewing time and method. *Food Res. Int.* **2012**, *48*, 57–64. [CrossRef]
52. Číž, M.; Čížová, H.; Denev, P.; Kratchanova, M.; Slavov, A.; Lojek, A. Different methods for control and comparison of the antioxidant properties of vegetables. *Food Control* **2010**, *21*, 518–523. [CrossRef]
53. Yashin, A.; Yashin, Y.; Wang, J.Y.; Nemzer, B. Antioxidant and Antiradical Activity of Coffee. *Antioxidants* **2013**, *2*, 230–245. [CrossRef]
54. Prior, R.L.; Wu, X.; Schaich, K. Standardized methods for the determination of antioxidant capacity and phenolics in foods and dietary supplements. *J. Agric. Food Chem.* **2005**, *53*, 4290–4302. [CrossRef] [PubMed]
55. Dudonné, S.; Vitrac, X.; Coutiére, P.; Woillez, M.; Mérillon, J.M. Comparative study of antioxidant properties and total phenolic content of 30 plant extracts of industrial interest using DPPH, ABTS, FRAP, SOD, and ORAC assays. *J. Agric. Food Chem.* **2009**, *57*, 1768–1774. [CrossRef] [PubMed]
56. Lizcano, L.J.; Bakkali, F.; Ruiz-Larrea, M.B.; Ruiz-Sanz, J.I. Antioxidant activity and polyphenol content of aqueous extracts from Colombian Amazonian plants with medicinal use. *Food Chem.* **2010**, *119*, 1566–1570. [CrossRef]
57. Wong, S.P.; Leong, L.P.; Koh, J.H. Antioxidant activities of aqueous extracts of selected plants. *Food Chem.* **2006**, *99*, 775–783. [CrossRef]
58. Vargas-Huertas, L.F.; Alvarado-Corella, L.D.; Sánchez-Kopper, A.; Araya-Sibaja, A.M.; Navarro-Hoyos, M. Characterization and Isolation of Piperamides from *Piper nigrum* Cultivated in Costa Rica. *Horticulturae* **2023**, *9*, 1323. [CrossRef]
59. Janissen, B.; Huynh, T. Chemical composition and value-adding applications of coffee industry by-products: A review. *Resour. Conserv. Recycl.* **2018**, *128*, 110–117. [CrossRef]

foods

MDPI

Article

What Drives Generation Z to Avoid Food Waste in China? An Empirical Investigation

Xin Qi [1,2,3], Muyuan Li [1], Jiayi Chen [1], Guohua Zhan [1] and Lu Niu [1,*]

1 Management College, Ocean University of China, Qingdao 266100, China; x.qi@ouc.edu.cn (X.Q.); limuyuan@stu.ouc.edu.cn (M.L.); 22110031002@stu.ouc.edu.cn (J.C.); zghzhanguohua@163.com (G.Z.)
2 Institute of Marine Development, Ocean University of China, Qingdao 266100, China
3 Innovation and Entrepreneurship Research Center, Ocean University of China, Qingdao 266100, China
* Correspondence: niulu1003@ouc.edu.cn

Abstract: Avoiding food waste has become an important global issue. Given the global impact of food waste and the profound influence of Generation Z on future development, it is crucial to guide them in cultivating awareness and behaviors to reduce food waste, thereby promoting sustainable development. Considering young consumers' specific characteristics and consumption environment, this study extended the Theory of Planned Behavior (TPB) framework by adding two constructs of moral self-identity and scarcity mindset. An online survey was conducted, receiving 417 valid responses, and the data were analyzed using structural equation modeling. This study shows that subjective norms, attitudes, and perceived behavioral control positively influence Generation Z's intentions to avoid food waste. Meanwhile, moral self-identity remarkably positively influences attitudes and perceived behavioral control, which in turn affects intention to avoid food waste. Moreover, the positive moderating role of scarcity mindset is verified. This study refines the exploration of food waste within the realm of the Generation Z group, and the findings are beneficial for relevant stakeholders to further develop personalized promotion strategies for Generation Z.

Keywords: avoid; food waste; Generation Z; scarcity mindset; TPB

1. Introduction

Food waste has become a critical global issue, with far-reaching implications for resource depletion and environmental pollution [1,2]. The United Nations Environment Programme (UNEP) and WRAP's 2021 Food Waste Index Report highlight that approximately 17% of food available to consumers worldwide is wasted [3]. This phenomenon not only reflects an enormous waste of resources, but also underscores the critical issue of global food insecurity. According to the 2023 Global Food Crisis Report, over 250 million people faced severe hunger in 2022, with seven countries at risk of famine [4]. Furthermore, UNEP's 2024 report points out that more than one billion tons of food are wasted annually, while 783 million people continue to suffer from hunger [5]. This stark contrast between waste and need drives home the urgent need for sustained efforts to combat food waste and its impact on global sustainability [6].

As the world's largest developing country, China faces rising economic growth alongside growing food production [7]. However, food waste remains a significant issue that cannot be overlooked [8]. According to the 2023 China Food and Nutrition Development Report, China's food loss and waste rate stands at 22.7%, with an economic loss of RMB 1.88 trillion in 2022, accounting for nearly 22.3% of the country's agricultural output [9].

Notably, food waste behaviors are particularly pronounced in urban areas, where rising incomes, large portion sizes, and evolving consumption habits contribute to significant waste. In fact, approximately 9.4% of food waste in China occurs at the consumption level, making it the largest contributor compared to food production or retail waste [9]. The expansion of the middle class and urban population has driven higher consumption, leading to challenges in managing food waste effectively. To mitigate global food shortages and reduce environmental impact, China must focus on addressing waste at the consumer level, offering valuable insights for other nations striving to meet sustainable development goals. Addressing food waste behavior at the consumer level, particularly among young, urban populations, could serve as a critical strategy for curbing waste and promoting sustainability in China.

Generation Z (born between 1995 and 2009) is rapidly becoming a key consumer group, and their behaviors related to food waste are crucial for both global and Chinese sustainability efforts [10]. Compared to previous generations, Generation Z is more socially responsible and environmentally conscious, often choosing brands that emphasize sustainability and ethical practices [11,12]. According to the 2023 China Food and Nutrition Development Report, a significant portion of food waste in China occurs at the consumption level, and Generation Z is the primary demographic group involved in this behavior [9]. With China's rapid urbanization and rising income levels, the consumption behaviors of Generation Z are expected to have a profound impact on national strategies to reduce food waste and promote sustainability. Recent studies show that younger generations, particularly Generation Z, are increasingly focused on sustainability and the principles of the circular economy. This growing concern has been seen in multiple sectors, with Generation Z demonstrating a preference for eco-friendly products and practices, making them more likely to support brands that prioritize environmental sustainability. These emerging trends reflect the increasing environmental awareness that drives Generation Z's attitudes and behaviors in both food waste and consumption patterns [13,14]. Existing studies have paid limited attention to the food waste behavior of Generation Z. Only a few studies [15,16] have explored their food waste behaviors, and there is still a lack of sufficient in-depth research on the underlying mechanisms and interventions that influence the food waste behaviors of Generation Z. This research gap is particularly prominent in China. Research on effective strategies to prevent food waste is of significant importance. Not only will it help reduce resource waste and alleviate environmental pressure, but it will also contribute to achieving sustainable development goals [17]. By exploring the psychological mechanisms, attitudes, and intentional behaviors of Generation Z that prevent food waste, this research can provide a theoretical basis for more targeted policies and social interventions, ultimately resulting in long-term social and ecological benefits. Therefore, there is an urgent need for research to understand how to prevent food waste behavioral intentions among Generation Z.

Various theories can be applied to explain individual behavior. Among these, the Theory of Planned Behavior (TPB), introduced by Ajzen [18], has been widely and successfully applied in food consumption studies [19–21]. However, the TPB has faced criticism for several reasons, particularly regarding its insufficient consideration of other important variables. It has been suggested that these variables evolve and adapt in different contexts and behavioral scenarios [22]. To better understand Generation Z's behavior of avoiding food waste, it is crucial to consider potential factors that could play an interesting role. Given the complexity of the ethical issues surrounding the problem of food waste [23], there is a clear need for more in-depth exploration. In addition, the impact of scarcity cannot be ignored, whether in the form of the actual availability of food resources in certain regions or in the sense of scarcity created by factors such as population growth and changing consumption patterns. Scarcity fundamentally changes the way people view food waste. It instills a sense of urgency and value and makes people more aware of the consequences of their actions [24,25]. Moral self-identity, reflecting a strong sense of responsibility, is particularly prominent among Generation Z, who are more socially conscious and environmentally aware [26]. This is because they are more exposed to social and environmental education in school and through various media platforms, which has cultivated their sense of moral obligation. They are more likely to internalize moral values related to sustainability. This identity could influence their attitudes and behaviors toward food waste [27,28]. Meanwhile, scarcity mindset, which is the belief in the limited availability of resources, also plays a crucial role for Generation Z. Growing up in an era of digitalization and resource scarcity, Generation Z tends to be more concerned about resource sustainability [29,30]. Therefore, this study integrates these two factors into the TPB framework to better understand Generation Z's behavior of avoiding food waste and inform targeted interventions to promote sustainable consumption.

This study aims to construct a suitable framework to explain the key factors influencing Generation Z consumers' intentions to avoid food waste. To achieve this, the study is based on the TPB, extracting variables from the TPB model (i.e., intention, attitude, subjective norm, perceived behavioral control), with a special consideration of moral self-identity and scarcity mindset, to build an extended framework. This study conducts a fit analysis to verify the applicability of the extended model in the context of food waste among Generation Z in China and uses path analysis and moderation tests to examine the implementation of each construct. This research fills the gap in the study of food waste prevention among Generation Z, a specific group. The findings provide valuable insights for psychological interventions to reduce food waste behaviors in Generation Z, contributing to the development of food waste prevention strategies. Moreover, the significance of this study also lies in providing empirical support for social and policy-level interventions, helping policymakers to design more effective public education approaches and interventions to encourage Generation Z to adopt sustainable consumption habits. By gaining a deeper understanding of the psychological motivations and behavior patterns of Generation Z, this research offers valuable insights for the food industry and environmental organizations, contributing to the global goals of promoting green consumption and sustainable development.

2. Theoretical Framework and Development of Hypotheses

2.1. Development of Hypotheses

2.1.1. Variables from the TPB Model

The Theory of Planned Behavior (TPB) posits that behavioral intention directly influences individual behavior, and that the three elements of attitude, subjective norms, and perceived behavioral control (PBC) collectively affect behavioral intention [18]. As a well-established social psychological theory, the TPB effectively explains rational behaviors that are not fully controlled by individual will, and has broad applicability in the field of individual behavior research [31–34]. The strengths of the TPB include the following: identifying perceptions of the behavior under study and continuously assessing the importance of its components; being considered as one of the most effective models for behavior intervention; and offering a flexible structure, where additional variables can be added to extend the theory [31]. In food waste research, scholars have utilized the TPB to explore behavior, incorporating contextual factors like motivation and financial attitudes [35], or integrating it with other behavioral theories for a comprehensive approach [36].

- Attitude

Attitude refers to a persistent predisposition of liking or disliking that an individual holds toward a particular object or behavior. It can also be understood as the positive or negative feelings or evaluations that an individual has about a specific behavior. According to the Theory of Planned Behavior (TPB), a person's attitude toward a particular behavior influences their behavioral intentions. It has also been suggested that attitude is often the most powerful predictor of behavioral intention [37]. Generation Z, known for their strong environmental awareness [38], tends to embrace and practice a zero-waste lifestyle [39], which suggests that they hold a more positive attitude toward reducing food waste. The intention to avoid food waste is largely influenced by Generation Z's attitude toward it. When Generation Z holds a more positive attitude toward avoiding food waste, their intention to do so becomes stronger. Thus, the following hypothesis is proposed in this study:

Hypothesis 1 (H1): *Attitude has a significant positive effect on intention to avoid food waste.*

- Subjective Norms

Subjective norms refer to the perceived pressure from significant others to support or oppose a behavior, shaped by normative beliefs and motivations to conform. Preferences of family [40], peers [41], and social media influencers [42] can subtly guide consumer decisions. For example, in China, family members often emphasize cultural values like frugality and respect for food, encouraging reductions in food waste. Similarly, peers and influencers on platforms like Weibo and Douyin (TikTok) have become key sources of normative influence. Previous research highlights the significant role of subjective norms in shaping behavioral intentions, such as promoting eco-friendly purchases [43], though some studies suggest their impact on green behaviors may vary [44]. In China, Generation Z, influenced by campaigns like "empty plate," experiences greater pressure to adopt sustainable behaviors, strengthening their intention to avoid food waste. Based on this, the following hypothesis is proposed in this study:

Hypothesis 2 (H2): *Subjective norms have a significant positive effect on intention to avoid food waste.*

- Perceived Behavioral Control (PBC)

Perceived behavioral control (PBC) refers to an individual's perception of the ease or difficulty in performing a behavior, influenced by control beliefs and their strength. Unlike attitude and subjective norms, PBC reflects objective factors such as experience, time, and available resources. For Generation Z, elements like food storage facilities, convenience of food takeout services, and time management skills influence their PBC. For instance, limited storage space or busy schedules can hinder their ability to avoid food waste. When individuals perceive fewer obstacles and more resources, their confidence in reducing food waste increases, thus enhancing their intention to do so [45]. Previous research, such as Jia Li et al.'s study on college students, has identified PBC as a key determinant of food waste reduction intentions [46]. Accordingly, the following hypothesis is proposed in this study:

Hypothesis 3 (H3): *PBC has a significant positive effect on intention to avoid food waste.*

2.1.2. Incorporating Additional Constructs in the TPB

- Moral Self-identity

Moral self-identity refers to an individual's recognition of their moral qualities, reflecting how central moral values are to their self-concept [26]. Those with a strong moral self-identity feel a responsibility to act in alignment with their values [47], making them more likely to view reducing food waste as a personal moral responsibility [48]. They experience guilt and responsibility toward wasteful behaviors, which motivates them to adopt positive attitudes and enhances their perceived behavioral control (PBC) in food waste avoidance [49]. Conversely, weaker moral self-identity may result in less motivation to act responsibly. This concept has been widely recognized as a reliable predictor of pro-social behavior [50], influencing both intentions and actions [51]. Additionally, individuals with a strong moral self-identity prioritize their internal moral code over external influences like convenience, which drives them to reject wasteful behaviors and feel they have the responsibility and ability to make the right choices [52,53]. This moral self-discipline enhances PBC, reinforcing their motivation to act in ways consistent with their values [54]. Based on this, the following hypotheses are proposed in this study:

Hypothesis 4 (H4): *Moral self-identity has a significant positive effect on attitude.*

Hypothesis 5 (H5): *Moral self-identity has a significant positive effect on PBC.*

- Scarcity Mindset

In traditional economics, resource scarcity refers to objective actual scarcity, but Mullainathan and others suggest it is a relative concept, where individuals feel they have fewer resources than needed for survival and development, which is known as the scarcity mindset [29]. Previous studies offer various definitions of the scarcity mindset. For instance, Hamilton et al. (2019) describe it as a state where an individual's ability to meet their needs is threatened due to limited resources [55]. Tripathi et al. define it as a perception of having a fixed and finite amount of resources to meet demands [56]. In the context of food waste avoidance, this study adopts the definition by Mehta et al., where individuals perceive that the supply of resources is insufficient to meet their normal growth needs [57]. The scarcity mindset significantly influences individuals' affective, cognitive, and behavioral styles. Research shows that a scarcity mindset leads to valuing behaviors, compensatory behaviors, and convergent behaviors [58]. Specifically, individuals with a scarcity mindset are more likely to increase their effective utilization of resources. Gao et al. demonstrated that activating the scarcity mindset can reduce food waste behavior [59]. Based on this, the following hypotheses are proposed in this study:

Hypothesis 6 (H6): *Scarcity mindset plays a positive moderating role in the influence of attitudes* (H6a), *subjective norms* (H6b), *and PBC* (H6c) *on the intention to avoid food waste.*

2.2. Theoretical Framework

Based on the TPB model, this study proposes a new framework by adding two variables: moral self-identity and scarcity mindset. To better present the hypotheses proposed in this study, the research structure is illustrated in Figure 1.

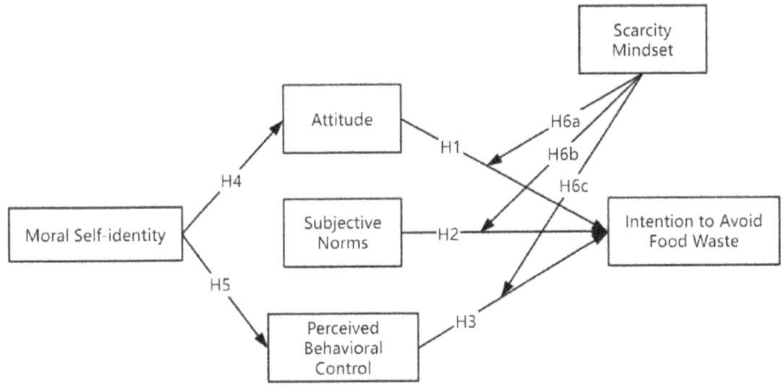

Figure 1. Conceptual model.

3. Materials and Methods

3.1. Data Collection

This study collected data to analyze the proposed conceptual framework through an online questionnaire posted on a professional survey platform (www.wenjuan.com). Recognizing that living habits may vary across regions, which could lead to differences within Generation Z, the survey was conducted in China, utilizing online platforms and social media channels popular among Generation Z. This approach helps mitigate the potential bias caused by geographical factors, ensuring the sample's representativeness and universality. Before the official launch, a pre-test was conducted with 30 randomly selected young consumers to refine the questionnaire and improve its clarity. This study specifically targeted Chinese Generation Z, and thus, an age selection question was included at the beginning of the survey to ensure only participants meeting the age criterion could proceed. In total, 470 questionnaires were collected, and 53 invalid responses were removed after data cleaning and logical consistency checks, leaving 417 valid responses (a response rate of 88.7%). According to Kline's recommendation [60], at least 10 valid responses per parameter are required for valid empirical analysis. With 25 measurement items, the minimum sample size needed was 250 usable responses, making the 417 valid responses sufficient for the subsequent data analysis.

3.2. Measures

This study is based on reliable previous research scales and selected measurement items from them. The content of the scale refers to established Chinese–English bilingual scales to ensure the accuracy of the expression to the content of the measurement items. The questionnaire includes two parts. The first part is questions on attitude, subjective norms, PBC, moral self-identity, and scarcity mindset. Items were measured using a five-point Likert-scale approach, divided into five levels (level 1 for strongly disagree and level 5 for strongly agree) for measurement. The specific items of the questionnaire and the sources used for them are shown in Table 1. The second part counted the demographic characteristics of the respondents to the questionnaire with five items.

Table 1. Questionnaire items and their source of adoption.

Variable	Item	Measurement Item	Adopted From
Intention to Avoid Food Waste (IAFW)	IAFW1	I try to avoid wasting food in my life.	[46,61]
	IAFW2	To avoid food waste, I will be careful to order in moderation.	
	IAFW3	I'll make a point of packing my leftovers.	

Table 1. *Cont.*

Variable	Item	Measurement Item	Adopted From
Subjective Norms (SN)	SN1	No one important to me is in favor of over-ordering.	[46,61,62]
	SN2	Many of my friends around me have responded to the "empty plate" campaign to reduce food waste.	
	SN3	Most people in my family would be careful to avoid wasting food.	
Attitude (AT)	AT1	There are still many hungry people in the world and it is immoral if we waste food.	[39,62,63]
	AT2	Avoiding food waste is a smart move.	
	AT3	It makes me sad when I see food being wasted.	
Perceived Behavioral Control (PBC)	PBC1	I think food waste can be avoided.	[61]
	PBC2	If I want to save food, I can always reach my goal.	
	PBC3	It's not difficult to order as much as I eat when I eat out.	
Moral Self-identity (MS)	MS1	It makes me feel good to be a person who has qualities like these.	[64]
	MS2	I am actively engaged in activities that reflect these qualities.	
	MS3	It's important to me to be a person who possesses these qualities.	
Scarcity Mindset (SM)	SM1	In our modern society with increasingly scarce resources, I tend to avoid food waste.	[65]
	SM2	I'll try to avoid food waste due to fear of food shortages.	
	SM3	When my money is tight, I tend to try to avoid food waste as much as possible.	

3.3. Data Analysis

In this study, the data were analyzed using Analysis of Moment Structures (AMOS) version 29 and Statistical Package for the Social Sciences (SPSS) version 29. Initially, we applied SPSS for descriptive analysis to observe the characteristics of the participants. Then, we used AMOS to conduct structural equation modeling (SEM) analysis in two steps. Firstly, validated factor analysis (CFA) was used to test the reliability and validity of the measurement model. Secondly, p-value, t-value, and standardized regression coefficient (β) were used to measure the complete structural model. Thus, model fit effects and hypothesized relationships could be assessed.

4. Results

4.1. Profile of the Respondents

The demographic characteristics of the respondents are presented in Table 2. The platform data indicated that the respondents came from 29 different provinces and regions across China, with the largest numbers originating from Shandong, Guangdong, and Henan. In terms of gender distribution, 67.8% of the respondents were female and 32.2% were male. The majority of participants were over 20 years old and had a high level of education. Specifically, 32.2% had completed high school or vocational education, while 64.1% held advanced degrees, including bachelor's or master's degrees. The level of monthly disposable income was relatively consistent across different subregions.

Table 2. Demographic characteristics of the sample (n = 417).

Demographic	Variables	Frequency	Percent (%)
Gender	Female	283	67.8
	Male	134	32.2
Age	15–19	50	11.9
	20–24	194	46.6
	25–29	173	41.5

Table 2. *Cont.*

Demographic	Variables	Frequency	Percent (%)
Education	Elementary school and below	3	0.7
	Junior high school	13	3.0
	Senior high school	84	20.2
	Technical secondary school	50	12.0
	University	239	57.3
	Master's or above	28	6.8
Monthly Disposable Income	<RMB 2000	109	26.1
	RMB 2000–3000	65	15.7
	RMB 3001–5000	77	18.5
	RMB 5001–8000	78	18.8
	RMB 8001–10,000	52	12.4
	>RMB 10,000	35	8.5

4.2. Measurement Model: Reliability and Validity

Table 3 presents the reliability and validity analysis of each measure. The results of the analysis showed that the Cronbach's coefficients of intention to avoid food waste, subjective norms, attitudes, PBC, moral self-identity, and scarcity mindset were 0.785, 0.809, 0.783, 0.786, 0.794, and 0.823. Cronbach's coefficients of the whole questionnaire reached above 0.7, which indicated that the data of the sample had good reliability. The KMO test assessment value (0.91 > 0.5) showed the adequacy of the sample taken, and Bartlett's spherical test determination ($p < 0.05$) confirmed the statistical significance of the correlation matrix between the variables, indicating that the data collected were suitable for use in conducting factor analysis. The results of the confirmatory factor analysis showed that the loadings of the items ranged from 0.714 to 0.804; these results were better than the 5% significance level and the benchmark of 0.4 for a sample size of 200 or more. The test items all had significant unidimensionality and good model fit. Table 4 demonstrates the square root of AVE for each construct, all of which are greater than the Pearson correlation coefficient values for each latent variable, indicating that the discriminant validity of the data is good.

Table 3. Measurement model: reliability and validity.

Constructs	Factor Loadings	Cronbach's α	AVE	CR
IAFW1	0.737	0.785	0.551	0.786
IAFW2	0.768			
IAFW3	0.721			
SN1	0.74	0.809	0.588	0.810
SN2	0.764			
SN3	0.792			
AT1	0.754	0.783	0.546	0.783
AT2	0.732			
AT3	0.733			
PBC1	0.745	0.786	0.552	0.787
PBC2	0.717			
PBC3	0.765			
MS1	0.714	0.794	0.564	0.795
MS2	0.767			
MS3	0.77			
SM1	0.751	0.823	0.608	0.823
SM2	0.804			
SM3	0.784			

Note. IAFW: intention to avoid food waste; SN: subjective norms; AT: attitude; PBC: perceived behavioral control; MS: moral self-identity; SM: scarcity mindset; AVE = Average Variance Extracted (AVE = ∑SMC/(∑SMC + ∑standard measurement error); SMC = Squared Multiple Correlation (i.e., squared value of correlation between the constructs); CR = Composite Reliability.

Table 4. Correlation matrix for discriminant validity.

	INT	N	ATT	PBC	SE	SRM
IAFW	0.742					
SN	0.487 ***	0.767				
AT	0.444 ***	0.499 ***	0.739			
PBC	0.395 ***	0.411 ***	0.452 ***	0.743		
MS	0.34 ***	0.433 **	0.460 ***	0.396 **	0.751	
SM	0.216 ***	0.244 ***	0.253 ***	0.230 ***	0.209 ***	0.78

Note. *** and **, respectively, represent 0.001 and 0.01 levels of significance. Diagonal elements show square roots of AVE; off-diagonal elements represent correlations between constructs; IAFW: intention to avoid food waste; SN: subjective norms; AT: attitude; PBC: perceived behavioral control; MS: moral self-identity; SM: scarcity mindset.

4.3. Structural Model: Goodness-of-Fit Statistics

Table 5 shows the goodness-of-fit indicators of the structural model. The data analysis shows that the theoretical framework proposed by the study has a good model fit ($X^2/df = 1.281$; RMSEA = 0.026; RMR = 0.042; GFI = 0.95; NFI = 0.95; CFI = 0.985) and all the indices meet the criteria. The model is acceptable and has a satisfied ability to explain behavioral intentions.

Table 5. Summary of fit indices from confirmatory factor analysis.

Fit Indices	Model	Recommended Value	Results
X^2/df	1.281	>1 and <5 *	Satisfactory
RMSEA	0.026	≤0.08 *	Satisfactory
RMR	0.042	≤0.05 *	Satisfactory
GFI	0.95	≥0.9 *	Satisfactory
NFI	0.95	≥0.9 *	Satisfactory
CFI	0.985	≥0.9 *	Satisfactory

Note. * Source: Bagozzi and Yi [66]; RMSEA: Root Mean Square Error Approximation; RMR: Root Mean Square Residual; GFI: Goodness-of-Fit Index; NFI: Normative Fit Index; CFI: Comparative Fit Index.

4.4. Hypothesis Testing

Table 6 tests and demonstrates the proposed substantive impacts of the hypothetical paths. The results showed that attitudes ($\beta = 0.248$, t = 3.703, $p < 0.001$), subjective norms ($\beta = 0.382$, t = 5.640, $p < 0.001$), and PBC ($\beta = 0.197$, t = 3.095, $p < 0.01$) towards Generation Z avoiding food waste in China significantly and positively influenced the extended TPB modeling on the intention to avoid food waste, as supported by H1 and H3. The moral self-identification of Generation Z significantly and positively affects attitudes and PBC ($\beta = 0.681$, t = 9.517, $p < 0.001$; $\beta = 0.596$, t = 8.532, $p < 0.001$), as suggested by H4 and H5; thus, H4 and H5 are supported.

Table 6. Hypotheses test results.

Hypothesized Path	Standardized Path Coefficients	t-Value	Result
H1: AT→IAFW	0.248	3.703 ***	Support
H2: SN→IAFW	0.382	5.640 ***	Support
H3: PBC→IAFW	0.197	3.095 **	Support
H4: MS→AT	0.681	9.517 ***	Support
H5: MS→PBC	0.596	8.532 ***	Support

Note. IAFW: intention to avoid food waste; SN: subjective norms; AT: attitude; PBC: perceived behavioral control; MS: moral self-identity; ** $p < 0.01$; *** $p < 0.001$.

Hypothesis H6 of this paper proposes that scarcity mindset plays a positive moderating role in the relationship between attitudes (H6a), subjective norms (H6b), PBC (H6c), and the intention to avoid food waste, i.e., scarcity mindset strengthens the effects of subjective norms, attitudes, and PBC on the intention to avoid food waste. In order to further test the hypothesis, this paper adopts a three-step test of hierarchical regression analysis and uses the interaction term of the variables to test the moderating effect of information cognition, which means that if ΔR^2 is significant when the interaction term is added, then a moderating effect exists.

The specific results of the test are shown in Tables 7–9. The empirical results of Model 1 in Table 7 show that attitude has a significant positive effect on the intention to avoid food waste, which further validates hypothesis H1. Model 2 in Table 7 shows that both attitude and scarcity mindset have a positive effect on the intention to avoid food waste, with regression coefficients of 0.41 ($p < 0.001$) and 0.104 ($p < 0.05$), respectively. Meanwhile, the results of Model 3 in Table 7 show that the regression coefficient of the interaction term is 0.246 ($p < 0.001$) and $\Delta R^2 = 0.063$ ($p < 0.001$). Thus, scarcity mindset plays a positive moderating role in the effect of attitude on the intention to avoid food waste. Similarly, we analyzed the empirical results in Tables 8 and 9, and all regression coefficients and ΔR^2 results met the test requirements. So, Hypotheses H6a, H6b, and H6c are established.

Table 7. H6a test results.

Variables		IAFW		
		Model 1	**Model 2**	**Model 3**
Constant Variables	Constant Term	2.222 ***	1.888 ***	1.017 ***
	Gender	0.08	0.076	0.056
	Age	0.137	0.133	0.105
	Education	−0.094 *	−0.089 *	−0.083 *
	Career	−0.067	−0.06	−0.043
	Monthly Disposable Income	0.043	0.04	0.044
Independent Variables	AT	0.436 ***	0.41 ***	0.475 ***
Moderator Variables	SM		0.104 *	0.241 ***
Interaction Terms	AT × SM			0.246 ***
	R^2	0.215	0.225	0.288
	ΔR^2	0.215 ***	0.01 *	0.063 ***

Note. * $p < 0.05$; *** $p < 0.001$; IAFW: intention to avoid food waste; AT: attitude; SM: scarcity mindset.

Table 8. H6b test results.

Variables		IAFW		
		Model 1	**Model 2**	**Model 3**
Constant Variables	Constant Term	2.247 ***	1.923 ***	1.113 ***
	Gender	0.067	0.064	0.038
	Age	0.106	0.104	0.112
	Education	−0.122 **	−0.115 **	−0.116 **
	Career	−0.019	−0.015	−0.021
	Monthly Disposable Income	0.021	0.02	0.032
Independent Variables	SN	0.466 ***	0.444 ***	0.547 ***
Moderator Variables	SM		0.097 *	0.210 ***
Interaction Terms	SN × SM			0.321 ***
	R^2	0.257	0.265	0.345
	ΔR^2	0.257 ***	0.009 *	0.080 ***

Note. * $p < 0.05$; ** $p < 0.01$; *** $p < 0.001$; IAFW: intention to avoid food waste; SN: subjective norms; SM: scarcity mindset.

Table 9. H6c test results.

Variables		IAFW		
		Model 1	**Model 2**	**Model 3**
Constant Variables	Constant Term	2.377 ***	1.953 ***	1.410 ***
	Gender	−0.016	−0.014	0.023
	Age	0.077	0.076	0.074
	Education	−0.087	−0.082	−0.084
	Career	−0.044	−0.037	−0.045
	Monthly Disposable Income	0.047	0.044	0.041

Table 9. *Cont.*

Variables		IAFW		
		Model 1	Model 2	Model 3
Independent Variables	PBC	0.418 ***	0.387 ***	0.449 ***
Moderator Variables	SM		0.130 **	0.188 ***
Interaction Terms	PBC × SM			0.227 ***
	R^2	0.168	0.183	0.225
	ΔR^2	0.168 ***	0.015 **	0.041 ***

Note. ** $p < 0.01$; *** $p < 0.001$; IAFW: intention to avoid food waste; PBC: perceived behavioral control; SM: scarcity mindset.

5. Discussion

The issue of food waste has become increasingly critical in contemporary society, with urgent attention needed to address its consequences. In particular, Generation Z is emerging as a significant force driving social change, particularly in terms of consumption habits, environmental awareness, and social responsibility. This study explores the factors influencing the food waste avoidance intentions of Chinese Generation Z from a psychosocial perspective. Building on the TPB, two additional constructs—moral self-identity and scarcity mindset—were incorporated into the framework, offering a comprehensive explanation of the factors avoiding Generation Z's food waste behavior.

Through questionnaire research and data analysis, we found that attitudes, subjective norms, and perceived behavioral control (PBC) derived from the TPB model had a positive and significant effect on Generation Z's intentions to avoid food waste, supporting findings from previous studies [35,36,67]. Notably, subjective norms had a stronger influence than both attitudes and PBC. This aligns with the findings of Russell et al. [36], who also observed a significant role of social influence in shaping pro-environmental behaviors. One possible explanation for this is the unique socio-cultural context and group characteristics of Generation Z, which amplify the role of subjective norms. In the unique socio-cultural context of China, collective values and peer influence are deeply rooted in decision-making processes [68,69]. This is especially true for Generation Z, who grew up in a highly socialized and digitalized environment and are therefore more easily driven by social influences and external expectations [70]. Online media and national campaigns, such as the "empty plate" initiative, serve as key channels reinforcing the social norms surrounding food waste avoidance, amplifying their influence on Generation Z's behavioral intentions [71,72]. Additionally, Generation Z values group identity and belonging, driven by a stable material life and life stage characteristics that heighten their need for social connection, respect, and self-fulfillment. As a result, they are more inclined to conform to social behavioral norms [73,74]. In contrast, attitudes and PBC are more shaped by personal experience and internal beliefs, making their influence comparatively smaller. Attitudes are more shaped by personal beliefs and experiences, which may vary greatly among individuals. Previous research has shown that PBC is influenced by environmental constraints [75]. For Generation Z in China, many individuals reside in dormitories or shared apartments, where limited access to food storage and reuse facilities is common. This living situation reduces their perceived control over food waste, as practical constraints hinder their ability to implement waste-avoidance strategies effectively. However, it should be noted that although the influence of attitudes and PBC is relatively weaker, they still play an important role in the overall behavior intention formation process.

Meanwhile, the results of this study indicate that moral self-identity has a significant positive impact on attitudes and PBC, which is consistent with the results of previous human behavioral studies [52,53,71]. Therefore, relevant stakeholders, such as policymakers and educators, can consider promoting food waste avoidance by strengthening Generation

Z's moral self-identity. By reinforcing moral education and public awareness, various sectors of society can provide opportunities for Generation Z to engage in moral practices, thereby stimulating their sense of moral responsibility to reduce food waste. Additionally, businesses can build a positive brand image by actively participating in food waste reduction initiatives and advocating for environmental protection, which may help gain the trust and goodwill of Generation Z [72]. Furthermore, this study finds that scarcity mindset positively moderates the relationship between subjective norms, attitudes, PBC, and intention to avoid food waste [68]. This suggests that Generation Z's inclination to avoid food waste may grow significantly in the future. As a result, governments and marketers should capitalize on this opportunity to enhance Generation Z's awareness of conservation. Through targeted advertisements and promotional sales of food nearing expiration, marketers can raise awareness about the food waste issue, guide consumers in developing a food conservation mindset, and attract greater consumer attention and support.

6. Conclusions

This study explores the food waste avoidance behaviors of Chinese Generation Z, a key demographic shaping the future, through an extended TPB model. By integrating moral self-identity and scarcity mindset into the TPB framework, this research highlights the significant roles of attitude, subjective norms, PBC, moral self-identity, and scarcity mindset in influencing food waste avoidance intentions. The contributions of this study, both theoretical and practical, can be summarized in three key areas. First, the introduction of moral self-identity and scarcity mindset expands the TPB model and deepens our understanding of the psychological mechanisms underlying food waste avoidance among Chinese Generation Z. The findings underscore the growing importance of these variables, particularly in the post-pandemic era, where their positive impact on food-saving behaviors is especially pronounced. Second, our empirical analysis confirms the robustness of the TPB model, revealing complex relationships between the variables and demonstrating the significant moderating effects of moral self-identity and scarcity mindset. These findings not only validate the TPB model, but also highlight the importance of considering emerging social influences on Generation Z's behavioral intentions in specific contexts.

In addition, policymakers and educators can play a key role in enhancing the moral self-identity of Generation Z by incorporating food waste awareness into moral education curricula and public events. For example, community service events such as food drives or educational seminars can encourage young people to engage in practices that reinforce the moral responsibility to reduce food waste. Businesses can also utilize ethical messages in their marketing campaigns to emphasize shared social and environmental responsibility. In addition, the moderating effect of the scarcity mentality suggests that raising awareness of resource limitations can increase the willingness of Generation Z to avoid food waste. Governments and organizations can design targeted interventions, such as food expiration awareness campaigns, to highlight the consequences of food waste and encourage behaviors such as purchasing discounted items that are nearing their expiration date. Grocery stores and food delivery platforms can also offer incentives for choosing imperfect or excess food, aligning individual behaviors with conservation goals. Finally, addressing practical barriers is critical, as PBC is often influenced by environmental constraints. For example, equipping dormitories and apartments with better food storage facilities, such as community refrigerators or portioned food sharing platforms, could allow Gen Z to manage food more effectively.

7. Limitations

This study has several limitations that warrant further exploration in future research.

First, it focused only on Generation Z's intentions to avoid food waste, rather than their actual food waste avoidance behavior. Given the well-documented intention–behavior gap [76], this study's findings may not fully reflect real-world practices. Future research should extend this model to examine actual food waste behaviors, using observational methods or longitudinal designs to better capture the relationship between intentions and actions.

Second, the study was conducted solely within the Chinese Generation Z population. Due to China's unique socio-cultural conditions, including the prevalence of only-child families and the resulting heightened expectations on this generation [77], the findings may have limited generalizability. Research has shown that subjective norms' impact on behavior varies significantly across national contexts. Therefore, cross-national comparisons involving Generation Z from diverse cultural backgrounds could uncover important differences in food waste attitudes and behaviors, offering valuable insights for developing globally effective policies and interventions.

Third, the inclusion of a pathway from TPB constructs to moral self-identity (MSI) could complicate the model without a strong theoretical foundation, potentially diluting the study's focus. While this pathway was not incorporated in the present study, we recognize its potential to provide valuable insights into behavioral mechanisms. Future research could consider exploring these interactions to further enrich the understanding of Generation Z's food waste behaviors.

Fourth, this study relied on self-reported data, which introduces potential biases such as social desirability bias and recall errors. Future research could incorporate objective measures, such as food diaries or waste tracking technologies, to validate and supplement self-reported findings. These methods would enhance the scientific rigor of studies on food waste behaviors.

The insights gained from addressing these limitations can contribute significantly to the ongoing efforts to promote sustainable consumption and global food conservation.

Author Contributions: Conceptualization, X.Q. and M.L.; Methodology, X.Q. and M.L.; Formal Analysis, X.Q., M.L. and G.Z.; Investigation, X.Q., M.L. and J.C.; Resources, L.N.; Writing—Original Draft Preparation, X.Q., M.L. and J.C.; Writing—Review and Editing, X.Q., M.L., J.C. and L.N.; Visualization, X.Q., G.Z. and L.N.; Supervision, X.Q. and L.N.; Validation, X.Q., G.Z. and L.N. All authors have read and agreed to the published version of the manuscript.

Funding: This research was supported by the National Social Science Fund of China (23CGL008).

Institutional Review Board Statement: The study was conducted in accordance with the Declaration of Helsinki, and approved on 11 October 2024 by the School of Management, Ocean University of China (Reference No. 20241011).

Informed Consent Statement: Informed consent was obtained from all subjects involved in the study.

Data Availability Statement: The raw data supporting the conclusions of this article will be made available by the authors on request.

Acknowledgments: The authors are thankful to the anonymous reviewers and the editor for their valuable comments.

Conflicts of Interest: The authors declare no conflicts of interest.

References

1. Berry, B.; Blackmer, T.; Haedicke, M.; Lee, S.; MacRae, J.D.; Miller, T.R.; Nayak, B.; Rivet-Prefontaine, L.; Saber, D.; Silka, L.; et al. Safe Circular Food Systems: A Transdisciplinary Approach to Identify Emergent Risks in Food Waste Nutrient Cycling. *Foods* **2024**, *13*, 2374. [CrossRef] [PubMed]
2. Barrera, E.L.; Hertel, T. Global food waste across the income spectrum: Implications for food prices, production and resource use. *Food Policy* **2021**, *98*, 101874. [CrossRef]
3. United Nations Environment Programme (UNEP), & WRAP. Food Waste Index Report 2021. Available online: https://wedocs.unep.org/bitstream/handle/20.500.11822/37946/UNEP_AR2021.pdf (accessed on 4 March 2021).
4. Global Network to Fight the Food Crisis. The Global Food Crisis Report 2023. Available online: https://www.fsinplatform.org/global-report-food-crises-2023 (accessed on 3 May 2023).
5. United Nations Environment Programme. Food Waste Index Report 2024. Available online: https://wedocs.unep.org/handle/20.500.11822/45230 (accessed on 27 March 2024).
6. Al-Obadi, M.; Ayad, H.; Pokharel, S.; Ayari, M.A. Perspectives on food waste management: Prevention and social innovations. *Sustain. Prod. Consump.* **2022**, *31*, 190–208. [CrossRef]
7. Xin, X.Y. Meeting People's Aspirations to Live a Better Life with a Mature and Established System in China. *Int. Crit. Thought.* **2021**, *11*, 157–170. [CrossRef]
8. Lan, Y.; Xu, B.J.; Huan, Y.Z.; Guo, J.H.; Liu, X.J.; Han, J.W.; Li, K.R. Food Security and Land Use under Sustainable Development Goals: Insights from Food Supply to Demand Side and Limited Arable Land in China. *Foods* **2023**, *12*, 4168. [CrossRef]
9. Chinese Academy of Agricultural Sciences. *China Food and Nutrition Development Report*; Chinese Academy of Agricultural Sciences: Beijing, China, 2023.
10. Yilmaz, B.; Kisaçtutan, E.D.; Karatepe, S.G. Digital natives of the labor market: Generation Z as future leaders and their perspectives on leadership. *Front. Psychol.* **2024**, *15*, 1378982. [CrossRef]
11. Robichaud, Z.; Yu, H. Do young consumers care about ethical consumption? Modelling Gen Z's purchase intention towards fair trade coffee. *Brit. Food J.* **2022**, *124*, 2740–2760. [CrossRef]
12. Thangavel, P.; Pathak, P.; Chandra, B. Consumer Decision-making Style of Gen Z: A Generational Cohort Analysis. *Glob. Bus. Rev.* **2022**, *23*, 710–728. [CrossRef]
13. Burlea-Schiopoiu, A.; Ogarca, R.F.; Barbu, C.M.; Craciun, L.; Baloi, I.C.; Mihai, L.S. The impact of COVID-19 pandemic on food waste behaviour of young people. *J. Clean. Prod.* **2021**, *294*, 126333. [CrossRef]
14. Goh, E.; Jie, F. To waste or not to waste: Exploring motivational factors of Generation Z hospitality employees towards food wastage in the hospitality industry. *Int. J. Hosp. Manag.* **2019**, *80*, 126–135. [CrossRef]
15. Aldaihani, F.M.F.; Islam, M.A.; Saatchi, S.G.; Haque, M.A. Harnessing green purchase intention of generation Z consumers through green marketing strategies. *Bus. Strategy Dev.* **2024**, *7*, 14. [CrossRef]
16. Gazzola, P.; Pavione, E.; Pezzetti, R.; Grechi, D. Trends in the Fashion Industry. The Perception of Sustainability and Circular Economy: A Gender/Generation Quantitative Approach. *Sustainability* **2020**, *12*, 2809. [CrossRef]
17. Urugo, M.M.; Teka, T.A.; Gemede, H.F.; Mersha, S.; Tessema, A.; Woldemariam, H.W.; Admassu, H. A comprehensive review of current approaches on food waste reduction strategies. *Compr. Rev. Food Sci. Food Saf.* **2024**, *23*, 70011. [CrossRef] [PubMed]
18. Ajzen, I. The theory of planned behaviour: Reactions and reflections. *Psychol. Health* **2011**, *26*, 1113–1127. [CrossRef] [PubMed]
19. Coşkun, A.; Özbük, R.M.Y. What influences consumer food waste behavior in restaurants? An application of the extended theory of planned behavior. *Waste Manag.* **2020**, *117*, 170–178. [CrossRef]
20. Qi, X.; Ploeger, A. Explaining Consumers' Intentions towards Purchasing Green Food in Qingdao, China: The Amendment and Extension of the Theory of Planned Behavior. *Appetite* **2019**, *133*, 414–422. [CrossRef]
21. Zhang, X.-Y.; Chao, C.-T.; Chiu, Y.-T.; Chen, H.-S. Study of the Correlation between Streaming Video Platform Content on Food Production Processes and the Behavioral Intentions of Generation Z. *Foods* **2024**, *13*, 1537. [CrossRef]
22. Armitage, C.J.; Conner, M. Efficacy of the theory of planned behaviour: A meta-analytic review. *Br. J. Soc. Psychol.* **2001**, *40*, 471–499. [CrossRef]
23. Chang, H.H. Is it unethical to waste food? exploring consumer's ethical perspectives and waste intentions. *Curr. Psychol.* **2022**, *41*, 8434–8448. [CrossRef]
24. Simpson, B.; Mesler, R.M.; White, K. Perceiving less but wasting more: The relationship between perceived resource scarcity and consumer food waste. *Appetite* **2024**, *202*, 15. [CrossRef]
25. Jaiswal, J.; Aagja, J. Influence of scarcity and environmental consciousness on food waste behaviour. *J. Consum. Behav.* **2024**, *23*, 1057–1069. [CrossRef]
26. Aquino, K.; Reed, A., 2nd. The self-importance of moral identity. *J. Pers. Soc. Psychol.* **2002**, *83*, 1423–1440. [CrossRef] [PubMed]
27. Smith, A. Moral Identity and Environmental Attitudes in Generation Z. *J. Consum. Res.* **2022**, *45*, 384–398.
28. Turner, M.; Harris, J. The Role of Moral Self-Identity in Food Waste Reduction. *Sustainability* **2020**, *12*, 3472.

29. Mullainathan, S.; Shafir, E. *Scarcity: Why Having too Little Means So Much*; Times Books/Henry Holt and Co: New York, NY, USA, 2013.
30. Lee, R.; Nguyen, J. Resource Scarcity and the Consumer Behavior of Generation Z: The Role of Scarcity Mindset. *J. Consum. Behav.* **2021**, *21*, 527–539.
31. Yuriev, A.; Dahmen, M.; Paillé, P.; Boiral, O.; Guillaumie, L. Pro-environmental behaviors through the lens of the theory of planned behavior: A scoping review. *Resour. Conserv. Recy.* **2020**, *155*, 104660. [CrossRef]
32. Kim, S.H.; Kuo, W.Y. The Role of Beliefs, Pride, and Perceived Barriers in Decision-Making Regarding Purchasing Value-Added Pulse Products among US Consumers. *Foods* **2022**, *11*, 824. [CrossRef]
33. Wu, L.Q.; Zhu, Y.; Zhai, J.Q. Understanding Waste Management Behavior Among University Students in China: Environmental Knowledge, Personal Norms, and the Theory of Planned Behavior. *Front. Psychol.* **2022**, *12*, 771723. [CrossRef]
34. Thanki, H.; Shah, S.; Oza, A.; Vizureanu, P.; Burduhos-Nergis, D.D. Sustainable Consumption: Will They Buy It Again? Factors Influencing the Intention to Repurchase Organic Food Grain. *Foods* **2022**, *11*, 3046. [CrossRef]
35. Aktas, E.; Sahin, H.; Topaloglu, Z.; Oledinma, A.; Huda, A.K.S.; Irani, Z.; Sharif, A.M.; van't Wout, T.; Kamrava, M. A consumer behavioural approach to food waste. *J. Enterp. Inf. Manag.* **2018**, *31*, 658–673. [CrossRef]
36. Russell, S.V.; Young, C.W.; Unsworth, K.L.; Robinson, C. Bringing habits and emotions into food waste behaviour. *Resour. Conserv. Recy.* **2017**, *125*, 107–114. [CrossRef]
37. Bosnjak, M.; Ajzen, I.; Schmidt, P. The Theory of Planned Behavior: Selected Recent Advances and Applications. *Eur. J. Psychol.* **2020**, *16*, 352–356. [CrossRef] [PubMed]
38. Gomes, S.; Lopes, J.M.; Nogueira, S. Willingness to pay more for green products: A critical challenge for Gen Z. *J. Clean. Prod.* **2023**, *390*, 136092. [CrossRef]
39. Panopoulos, A.; Poulis, A.; Theodoridis, P.; Kalampakas, A. Influencing Green Purchase Intention through Eco Labels and User-Generated Content. *Sustainability* **2023**, *15*, 764. [CrossRef]
40. Lin, J.; Zhang, Z. Research progress and implications of household food waste studies:New contents, new methods and new perspectives. *J. Nat. Resour.* **2023**, *38*, 1178–1193.
41. Teng, C.C.; Wang, Y.C.; Chuang, C.J. Food choice motives and dining-out leftover prevention behavior: Integrated perspectives of planned behavior and norm activation. *Int. J. Hosp. Manag.* **2022**, *107*, 10. [CrossRef]
42. Salková, D.; Maierová, O.; Stanislavská, L.K.; Pilar, L. The Relationship between "Zero Waste" and Food: Insights from Social Media Trends. *Foods* **2023**, *12*, 3280. [CrossRef]
43. Li, L.; Long, X.L.; Laubayeva, A.; Cai, X.; Zhu, B.Z. Behavioral intention of environmentally friendly agricultural food: The role of policy, perceived value, subjective norm. *Environ. Sci. Pollut. R* **2020**, *27*, 18949–18961. [CrossRef]
44. Paul, J.; Modi, A.; Patel, J. Predicting green product consumption using theory of planned behavior and reasoned action. *J. Retail. Consum. Serv.* **2016**, *29*, 123–134. [CrossRef]
45. Hagger, M.S.; Cheung, M.W.L.; Ajzen, I.; Hamilton, K. Perceived Behavioral Control Moderating Effects in the Theory of Planned Behavior: A Meta-Analysis. *Health Psychol.* **2022**, *41*, 155–167. [CrossRef]
46. Jia, L.; Qiao, G. Determinants of university student's food waste prevention behaviour-An analysis based on the theory of planned behavior and norm activation model. *J. Arid Land Resour. Environ.* **2022**, *36*, 38–46.
47. Hardy, S.A.; Carlo, G. Moral Identity: What Is It, How Does It Develop, and Is It Linked to Moral Action? *Child Dev. Perspect.* **2011**, *5*, 212–218. [CrossRef]
48. Wu, B.; Yang, Z.Y. The impact of moral identity on consumers' green consumption tendency: The role of perceived responsibility for environmental damage. *J. Environ. Psychol.* **2018**, *59*, 74–84. [CrossRef]
49. Patrick, R.B.; Bodine, A.J.; Gibbs, J.C.; Basinger, K.S. What Accounts for Prosocial Behavior? Roles of Moral Identity, Moral Judgment, and Self-Efficacy Beliefs. *J. Genet. Psychol.* **2018**, *179*, 231–245. [CrossRef]
50. Huang, Y.H.; Ma, E.; Yen, T.H. Generation Z diners' moral judgements of restaurant food waste in the United States: A qualitative inquiry. *J. Sustain. Tour.* **2022**, 1–20. [CrossRef]
51. Yang, L.H.; Cai, G.Y.; Yong, S.H.; Shi, H.H. Moral identity: A mediation model of moral disengagement and altruistic attitude. *Soc. Behav. Personal.* **2020**, *48*, 1–13. [CrossRef]
52. Vadera, A.K.; Pathki, C.S. Competition and cheating: Investigating the role of moral awareness, moral identity, and moral elevation. *J. Organ. Behav.* **2021**, *42*, 1060–1081. [CrossRef]
53. Li, D.L.; Cheng, G.; Wang, C.Y. The influence of moral identity on green consumption. *Front. Psychol.* **2022**, *13*, 1020333. [CrossRef]
54. Dowd, K.; Burke, K.J. The influence of ethical values and food choice motivations on intentions to purchase sustainably sourced foods. *Appetite* **2013**, *69*, 137–144. [CrossRef]
55. Hamilton, R.; Thompson, D.; Bone, S.; Chaplin, L.N.; Griskevicius, V.; Goldsmith, K.; Hill, R.; John, D.R.; Mittal, C.; O'Guinn, T.; et al. The effects of scarcity on consumer decision journeys. *J. Acad. Market. Sci.* **2019**, *47*, 532–550. [CrossRef]
56. Tripathi, R.C.; Pande, N.; Tripathi, V.N.; Shankar, S.; Pande, A.; Bahadur, R. Scarcity Mindsets and Generational Differences in India: The Role of Cognitive and Affective Factors. *Psychol. Dev. Soc.* **2022**, *34*, 45–78. [CrossRef]

57. Mehta, R.; Zhu, M. Creating When You Have Less: The Impact of Resource Scarcity on Product Use Creativity. *J. Consum. Res.* **2016**, *42*, 767–782. [CrossRef]
58. Lei, L.; Wang, J.; Liu, W. The effect of scarcity on individuals' psychology and behavior:An explanation from a more integrated perspective. *Adv. Psychol. Sci.* **2020**, *28*, 833–843. [CrossRef]
59. Gao, H.; Jia, H.; Guo, B. Resources Available for Me Versus Us: Implications for Mitigating Consumer Food Waste. *J. Marketing. Res.* **2024**, *61*, 619–637. [CrossRef]
60. Kline, R.B. *Principals and Practice of Structural Equation Modeling*, 3rd ed.; The Guilford Press: New York, NY, USA, 2011.
61. Jia, L.; Zhang, Y.; Qiao, G. Consumer-Related Antecedents of Waste Behavior in Online Food Ordering: A Study among Young Adults in China. *Foods* **2022**, *11*, 3098. [CrossRef] [PubMed]
62. Visschers, V.H.M.; Wickli, N.; Siegrist, M. Sorting out food waste behaviour: A survey on the motivators and barriers of self-reported amounts of food waste in households. *J. Environ. Psychol.* **2016**, *45*, 66–78. [CrossRef]
63. Liao, F.; Qing, P.; Hou, M. Analysis on the influencing factors of consumers wasting food behaviors: Based on the theory of planned behaviors. *Res. Agric. Modern.* **2020**, *41*, 115–124.
64. Zengkui, W. The Revision of Moral self-identity Scale for Chinese Adolescents. *Soc. Psychol. Pers. Sci.* **2008**, *23*, 41–45.
65. DeSousa, M.; Reeve, C.L.; Peterman, A.H. Development and initial validation of the Perceived Scarcity Scale. *Stress. Health* **2020**, *36*, 131–146. [CrossRef]
66. Bagozzi, R.P.; Yi, Y. On the Evaluation of Structural Equation Models. *J. Acad. Market. Sci.* **1988**, *16*, 74–94. [CrossRef]
67. Amato, M.; Verneau, F.; Coppola, A.; La Barbera, F. Domestic Food Waste and Covid-19 Concern: An Application of the Theory of Planned Behaviour. *Sustainability* **2021**, *13*, 8366. [CrossRef]
68. Zhao, S.; Balabaskar, R. The Influence of Social Media on Generation Z: An Exploration of the Role of Digital Technologies in Consumer Behavior. *J. Mark. Dev. Compet.* **2020**, *14*, 74–86.
69. Chen, C.Q.; Wang, X.K.; Shen, T.H.; Tang, S.; Tao, S.; Wang, D.Y. Unveiling the Heterogeneity of Sensation Seeking and Collectivism Development in Chinese Adolescents. *J. Pers.* **2024**, *13*. *Epub ahead of print*. [CrossRef] [PubMed]
70. Zhang, L.; Zhang, M.; Jia, J.; Peng, X.; Zhu, J.X.; You, S.B. Collectivist culture, environmental regulation and pollution emissions: Evidence from China. *Front. Psychol.* **2024**, *14*, 17. [CrossRef] [PubMed]
71. Williams, A. The Impact of Government Policies on the Behavior of Generation Z: The Role of Social and Environmental Influence. Environ. *Behav. Public Policy J.* **2019**, *22*, 123–135.
72. Tewari, P.; Malhotra, R. Social Media and Generation Z's Consumer Behavior: The Role of Social Influence. *Int. J. Consum. Stud.* **2020**, *44*, 628–635.
73. Hernandez, R.; Franco, J.; Rodríguez, S. The Role of Group Identity and Belonging in Shaping Generation Z's Social Behaviors. *J. Soc. Psychol.* **2020**, *52*, 456–465.
74. Smith, J.; Brown, L. Generation Z and Their Sense of Social Belonging: The Influence of Material Stability and Life Stage. *J. Adolesc. Res.* **2021**, *40*, 300–312.
75. Wang, B.R.; Li, Y. Consumers' Intention to Bring a Reusable Bag for Shopping in China: Extending the Theory of Planned Behavior. *Int. J. Environ. Res. Public Health* **2022**, *19*, 13. [CrossRef]
76. Liu, Y.Q.; Wang, P.; Zhang, M.Y.; Chen, X.; Li, K.; Qu, J.Y. Bridging the Intention-Behavior Gap in Organic Food Consumption: Empirical Evidence from China. *Foods* **2024**, *13*, 2239. [CrossRef]
77. Lu, H.J.; Chang, L. Parenting and Socialization of Only Children in Urban China: An Example of Authoritative Parenting. *J. Genet. Psychol.* **2013**, *174*, 335–343. [CrossRef]

foods

MDPI

Review

Quality Properties of Bakery Products and Pasta Containing Spent Coffee Grounds (SCGs): A Review

Mitra Ahanchi [1], Elizabeth Christie Sugianto [2], Amy Chau [2] and Ali Khoddami [2,*]

[1] Faculty of Management, University of Warsaw, ul. Szturmowa 1/3, 02-678 Warszawa, Poland
[2] School of Life and Environmental Sciences, Faculty of Science, The University of Sydney, Sydney, NSW 2006, Australia
[*] Correspondence: ali.khoddami@sydney.edu.au

Abstract: Coffee is one of the most consumed and popular beverages worldwide, and it produces a significant quantity of waste. Spent coffee grounds (SCGs) are one of the major waste products that can be used as an ingredient for creating novel foods. Therefore, the effect of incorporating varying percentages of spent coffee grounds (SCGs) on the quality properties of bakery products and pasta is reviewed. Chemically, SCGs alter protein, fat, fiber, ash, and bioactive compound levels in bakery and pasta products, improving nutritional value and promoting health benefits. The impact of SCGs on the physical characteristics of baked goods depends on factors like SCG concentration and processing methods, which influence product texture and structure. Sensory properties are vital for consumer acceptance. SCGs can add unique flavors and colors to baked goods, but more attention is needed to optimize the SCGs' incorporation concentration for a better consumer appeal. In conclusion, integrating SCGs into bakery products and pasta offers nutritional enhancement, sustainability, and sensory improvement opportunities. Optimizing product quality allows manufacturers to leverage SCGs' potential in the food industry.

Keywords: coffee; spent coffee grounds; bioactive components; quality properties; bakery products

1. Introduction

Coffee stands out as one of the most widely consumed and favored beverages globally, the second most traded commodity globally, just after petroleum, highlighting its significant influence and dominance in the global market [1–3]. The latest data from the International Coffee Organization (ICO) reveal that global coffee consumption reached 170.3 million 60 kg bags in 2021/2022 [4]. While coffee has energizing and uplifting benefits, it is crucial to consider the waste produced [5]. The amount of waste the coffee industry creates is anticipated to rise in the coming years due to the steady growth in coffee production and consumption [2]. The waste includes husks, pulp, mucilage, silverskins, and SCGs, which are produced on the farm and finished in the coffee shops [3]. SCGs are the primary waste material produced during processing, accounting for 0.6 tons per ton of coffee beans [1,6,7]. To adhere to current national regulations, waste management programs must be effective. The SCGs produced are often gathered by specialist organizations and sold for various uses, such as gardening, composting, bioenergy production, and mushroom cultivation [7].

There is much evidence of coffee's health-promoting properties, which include anti-allergic, anti-inflammatory, anti-diabetic, anti-mutagenic, anti-cancer, antioxidant, and antimicrobial properties [2,8–11]. In addition, there is an inverse relationship with several illnesses, including Parkinson's, Alzheimer's, depression, fatty liver, and metabolic disorders [6,9,12]. Protective effects on various nervous, skeletal, reproductive, immune, and cardiovascular systems have also been reported [9,12]. The rich phytochemistry of coffee and its by-products, which comprise elements like caffeine and other physiologically active components primarily from the polyphenol and alkaloid groups, is responsible for these health advantages [2,6,8,9,12].

A key initiative in enhancing nutrition is raising the production of widely consumed items with substantial nutritional and biological significance. Proper nutrition plays an important role in treating and preventing many diseases [13]. SCGs are a source of macro-nutrients (carbohydrates, insoluble fiber, lipids, and protein), micro-nutrients (magnesium, sodium, potassium, and iron), and bioactive compounds (tannins, caffeine, and polyphenols) [5,6]. The upcycling of SCGs supports the development of functional foods and helps create a circular economy by repurposing waste materials. However, the impact of reducing environmental pollution may be limited for a few reasons. First, the global consumption of coffee results in a massive volume of SCGs, and while their application in food production can mitigate some waste, the scale at which SCGs can be incorporated into food products is relatively small compared to the total waste generated. Furthermore, there are challenges related to the sensory qualities (e.g., taste and texture) of SCG-enriched products, which may limit consumer acceptance and market scalability. In addition, there is a concern regarding the potential contamination by mycotoxins produced by certain molds, particularly under improper storage conditions [14–17].

Nonetheless, integrating SCGs into food production aligns with sustainable waste management practices by promoting the reuse of by-products, reducing landfill waste, and potentially lowering the demand for virgin agricultural inputs. This is particularly significant as it offers a more environmentally friendly alternative to conventional waste disposal methods, such as incineration or landfilling, which contribute to greenhouse gas emissions. While it may not drastically lower overall pollution, this approach supports broader sustainability goals by contributing to resource efficiency and waste minimization [6,7,11,18,19].

Bakery products are widely consumed and accepted [20,21], appealing to a broad range of age groups due to their widespread availability, convenience, and universal enjoyment [20]. Also, they are an essential source of nutrients (carbohydrates, proteins, vitamins, and minerals) [22–24]. However, these products have come into several issues in today's world. Firstly, bakery products contain excessive levels of sugar and fat, which can lead to weight gain as well as the development of non-communicable diseases like diabetes, cardiovascular disease, and obesity [25]. The second issue is product deterioration, encompassing chemical, physical, and microbiological changes in baked goods [25,26].

Since SCGs contain bioactive compounds, including antioxidant dietary fibers and phenolic compounds, it can be expected that using this material in the formulation of bakery and pasta products assists in overcoming these issues [7,27]. On the other hand, using SCGs in baking products has been shown to improve the sensory, physical, and chemical properties of these products [18,28–32].

The innovation of this review lies in the comprehensive evaluation of SCGs as a sustainable food ingredient. While previous studies have explored various aspects of SCG utilization, this review focuses on their specific impact on bakery products and pasta, addressing gaps in the existing literature by integrating findings from different fields. This review is new in compiling and presenting past research data regarding the effects of incorporating SCGs into the chemical, physical, microbiological, and sensory properties of bakery products. Moreover, it promotes the environmental benefits of incorporating SCGs into food products, such as reducing food waste and promoting circular economy practices. By exploring these dual benefits, we aim to contribute to the growing body of research that seeks sustainable solutions within the food industry.

2. Methods of Review

A comprehensive search was conducted using the following reputable databases: Google Scholar, ScienceDirect, PubMed, Springer, and Wiley, to find related peer-reviewed published research. Studies were selected using the keywords "spent coffee grounds" (SCGs), "grain-based products", "bakery products", "bread", "cakes", "biscuits", "cookies", "wafers", "muffins", "crackers", and "pasta". They were thoroughly read to ensure that the chosen papers addressed the research topic. The authors, the year of publication, the goals, the methodology, and the findings were among the data assessed. Compatibility with the primary topic and readability were the two requirements for inclusion. The exclusion criteria included animal studies, literature reviews without full text, and incompatibility with the core theme.

After searching the databases for the literature, classifying and analyzing the outcomes, and investigating the studies, 21 studies were found to integrate this review. Then, the papers' full texts were examined to ensure they addressed the topic of the current investigation. The chosen literature was published from 2012 onward.

3. Results and Discussion

Out of the 21 works of literature that were assessed, 47.61% of the studies produced cookies (n = 10), 14.28% made bread (n = 3), 14.28% made biscuits (n = 3), 9.52% (n = 2) made cakes, 9.52% (n = 2) made muffins, and 4.76% (n = 1) made pasta that contained SCGs. Among these studies (n = 21), 95.23% conducted chemical analyses (n = 20), 80.95% conducted physical analyses (n = 17), 19.04% conducted microbial analyses (n = 4), and 76.19% conducted sensory analyses (n = 16).

The research that examined bakery products and pasta prepared with SCGs is displayed in Table 1. A massive variation was seen in the amount of SCGs incorporated into baked products. Studies involving cookies found a minimum of 0.05% and a maximum of 30.0% SCGs incorporation. The amounts for breads and biscuits ranged from 2.0% to 11.0%, while the contents for pasta and cakes varied from 1.0% to 10.0%. Muffins showed the greatest variation in SCGs incorporation, with the minimum amount being 1.0% and the maximum being 61.0%. Of the studies, 66.66% (n = 14) also mentioned the variety of coffee they had used: 'Arabica' and/or 'Robusta'.

Table 1 is instructed to precisely present the impact of the inclusion of SCGs or SCG derivatives on the chemical, physical, and sensory properties of baked products.

3.1. Bakery Products' Chemical Properties

3.1.1. Cookies

In terms of moisture content, Aguilar-Raymundo et al. [27] noted that cookies containing 10.0%, 17.5%, and 25.0% SCGs demonstrated a reduced moisture content compared to the control cookies that did not include SCGs. The decrease in moisture levels may be attributed to the elevated fiber content present in SCGs, which can absorb water. This phenomenon could contribute to an enhanced shelf life of cookies by lowering their water activity. Similarly, Oliviera Batista et al. [32] reported a decrease in moisture in gluten-free cookies due to the inclusion of SCGs. Azuan et al. [29] presented divergent results, indicating that cookies supplemented with SCG extract from Arabica coffee beans exhibited a higher moisture content. This suggests that the impact of SCGs on moisture may depend on the type of coffee bean and processing method. Desai et al. [33] also observed that adding roasted green spent coffee grounds flour (RGCSF) and unroasted green spent coffee grounds flour (UGCSF) to cookies increases the moisture content of this product. Trà et al. [18] conducted a study on cookies incorporating SCGs derived from Robusta coffee beans, revealing a marginal rise in moisture content. However, the variations noted were minimal compared to the findings reported by Azuan et al. [29].

Regarding lipid stability, Koay et al. [7] found that adding SCG powder to shortbread improved lipid stability due to its high dietary fiber and antioxidant properties. The findings indicate that incorporating SCGs resulted in an enhancement of both, which has the potential to enhance the nutritional value of baked goods, in addition to prolonging their shelf life through lipid stabilization. Meerasri et al. [34] similarly illustrated that substituting butter with oil derived from SCGs in cookie formulations enhanced the total phenolic content (TPC) and antioxidant activity while preserving moisture levels. This suggests that SCG oil is a viable alternative to butter, offering enhanced lipid stability. Also, Sérioa et al. [35] investigated the antioxidant activity of cookies containing SCG extracts with 1.0%, 2.0%, and 3.0% concentrations. They demonstrated that the increase in the amount of SCG extract enhanced the antioxidant activities of cookies. SCGs contain a variety of bioactive compounds, including phenolics, chlorogenic acid, and caffeine, which play a significant role in their antioxidant capabilities. Sharma et al. [36] investigated the impact of roasted spent coffee grounds (RSCGs) and spent coffee ground residue (SCGR) on the composition of cookies, revealing that RSCGs provided notably greater concentrations of bioactive compounds in comparison to SCGR. Roasting SCGs appears to increase the accessibility of phenolic compounds and various antioxidants, thereby enhancing lipid stability and prolonging the shelf life of the goods.

In terms of nutritional quality, Aguilar-Raymundo et al. [27] observed that cookies fortified with SCGs demonstrated elevated concentrations of fat, crude fiber, and ash in comparison to the control cookies, whereas the protein content was diminished. The enhancement of fiber content, especially in the form of crude fiber, represents a significant nutritional advantage of cookies enriched with spent coffee grounds (SCGs). Fiber is crucial for supporting gastrointestinal health and may decrease the risk of chronic conditions, including cardiovascular disease and obesity [37,38]. Azuan et al. [29] also observed improvements in ash, fat, and protein content when extracts from SCGs were incorporated into cookies. Desai et al. [33] conducted a comparative analysis of the nutritional effects of roasted and unroasted spent coffee grounds incorporated into cookies. Their findings indicated that RGCSF significantly increased the protein, ash, and soluble dietary fiber levels, whereas UGCSF was more effective in enhancing the insoluble dietary fiber content. This suggests that roasting may alter the fiber profile and nutrient retention in SCGs, making roasted SCGs a more suitable choice for increasing protein and soluble fiber in cookies. Oliveira Batista et al. [32] conducted a study on the incorporation of SCGs into gluten-free cookies, revealing that elevated levels of SCGs corresponded with enhanced concentrations of protein, lipids, ash, and fiber. The findings align with Aguilar-Raymundo et al. [27] and Azuan et al. [29], who documented an increase in fiber and ash content in cookies supplemented with SCGs. Koay et al. [7] further illustrated that the incorporation of SCG powder into shortbread resulted in a decrease in the caloric value of the cookies while simultaneously enhancing their nutritional characteristics. The elevated dietary fiber content of SCGs contributed to a reduction in the glycemic index of the baked products. In addition, Castaldo et al. [39] reported higher antioxidant activity in cookies enriched with 5.0% SCGs, confirming the antioxidant benefits of SCGs.

The studies consistently show that SCGs, whether in extract, flour, or powder form, significantly improve cookies' antioxidant activity, TPC, and fiber levels [33–35,39]. There is a general trend of increased ash, fat, and protein content with SCGs inclusion, with variations in moisture content depending on the SCGs processing method and bean type [27,29,32]. The roasting of SCGs enhances bioactive compounds and antioxidants, while SCG oil can effectively replace butter in cookie formulations [33,34,36]. Overall, SCGs' high fiber content and antioxidant properties make them a valuable functional ingredient for improving the nutritional profile of cookies, potentially extending shelf life by reducing moisture and enhancing health benefits, such as improved gastrointestinal health, reduced risk of cardiovascular diseases, and weight management [37,38,40]. The findings indicate that SCGs offer versatile benefits, particularly in gluten-free and low-calorie cookie formulations [7,32,41].

3.1.2. Breads

Daniel et al. [42] assessed the effect of adding SCGs at concentrations of 2.0%, 4.0%, 6.0%, 8.0%, and 10.0% on the chemical features of bread. They observed that bread with 10.0% SCGs had the highest protein level. Fat content varied among samples containing coffee grounds, possibly influenced by the amount of oil used. Also, the coffee brewing process affects the fat content because a considerable portion of the unsaponifiable fat compounds is eliminated during this process [43]. Bread with 10.0% SCGs had the highest ash content, which indicates a higher amount of inorganic elements in SCGs than regular white wheat flour. The fiber level of all breads containing SCGs was also noticeably enhanced compared to white bread, with the highest level found for the sample containing 10.0%

SCGs. Total carbohydrates were higher in samples with 2.0% and 4.0% SCGs, suggesting that SCGs-enriched bread could be a beneficial energy source. SCG-supplemented breads also showed higher levels of total phenolics and flavonoids, likely due to the high content of these compounds in SCGs [42].

Chau [44] examined the chemical properties of white bread formulated with 2.0% and 4.0% SCG varieties of Robusta and Arabica (data are not published). The study found that adding both varieties of SCGs into bread enhanced the level of resistant starch and reduced digestible starch content. Both types of SCGs enhanced the total phenolic value and antioxidant capacity in breads. Bread with 2.0% Robusta SCGs had a slightly higher phenolic content and antioxidant capacity than bread with 2.0% Arabica SCGs, while 4.0% Robusta SCGs bread showed significantly higher values than 4.0% Arabica SCGs bread. Differences in phenolic content may be attributed to SCGs source and coffee extraction parameter variations [45]. The study suggests that SCGs could serve as a potential antioxidant source in bread, potentially protecting against oxidative damage and inhibiting starch-digesting enzymes [42].

Both studies provide valuable insights into the chemical properties of SCG-enriched bread, demonstrating improvements in protein, fiber, mineral content, phenolic compounds, and antioxidant capacity [42,44]. SCGs show promise as a functional ingredient, offering potential health benefits such as improved gut health, better glycemic control, and enhanced antioxidant activity [37,38,42]. However, further studies are needed to explore the sensory and consumer acceptability aspects, as well as the long-term health impacts of incorporating SCGs into bakery products. The variations in the chemical composition of bread due to differences in SCGs concentration and source emphasize the need for careful optimization when formulating SCG-enriched bread for commercial purposes [42,44].

3.1.3. Biscuits

Martinez Saez et al. [30] compared the level of amino acids in biscuits formulated with 3.50%, 3.64%, 3.77%, 3.94%, 4.24%, and 4.40% of SCGs from Robusta coffee beans with commercial biscuits. They reported a significant difference between biscuits containing 4.24% and 4.40% SCGs and commercial biscuits. The level of amino acids in commercial biscuits was significantly lower than in biscuits containing 4.24% and 4.40% coffee grounds. This finding suggests that the inclusion of SCGs at these concentrations enhances the nutritional profile of biscuits by increasing their amino acid content, which is essential for improving the nutritional value, particularly for protein quality. This could be linked to the bioactive compounds and residual proteins in SCGs, which enhance the overall protein quality in biscuits and could be valuable for creating more nutrient-dense bakery products.

Ali et al. [46] investigated the effect of SCGs with levels of 2.0%, 4.0%, and 6.0% inclusion on the chemical composition of biscuits. Their results indicated that although the inclusion of SCGs reduced protein content, other beneficial changes were observed. Specifically, the fiber, ash, and moisture content increased with higher concentrations of SCGs, enhancing the biscuits' overall nutritional value, particularly the dietary fiber, which plays a critical role in digestive health. The absence of significant differences in fat content suggests that SCGs do not substantially alter the lipid profile of biscuits, which could be an advantage for maintaining a consistent texture and mouthfeel. This study highlights that while there is a trade-off with reduced protein content, the increased fiber and mineral content from SCGs may offer functional and health benefits that make these biscuits appealing from a nutritional standpoint.

Campos-Vega. [47] evaluated the chemical composition of biscuits containing 7.77% SCGs (added with fructooligosaccharides; SCG-FOS) and 11.11% SCGs antioxidant dietary fiber (SCF). Their findings revealed that the SCF biscuits exhibited the highest protein levels, while the SCG-FOS biscuits had the highest lipid content. Notably, the SCF biscuits also had the most total dietary fiber, suggesting that SCGs, especially in combination with other dietary fiber sources, significantly improve the fiber content of biscuits. However, while the total phenolic content remained similar across all biscuits, the extractable antioxidant capacity varied, with traditional biscuits (TBs) exhibiting the highest antioxidant capacity. This discrepancy could be attributed to differences in the bioavailability of phenolic compounds in SCGs and their interactions with other biscuit ingredients. Interestingly, despite the enhanced nutritional properties, no significant differences in the carbohydrate content were noted, indicating that incorporating SCGs and other functional ingredients does not drastically alter the energy content of the biscuits. This study highlights the potential for SCGs to enhance specific nutritional attributes, such as protein and fiber content. However, further optimization is required to retain or boost the antioxidant capacity.

In summary, the inclusion of SCGs in biscuits presents both nutritional benefits and certain limitations [30,46,47]. Studies have shown that while SCGs can enhance dietary fiber and mineral content, they may reduce protein content, depending on the formulation [46,47]. Furthermore, the antioxidant capacity of SCG-enriched biscuits may vary, possibly due to the interactions between SCGs and other ingredients [47]. Overall, SCGs offer potential as a functional ingredient in biscuits, contributing to enhanced fiber and nutrient density [46,47]. However, further research is necessary to optimize formulations for improved antioxidant properties and protein retention.

3.1.4. Cakes

The protein, fat, dietary fiber, and ash levels of sponge cakes containing SCGs (2.0%, 4.0%, and 6.0%) were studied by Hussein et al. [31]. Their results revealed that adding SCGs did not significantly change the protein and fat content compared to the control sample, suggesting that SCGs do not interfere with the stability of protein or fat in baked products. However, a notable rise in the dietary fiber and ash content was observed as the SCG concentration increased. This suggests that SCGs can enhance the fiber content of cakes, improving their nutritional profile for consumers looking for fiber-enriched products. The increase in ash indicates a higher mineral content in the cakes, which could provide additional health benefits.

Similarly, Ahmad et al. [28] investigated the impact of incorporating SCGs (1.0%, 2.0%, and 3.0%) into sponge cakes. Their study emphasized the potential functional and nutritional improvements that SCGs offer. While the control sponge cake had the highest dry matter value, cakes enriched with SCGs showed increased ash, crude fiber, and minerals levels, especially at the 3.0% concentration. These results indicate that SCGs can enhance the mineral and fiber content in cakes, offering a more nutrient-dense product. Moreover, SCG-enriched cakes also demonstrated elevated levels of total phenolics, tannins, and antioxidant activity, highlighting SCGs' ability to impart bioactive compounds into the product. The higher antioxidant activity could be particularly appealing from a health perspective, as antioxidants play a crucial role in neutralizing free radicals and may contribute to reducing the risk of chronic diseases.

These studies collectively suggest that SCGs can be a valuable ingredient in cakes, maintaining their elemental macronutrient composition (e.g., protein and fat) and enhancing their functional and nutritional properties, especially regarding fiber, minerals, and antioxidants. The use of SCGs could cater to health-conscious consumers seeking cakes with added health benefits, particularly in terms of dietary fiber intake and antioxidant capacity [28,31].

3.1.5. Muffins

Severini et al. [48] evaluated the chemical properties of muffins supplemented with 15.0% and 30.0% SCGs from Arabica coffee beans. They observed that adding SCGs to muffins increased the antioxidant capacity, phenolic compounds, and dietary fiber. The authors suggest that SCGs supplementation could serve as a functional ingredient, providing added health benefits.

Benincá et al. [49] examined the antioxidant properties of muffins supplemented with 1.0%, 16.0%, 31.0%, 46.0%, and 61.0% of SCGs from Arabica coffee. They reported that adding SCGs increased antioxidant activity, total phenolics, caffeine, chlorogenic acid

(5-CGA), and trigonelline contents. These bioactive compounds are well known for their potential health benefits, such as reducing oxidative stress and improving cardiovascular health. The dose-dependent relationship observed in this study emphasizes the potential of SCGs as a valuable addition to muffins.

These studies highlighted that incorporating SCGs into muffins is a promising strategy for enhancing the nutritional profile of bakery products. The increase in antioxidant capacity, phenolic compounds, and dietary fiber suggests that SCGs could be used to develop functional foods with added health benefits [48,49].

3.1.6. Pasta

Sugianto [19] studied the moisture value of pasta containing 0.0%, 4.0%, 8.0%, and 10.0% SCGs from Arabica coffee beans. They did not observe any noticeable differences in the moisture value between the samples enriched with 4.0%, 8.0%, and 10.0% SCGs compared to the 0.0% SCGs pasta. One possible explanation for the lack of differences in moisture content could be the relatively small quantity of added SCGs. The fibrous nature of SCGs may also limit their interaction with water molecules compared to other ingredients. Furthermore, the drying temperature is crucial in regulating moisture loss, and it is possible that the SCG-enriched pasta behaved similarly to the control sample under the same drying conditions. This suggests that the drying process may have a more dominant effect on moisture content than the addition of SCGs. In addition to moisture content, Sugianto noted the potential impact of SCGs on non-enzymatic browning reactions. Moisture levels play a crucial role in browning processes, where a higher moisture content generally limits the extent of Maillard reactions and caramelization during drying.

This study provides valuable insight into the moisture behavior of SCG-enriched pasta. The results indicate that the inclusion of up to 10.0% SCGs does not significantly affect moisture retention during drying [19].

Table 1. The impact of SCGs or SCG's derivative inclusion on chemical, physical, and sensory properties of baked products.

Product	SCGs or SCGs Derivatives Concentrations (%)	Chemical	Physical	Sensory	References
	10.0, 17.5, and 25.0	Increased fat, crude fiber, and ash. Decreased moisture and protein.	All the treatments were soft, crumbly, and lacked elasticity.	The 17.5% SCGs incorporation is preferred in terms of impact on hardness, granularity, flavor, and taste.	[27]
	10.0, 20.0, and 30.0	Increased protein, ash, lipid, and fiber contents. Decreased moisture content.	Improved spread ratio and specific volume. Led to higher mass loss at q 10% inclusion rate. Decreased hardness at a 20% and 30% inclusion rate. Color shifted to dark brown.	The 10% and 20% inclusions were acceptable in terms of flavor and general acceptance.	[32]
	0.27, 0.53, 0.80, 1.07, and 1.33	Increased moisture, ash, fat, protein, crude fiber, calories, antioxidant activity, and TPC.	Enhanced a^* and decreased L^* and b^*. Significant impact with the inclusion of 0.80%. Enhanced fracturability.	The highest appearance scores were related to the 0% extract. The brightest score was related to the 0.53% extract.	[29]
	3.0, 6.0, 10.0, and 12.0	Increased water activity and moisture content with UGCSF and RGCSF.	Increased diameter and spread ratio with RGCSF. Reduced L^*, a^*, and b^* with UGCSF and RGCSF.	The 10% and 12% were preferred in color, appearance, aroma, taste, crispiness, and general acceptability.	[33]
	0.05, 0.10, 0.15, 0.20, and 0.25	Increased moisture, fat, ash, total phenolic contents, and antioxidant activity. Decreased total carbohydrate and starch content.	Increased dough hardness. Reduced adhesiveness, cohesion, and springiness. Caused darker color.	Steady general consumer acceptance for concentration ranged from 0.05% to 0.20%.	[18]

Table 1. *Cont.*

Product	SCGs or SCGs Derivatives Concentrations (%)	Chemical	Physical	Sensory	References
Cookies/ shortbread	2.0, 4.0, 6.0, 8.0, and 10.0	Increased calorie, moisture, ash, crude protein, crude fiber, total phenolic contents, and antioxidant activity. Decreased carbohydrate and peroxide values.	Reduced water loss. Reduced L^* and b^* values but enhanced the a^* value. Increased hardness at lower concentrations.	Preferred 10% inclusion in terms of aroma and hardness.	[7]
	10.0, 20.0, and 30.0	Increased phenolic content and antioxidant activity.	Increased the diameter. Remained unchanged in the thickness.	Lower flavor rating scores for 20% and 30% inclusion. No significant differences in the attributes of color, hardness, crumbness, and odor scores.	[34]
	1.0, 2.0, and 3.0	Enhanced antioxidant activity.	-	Impact on color: preferred at lower levels.	[35]
	5.0, 7.0, and 10.0	Inclusion of RSCGs and SCGR, enhanced fiber level, total phenolics, and antioxidant activity.	The treatment with 7% RSCGs had a higher hardness, whereas the treatment with 10% SCGR had a higher hardness Both treatments had similar fracturability values. Enhanced a* and decreased b* color values.	The 7.0% SCGR treatment was preferred in terms of texture and taste.	[36]
	5.0	Enhanced antioxidant activity.	-	-	[39]
	2.0, 4.0, 6.0, 8.0, and 10.0	Increased protein, total phenolics, antioxidant activity, and total flavonoid levels. Highest ash, mineral, and fiber content for a 10% inclusion rate. Higher carbohydrate for 2.0 and 4.0% inclusion rates.	Reduced water loss. Higher loaf volume at a 2.0% inclusion rate.	Preferred color, aroma, texture, and general acceptability were for a 2.0% inclusion rate.	[42]
Bread	2.0 and 4.0	Enhanced resistant starch, total phenolics, and antioxidant activity	Darker and redder crumb and crust color.	-	[44]
	4.0	-	Decreased hardness, springiness, elasticity, and cohesiveness. Reduced resilience, brightness, redness, and yellowness. Heavier loaf and a lower specific volume.	Preferred in terms of aroma and taste.	[50]
	3.50, 3.64, 3.77, 3.94, 4.24, and 4.40	Increased amino acid contents.	No significant differences in texture.	Increased the color scores at 3.50%, 3.64%, and 4.24% inclusion. Similar overall acceptance for 3.50%, 3.64%, and 4.24% inclusion rates and control.	[30]
Biscuit	2.0, 4.0, and 6.0	Enhanced fiber, ash, and moisture content. Reduced protein content.	Reduced the L^* and b^* values, and Enhanced the a^* value.	No significant difference in aroma and taste at 2.0% and 4.0% inclusion rates. No significant difference in texture and overall acceptance.	[46]
	7.77, and 11.11	The highest protein and fiber were related to the SCF, while the highest lipid was related to the SC-FOS. Total phenolic content was higher in SCF.	-	-	[47]

Table 1. *Cont.*

Product	SCGs or SCGs Derivatives Concentrations (%)	Chemical	Physical	Sensory	References
Sponge Cake	2.0, 4.0, and 6.0	Increased dietary fiber and ash contents.	Intensified browning in both the crust and crumb. Increased a^* and b^* values. Increased dough water absorption and development time. Decreased dough stability. Improved mixing tolerance index. Increased weight and specific volume. Improved baking quality. Increased adhesiveness. Decreased hardness, gumminess, and chewiness.	Color, taste, smell, texture, appearance, and general acceptability were preferred for 2.0% and 4.0% inclusion rates.	[31]
	1.0, 2.0, and 3.0	Reduced the dry matter. Enhanced crude fiber, all minerals (except Na), ash, total phenolics, total tannin contents, and antioxidant activity.	Decreased L^* value of the crust and crumb. No notable change in a^* value. Decreased b^* value for both crust and crumb.	Negatively impacted the appearance, shell color, crumb color, texture, taste, and smell.	[28]
Muffins	5.25, and 10.5	Enhanced antioxidant activity, total phenolics, and total dietary fiber.	Reduced the specific gravity. Reduced the L^*, a^* and b^* color values.	Received high sensory scores in terms of appearance, color, odor, taste, softness, and overall acceptance.	[48]
	1.0, 16.0, 31.0, 46.0, and 61.0	Increased the total phenolic content, antioxidant activity, caffeine, chloro-genic acid (5-CGA), and trigonelline.	Increased the L^* value and hardness. Reduced the a^* and b^* color values. No significant difference in elasticity, chewiness, and cohesiveness across various concentrations.	-	[49]

3.2. Bakery Products' Physical Properties

3.2.1. Cookies

Aguilar-Raymundo et al. [27] studied the effect of adding SCGs at 10.0%, 17.5%, and 25.0% on the textural properties of cookies using a texture analyzer. Despite varying SCGs percentages, no significant differences were observed in the samples' cohesiveness, hardness, and springiness. Hardness values ranged from 33 to 51 N, while cohesiveness and springiness remained relatively low, ranging from 0.08 to 0.11 and 11.9% to 19.4%, respectively. Overall, the cookies were characterized as soft, crumbly, and lacking elasticity.

Azuan et al. [29] evaluated cookies' color and textural properties supplemented with 0.25%, 0.53%, 0.80%, 1.07%, and 1.33% SCG extracts. These researchers used the Konica Minolta color measuring system and the texture analyzer to check the color and texture properties of the cookies. The researchers observed changes in cookies' color and textural properties when incorporating SCGs extracts. The increase in the amount of added SCG extract in cookies resulted in decreasing lightness (L^*) and yellowness (b^*) while redness (a^*) increased. These color changes are attributed to the Maillard reaction during baking and the SCG extract's original dark color [31,37]. Regarding hardness, there was no notable variation between the samples enriched with the SCG extract and the control sample. However, in terms of fracturability, there was a notable variation between the samples enriched with the SCG extract and the control sample. The samples' fracturability rose when the SCG extract was incorporated; adding a 1.33% SCG extract gave the maximum fracturability value of 40.03 mm, while the control sample's fracturability value was 38.43 mm [29].

Desai et al. [33] investigated the effect of adding RGCSF and UGCSF (3.0%, 6.0%, 10.0%, and 12.0%) on the physical features of cookies. These researchers used the colorimeter and the texture analyzer to check the color and texture attributes of the cookies. They indicated that cookies with 10.0% UGCSF were heavier than control cookies, likely due to the higher dietary fibers in UGCSF that retain moisture during baking. Conversely, cookies with 12.0%

RGCSF were lighter than controls. RGCSF increased the cookie diameter and spread ratio, indicating desirable quality traits that consumers prefer. Cookies with 10.0% UGCs had the lowest weight and maximum spread ratio. Thickness varied, with cookies containing 10.0% and 12.0% UGCSF being the thickest, while those with 10.0% RGCSF were the thinnest. Increasing UGCSF and RGCSF concentrations led to reduced lightness (L^*), redness (a^*), and yellowness (b^*) parameters, attributed to the darker RGCSF and caramelization of sugars in UGCSF during baking.

Meerasri and Sothornvit [34] studied the physical characteristics of cookies made with 10.0%, 20.0%, or 30.0% SCG oil. These researchers used dynamic oscillatory tests to determine the rheological features of dough. The diameter was measured as an average value when placing four cookies edge to edge. Thickness was measured using a vernier caliper by taking the average of stacking four cookies on top of each other. The spread ratio was calculated by dividing the average value of the diameter by the average value of the thickness of the cookies [51]. They also used a texture analyzer and a spectro-guide spectrophotometer to measure the texture and color characteristics. They exhibited that cookies made with 10.0%, 20.0%, and 30.0% SCG oil exhibited reduced storage modulus (G') and loss modulus (G'') values as opposed to cookies made with butter. The addition of SCG oil weakened the dough microstructure and reduced the viscosity. The cookies' diameters increased when using SCG oil, likely due to the weaker microstructure, allowing easier spreading during baking. However, the thickness and spread ratio remained unaffected by the SCG oil substitution. A substantial reduction in hardness was observed in cookies with higher SCG oil content, suggesting altered texture due to increased softening of the dough microstructure. In addition, the color parameters showed minimal differences, but when 30.0% SCG oil was used, the lightness (L^*) and yellowness (b^*) notably decreased [39].

In another study, Sharma et al. [36] examined the effect of 5.0%, 7.0%, and 10.0% RSCGs and SCGR on the texture and color of cookies using a texture analyzer and a LabScan XE spectrophotometer, respectively. The control treatment had a fracturability of 14.99 mm and a hardness of 4.74 N. Cookies with 7.0% RSCGs and 10.0% SCGR showed texture profiles similar to the control regarding hardness and fracturability. Cookies with 7.0% RSCGs had a hardness of 3.15 N and a fracturability of 13.23 mm, while those with 10.0% SCGR had a hardness of 4.80 N and a fracturability of 13.2 mm. SCG-enriched cookies appeared darker due to the presence of melanoidins, unlike SCGR cookies. RSCGs and SCGR-added cookies exhibited increased redness but decreased yellowness compared to the control.

Trà et al. [18] studied cookies containing SCGs (0.0%, 0.10%, 0.15%, 0.20%, and 0.25%). They used a texture analyzer and Konica Minolta color system to evaluate texture and color features. These researchers observed that the hardness of cookie dough increased with higher concentrations of SCGs in the formulation. At the same time, adhesiveness, cohesion, and springiness decreased marginally. Higher SCGs concentrations reduced cookie brightness and increased ΔE values, indicating a darker color and reduced red and yellow tints.

Oliviera Batista et al. [32] assessed the physical characteristics of gluten-free cookies enriched with SCG concentrations of 10.0%, 20.0%, and 30.0%. The spreadability was determined using the ratio between the diameter and thickness values of the cookies after baking. The specific volume was expressed using the apparent volume (millet seed displacement method) and cookie mass after baking. The hardness of the gluten-free cookies was obtained using a TA.XT2 texture analyzer. Findings showed that incorporating 10.0% or 20.0% SCGs into gluten-free cookies improved their physical characteristics, including a better spread ratio and specific volume, compared to the control. However, the cookies with 10.0% SCGs exhibited a higher mass loss, while 30.0% SCGs did not affect these metrics. The higher fiber content in 20.0% and 30.0% SCGs cookies may have competed for water during baking, reducing the water availability for sugar dissolution and increasing viscosity, thereby decreasing the spread ratio [33,52]. SCGs' higher lipid level and larger particle size may have also contributed to the increased spread ratio [52]. Hardness remained unchanged at 10.0% SCGs but decreased at 20.0% and 30.0% SCGs [30], likely due to the enhanced water retention and softer texture from high fiber and lipid levels [27]. Moreover, items with higher lipid levels might have resulted in a softer texture [51]. SCGs improved textural characteristics, potentially making gluten-free cookies more robust than regular cookies [53]. In terms of color, the gluten-free cookies without SCGs were lighter in color, whereas the gluten-free cookies incorporated with SCGs presented a dark brown color [32]. These color shifts could be related to the dark hue of SCGs and the sugars' caramelization during baking [37].

Koay et al. [7] investigated the effect of 2.0%, 4.0%, 6.0%, 8.0%, and 10.0% SCGs on shortbread dimensions, color, and texture. Diameter and thickness were measured with a digital vernier caliper, and the spread ratio was calculated by dividing the diameter by the thickness. Additionally, the Konica Minolta color measuring system and the texture analyzer were applied to investigate the color and texture attributes of the cookies. They found noticeable changes in diameter and thickness before and after baking. These changes were consistent across samples with different SCGs incorporation levels. However, there was a significant reduction in water loss in SCG-containing samples, likely due to the water-retaining properties of SCGs fibers. Despite minimal differences in weight and spread, baking caused a slight decrease in both parameters for all samples due to moisture loss [37]. Regarding color, SCGs incorporation resulted in darker shortbread with lower brightness (L^*) and yellowness (b^*) values, attributed to SCG's dark color and enzymatic browning during baking. However, the redness (a^*) value increased with higher SCGs levels, likely due to non-enzymatic browning reactions during baking. Texturally, the hardness of shortbread varied with the SCGs content. While the hardness initially increased with low SCGs levels, it gradually decreased with higher SCGs contents. This was attributed to the substitution of flour with SCGs. Higher SCG contents increased moisture in the shortbread, leading to a softer texture over time [7].

3.2.2. Breads

Daniel et al. [42] evaluated the impact of adding SCGs in concentrations of 2.0%, 4.0%, 6.0%, 8.0%, and 10.0% on the physical features of bread. The highest loaf volume was associated with the sample containing 2.0% SCGs. Loaf volume was measured 50 min. after the loaves were removed from the oven using the rapeseed displacement method. The baking loss is obtained by subtracting the loaf weight (g) from the initial dough weight. In terms of baking loss, bread baked with a 10.0% SCGs concentration lost less water (9.29%), resulting in a higher weight for the bread. This could be due to fiber, a component of SCGs, having a high water retention capacity.

Rwubatse et al. [50] studied the physical attributes of whole wheat bread formulated with 4.0% SCGs. These researchers used a texture analyzer and a color reader to evaluate texture and color properties. Additionally, the specific volume and density were determined by an AACCI-approved technique 10-05-01 [54]. They stated that bread enriched with SCGs exhibited lower hardness compared to the control bread, indicating a softer texture. Prolonged dough fermentation combined with SCGs led to decreased springiness, or the bread's ability to rebound after compression, suggesting reduced elasticity [55]. Moreover, the addition of SCGs decreased cohesiveness, making the bread more prone to breaking or crumbling. Combining SCGs with whole wheat flour further reduced the resilience associated with bread staleness [50]. SCG-enriched bread also displayed lower values of lightness (L^*), redness (a^*), and yellowness (b^*) compared to the control, attributed to SCG's antioxidant properties mitigating non-enzymatic browning reactions [30]. SCG-enriched bread was heavier than the control, especially when fermented for extended periods, and had a lower specific volume due to SCGs fiber decreasing the gluten network essential for gas retention [50,56]. Furthermore, SCGs incorporation decreased bread density by enhancing dietary fiber content. Breads fermented for different durations exhibited variations in density, with SCG-enriched bread from shorter fermentation periods being denser compared to those from longer fermentation times [50].

Chau [44] evaluated the effect of incorporating SCGs from Arabica and Robusta varieties at 2.0% and 4.0% levels on the physical properties of bread (crust, crumb color, specific volume, density, and texture). A chromameter was used to assess the crust and crumb color. Specific volume and density were measured using the AACC 10-05.01 rapeseed displacement technique [54]. Furthermore, texture parameters were determined using a texture analyzer [44]. The crumb of SCG-enriched bread showed a darker and redder color due to the high polyphenol and melanoidin content resulting from the Maillard reaction during coffee roasting [7,32]. Higher SCGs concentrations led to darker crusts and crumbs [44]. Arabica SCGs had no significant effect on specific volume or density, whereas Robusta SCGs reduced specific volume and increased density. This was attributed to the phenolics and dietary fiber in the SCGs competing with wheat flour proteins for water, which hindered gluten network formation and reduced gas retention during fermentation [57,58]. Bread hardness increased with SCGs incorporation, alongside a trend toward increased gumminess, attributed to the reduced yeast and dough enzymatic activity [59,60]. Cohesiveness and resilience decreased with SCGs addition, resulting in a crumbly texture. However, the differences between the SCG-enriched and control bread diminished over storage time [44], suggesting potential benefits in extending shelf life by delaying staling [61,62].

3.2.3. Biscuits

Ali et al. [46] examined the impact of incorporating SCGs into the color characteristics of biscuits at concentrations of 2.0%, 4.0%, and 6.0%. A spectro-colorimeter was used to measure the biscuits' color. They observed that with the increase in the amount of dark SCGs, the L^* value indicated that the biscuits' lightness decreased. The a^* value, representing redness, increased with higher SCGs levels, while the b^* value, indicating yellowness, decreased compared to the control. This contradicted expectations based on the Maillard reaction, where increased a^* and b^* values typically correlate with heightened redness and yellowness [63]. The discrepancy suggests that the brown hue from SCGs might be too dark to be visually perceived, complicating interpretation [64]. Therefore, these findings do not indicate the browning reaction's progression during SCGs incorporation into biscuits [46].

3.2.4. Cakes

Hussein et al. [31] investigated the physical properties of sponge cakes containing 2.0%, 4.0%, and 6.0% SCGs. A tristimulus colorimeter with the CIELAB color space was used to measure color factors. The farinogram characteristics of dough blends were assessed to measure rheological attributes according to AACC guidelines [65]. The volume, weight, and specific volume of samples were measured using the AACC technique [65]. A testing system equipped with a cylinder probe was used to evaluate texture parameters. The findings indicated that adding SCGs intensified browning in both crust and crumb compared to the control cakes. Initial nonenzymic browning led to increased redness (a^*) and yellowness (b^*). Regarding the rheological features of cake dough, adding SCGs increased water absorption and the dough development time. In addition, dough stability decreased with SCGs inclusion, indicating a weaker dough. The primary component of SCGs, fibers, significantly influences the rheological characteristics of the resulting sponge cake. Regarding baking quality, cakes containing SCGs showed an increase in specific volume and weight compared to the control. In terms of textural characteristics, cakes with SCGs had lower hardness but higher adhesiveness and lower gumminess and chewiness compared to the control [31].

Ahmed et al. [28] studied the impact of supplementing sponge cakes with SCG powder at 1.0%, 2.0%, and 3.0% levels. Color properties were analyzed using the Minolta calorimeter. These researchers reported that these additions did not significantly affect the cakes' weight, height, and specific volume compared to the control sample. In terms of color properties, the crust's brightness (L^*) decreased in products with higher SCGs, except for sponge cakes with 1.0% SCGs. There was no notable change in redness (a^*) across samples. Yellowness (b^*) decreased significantly in all coffee-supplemented cakes. Additionally, the crumb's brightness (L^*) decreased in all samples, while the cake containing 1.0% SCGs showed no significant change in redness (a^*). The increase in redness (a^*) in other samples suggests the initiation of the Maillard reaction [66]. Yellowness (b^*) remained unchanged across all SCG-supplemented cakes.

3.2.5. Muffins

Severini et al. [48] evaluated the effect of adding SCGs at concentrations of 15.0% and 30.0% on the physical characteristics of muffins. A digit CAL SI caliper, colorimeter, and texture analyzer were used to assess the muffins' dimensions, color, and texture profile, respectively. The results indicated no significant difference in the overall dimensions of muffins containing SCGs and the control treatment. However, there were notable differences in how the pores were distributed throughout each portion. In terms of color attributes, the integration of SCGs led to a reduction in color factors (L^*, a^*, b^*). In terms of texture, adding SCGs to muffins did not cause significant changes in textural parameters such as hardness, springiness, cohesiveness, chewiness, and gumminess.

Benincá et al. [49] conducted a study investigating the impact of SCGs at various concentrations (1.0%, 16.0%, 31.0%, 46.0%, and 61.0%) on the color and texture of muffins. These researchers used a colorimeter and a texture analyzer, respectively, to check the color and texture attributes of the muffins. The addition of SCGs resulted in significant differences in color characteristics, with lower L^* values found in muffins with higher concentrations of SCGs, which could be due to the Maillard reaction producing melanoidins. SCG-enriched muffins had lower a^* and b^* values than control muffins. Texture analysis revealed greater hardness in SCG-enriched muffins. However, parameters like elasticity, chewiness, and cohesiveness did not significantly differ between control and SCG-enriched muffins up to 61.0%, indicating that SCGs substitution for wheat flour did not alter the textural aspects. The study concluded that the addition of SCGs of up to 61.0% maintained the muffin structure while enhancing hardness, suggesting potential applications in product development.

3.2.6. Pasta

Sugianto [19] investigated the cooking quality (optimal cooking time, water uptake, cooking loss, and swelling index), hardness, and pasta color, containing 4.0%, 8.0%, and 10.0% of SCGs. The optimal cooking time for pasta was measured by boiling 10 g of pasta in 100 mL of water. Starting at 5 min, a piece was removed every minute and compressed between glass plates to check for a white core. The time was recorded when no white core remained. Fifteen grams of pasta were placed in a stainless-steel strainer and submerged in 300 mL of boiling water for 9 min to assess the cooking quality, including water uptake, cooking loss, swelling index, and hardness. After cooking, the pasta was rinsed under cold water, shaken to remove excess water, and left to drain for 5 min before weighing. To measure the swelling index, cooked pasta was dried at 105 °C until a constant weight was reached [67]. The swelling index was calculated as the ratio of the cooked product weight to its dried weight. In terms of the optimal cooking time, pasta containing SCGs tended to break more easily than commercial pasta when cooked, suggesting that the optimal cooking time for SCG-containing pasta might be lower than what was instructed by the pasta manufacturer due to the fiber weakening the gluten matrix [68]. Pasta with SCGs showed a higher water uptake compared to the control, possibly due to the differences in drying techniques affecting the moisture content and higher fiber content disrupting the protein–starch matrix. SCG incorporation reduced the pasta swelling index [19], likely due to the presence of fiber and protein forming cohesive networks that impede water flow and starch gelatinization [66,69,70]. Uncooked pasta with SCGs, particularly at higher concentrations, exhibited greater hardness, indicating a potential impact of SCGs on pasta texture [19]. Cooked pasta with SCGs retained its shape and exhibited color differences, with SCGs contributing to darker coloration [19].

The findings indicate that including SCGs in baked goods affects various physical characteristics, including texture and color [19,28,29,36,44,46,49,50]. SCGs consistently improve features in cookies, bread, and cakes, primarily through the Maillard reaction, which darkens the color and alters the textural properties [7,29,31,32,44,50]. A common trend across products is the reduction in lightness and an increase in hardness, although specific effects vary by formulation [18,29,44,46,49]. For instance, SCGs increase cookie fracturability [29] but may decrease bread cohesiveness [50], illustrating product-specific differences. On the other hand, the impact of SCGs on the texture, color, and cooking quality of food products—while differing based on the specific product and the concentration of SCGs—underscores their potential to contribute to the development of healthier and more innovative food options [44,49].

3.3. *Bakery Products' Microbiological Properties*

Desai et al. [33] investigated the impact of adding UGCSF and RGCSF flour (3.0%, 6.0%, 10.0%, and 12.0%) on the microbiological characteristics of cookies. The study demonstrated that the UGCSF had a microbial load of less than 80 CFU/g for endospores, while yeast, mold, and aerobic bacteria were undetected. The absence of mold in UGCSF undoubtedly lowers the possibility of ochratoxin exposure. It is evident that the UGCSF has outstanding microbiological quality and, as a result, has the potential to be a food ingredient that is safe and rich in nutraceuticals. RGCSF had less than 10 CFU/g. The microbial burden was safe across all cookies examined for this research.

Sharma et al. [36] evaluated the probiotic features of cookies supplemented with 5.0, 7.0, and 10.0% RSCGs and SCGR. The results showed that both cookie types had a similar probiotic viable cell count.

Martinez Saez et al. [30] counted aerobic microorganisms, endospores, mold, and yeast in biscuits enriched with 3.50%, 3.64%, 3.77%, 3.94%, 4.24%, and 4.40% of SCGs. Findings showed that adding SCGs did not influence the number of microorganisms in the biscuit formulation and complied with the national microbiological requirements for biscuits (RD 1124/82, 1982). Also, the total aerobic microorganisms were reported to be 104 CFU/g, while endospores, mold, and yeasts were lower than 102 CFU/g for the biscuit containing 4.24% SCGs.

Ahmed et al. [28] investigated the microbial properties of sponge cakes containing SCGs at 1.0%, 2.0%, and 3.0% stored for 0, 7, and 14 days. The total bacterial count, mold, and yeast number elevated significantly in the control cake and in cakes supplemented with used SCGs at 7 and 14 days of storage compared to cakes stored for 0 days. However, sponge cakes supplemented with varying levels of SCGs demonstrated a noticeable reduction in the total bacterial count, total mold, and yeast compared to sponge cake without SCGs during each storage period, indicating the positive impact of SCGs on the microbial quality of the cakes.

Incorporating SCGs into baked goods does not compromise microbiological safety [28,33]. The studies indicate that SCGs can be safely utilized as a nutritious ingredient in various baked goods, ensuring their quality and safety for human consumption [28,30,33,36].

3.4. *Bakery Products' Sensory Properties*

3.4.1. Cookies

Aguilar-Raymundo et al. [27] used a nine-point hedonic scale to rate the sensory characteristics of cookies containing concentrations of 10.0%, 17.5%, and 25.0% SCGs. In addition, they studied the sensory parameters (hardness, granularity, flavor, and taste) of cookies using a Just-About-Right scale (JAR) method. The findings indicated that cookies with 17.5% SCGs were the favorites among consumers, followed by those with 10.0% SCGs. However, no significant difference in overall liking was noted across the formulations. While over 56% of participants found the cookies' coffee taste acceptable, most found the hardness low, and about 50% thought the coffee aroma was insufficient. The study suggests that SCGs concentrations between 17.5% and 25.0% might be optimal, as higher levels of SCGs could make the coffee taste overwhelming. Overall, cookies with SCGs were enjoyed, but those with 17.5% SCGs were preferred in terms of sensory attributes like hardness, granularity, flavor, and taste.

In another study by Sérioa et al. [35], the effect of adding 1.0%, 2.0%, and 3.0% of the SCG extract on the sensory properties of cookies was tested. The participants were reported to prefer samples with fewer incorporated extracts.

Azuan et al. [29] evaluated the sensory parameters of cookies supplemented with 0.27%, 0.53%, 0.80%, 1.07%, and 1.33% SCG extracts from Arabica coffee involving 40 untrained assessors using a nine-point hedonic scale. There was no discernible difference in the scores for the appearance of the SCG extract-supplemented and the control samples. Nonetheless, the control sample indicated the highest appearance score, followed by samples containing 0.53%, 0.80%, 1.07%, 1.33%, and 0.27% SCG extracts. The sample supplemented with 0.53% SCG extract demonstrated the highest color score in lightness, followed by samples containing 0.80%, 1.33%, and 1.07% SCG extracts, the control, and 0.27% SCG extract.

Desai et al. [33] assessed the effects of UGCSF and RGCSF in cookies at amounts of 3.0%, 6.0%, 10.0%, and 12.0% using a seven-point scale. They reported that cookies containing 10.0% and 12.0% RGCSF had more favorable sensory properties in terms of color, appearance, aroma, taste, crispiness, and overall acceptability.

Meerasri and Sothornvit [34] examined the impact of the inclusion of 10.0%, 20.0%, and 30.0% extracted oil from SCGs on the sensory attributes of cookies. These researchers used a nine-point hedonic scale to examine these attributes. No significant differences were found for color, hardness, crumbliness, and odor scores between cookies that contained SCG oil and thee control. On the other hand, the cookies with 20.0% or 30.0% SCG oil received lower flavor rating scores. The higher contents of SCG oil caused the cookies to smell like cooked coffee oil and impacted the mouthfeel when consumed.

Sharma et al. [36] investigated the sensory features of cookies supplemented with 5.0%, 7.0%, and 10.0% SCG residue (RSCG) using a nine-point hedonic scale. Cookies supplemented with 7.0% RSCG had the highest general acceptance rate in terms of texture and taste. At the same time, an enhancement in RSCG content gave the cookies an extra bitter taste and negatively impacted their texture and overall acceptance. SCG oil has a strong smoky smell associated with bitterness and astringency; thus, it cannot be utilized to bake cookies.

Trà et al. [18] studied the sensory characteristics of cookies enriched with 0.05%, 0.10%, 0.15%, 0.20%, and 0.25% of SCGs using a nine-point hedonic scale. They reported that the general acceptance of the SCG-enriched cookies remained unchanged when the SCGs ratio increased from 0.05% to 0.20%. However, the sensory score of cookies with 0.25% SCG dropped, which most likely resulted from the product's significantly greater hardness.

Oliveira Batista et al. [32] proposed developing gluten-free cookies with SCGs (10.0%, 20.0%, and 30.0%). These researchers used a nine-point hedonic scale to evaluate sensory properties. The flavor and general acceptability of the cookies declined when higher concentrations of SCGs (30.0%) were used in the cookies' formula. There were no variations in the assessed formulations' acceptance of texture and appearance. Scores of samples enriched with SCGs ranged from 5.95 to 6.87 across the analyzed qualities, indicating that consumers found some samples to be neither very appealing nor very objectionable. The cookies' acceptability parameters revealed that the formulations containing 10.0% and 20.0% SCGs were acceptable and suitable for commercialization.

Koay et al. [7] evaluated the effect of SCGs at levels of 2.0%, 4.0%, 6.0%, 8.0%, and 10.0% on the sensory parameters of cookies using a nine-point hedonic scale. In terms of color, the control group scored highest (7.15), with the acceptability decreasing as the SCGs levels increased due to the darker color of SCGs. Interestingly, the 10.0% SCGs shortbread ranked second in color preference, likely because panelists found the darker hue more prosperous. In terms of aroma, SCGs levels significantly influenced aroma, with 10.0% of SCGs receiving higher scores. However, there was no significant difference between the 6.0% and 8.0% SCGs shortbreads, likely due to the loss of volatile aromatics during coffee brewing. In terms of taste, no significant difference was observed in the taste of 2.0% to 8.0% of SCGs shortbreads. However, the control sample outperformed the 10.0% SCG shortbread, possibly due to the presence of bitter compounds like caffeine. In terms of hardness, the 10.0% SCGs shortbread was comparable to the control in hardness, while lower SCG concentrations showed significant differences from the control. Regarding fracturability and oiliness, SCGs incorporation did not affect the shortbread's crumbliness. The oiliness of the 10.0% SCGs inclusion in shortbread was comparable to the control in acceptability. The control sample scored highest in general acceptability, followed by 10.0% SCGs shortbread. Thus, SCGs can be incorporated into shortbread without negatively impacting sensory qualities.

3.4.2. Bread

Daniel et al. [42] investigated bread's sensory parameters (color, taste, aroma, and texture) enriched with 2.0%, 4.0%, 6.0%, 8.0%, and 10.0% SCGs using a nine-point hedonic scale. There were noticeable color variations between the control and SCG-enriched breads. The color acceptance of bread decreased as the amount of added SCGs increased. The aromas of white and bread with 6.0%, 8.0%, and 10.0% SCGs incorporations did not change significantly. The acceptability of the texture of the bread diminished with the addition of SCGs.

Rwubatse et al. [50] tested the sensory properties of whole wheat bread formulated with 4.0% SCGs using a seven-point hedonic scale. The findings indicated that customers were satisfied with the aroma and taste of whole wheat bread containing SCGs.

3.4.3. Biscuit

Martínez Saez et al. [30] studied the sensory characteristics of biscuits containing 3.50%, 3.64%, 3.77%, 3.94%, 4.24%, and 4.40% SCGs compared to commercial biscuits using a seven-point hedonic scale. The findings showed that biscuits containing 3.50%, 3.64%, and 4.24% of SCGs had the highest scores for color. Regarding the texture and taste, no significant difference was reported between biscuits containing 3.64% and 4.24% SCGs with commercial biscuits. Also, the overall acceptance was similar for biscuits, with 3.50%, 3.64%, and 4.24% SCGs compared to commercial biscuits.

Ali et al. [46] evaluated the effect of SCGs (2.0%, 4.0%, and 6.0%) on sensory parameters (aroma, taste, color, texture, and overall acceptance) of biscuits. The authors stated that no significant differences were found in the aroma and taste between biscuits containing 2.0% and 4.0% SCGs and the control. However, adding 6.0% SCGs caused a significant diminish in respected characteristics. In terms of color, a significant difference was observed between the samples containing SCGs and the control sample, and with the enhancement in the level of SCGs, the color score diminished. On the other hand, no substantial difference was observed between the samples containing SCGs and the control sample.

3.4.4. Cakes

Hussein et al. [31] examined the sensory parameters (color, taste, smell, texture, appearance, and general acceptability) of sponge cakes enriched with 2.0%, 4.0%, and 6.0% SCGs. The findings showed a noticeable difference between the sponge cakes enriched with SCGs and the control sponge cake. The color, taste, smell, texture, appearance, and general acceptability of sponge cake incorporated with 2.0% and 4.0% SCGs were significantly higher than those of the samples with 6.0% SCGs. Reduced glycemic sugars significantly influence the acceptability of smell and taste, presumably due to their roasted, sweet, and caramel flavors, which are regarded as the primary and most favorable sensory characteristics. Notably, the texture of samples containing SCGs seems to be almost equal to or even better than that of the control group. This could be attributed to the high emulsifying activity and emulsion stability of SCGs.

Ahmed et al. [28] investigated the sensory parameters (appearance, crust color, crumb color, texture, taste, smell, and overall) of sponge cakes supplemented with 1.0%, 2.0%, and 3.0% SCGs using a ten-point numeric scale. They stated that all the sensory parameters decreased significantly in all baked cakes that contained SCGs.

3.4.5. Muffins

Severini et al. [48] studied the sensory characteristics, including appearance, color, odor, off-odor, taste, softness, and overall acceptance, of muffins enriched with 5.25% and 10.50% SCGs using a five-point hedonic scale. In terms of general acceptance, appearance, and color, samples containing 10.50% SCGs scored higher than those containing 5.25% SCGs. There was no difference in the odor, taste, and softness of coffee samples, while the sample containing 10.50% SCGs had more off-odor than other samples. Overall, these researchers reported that the sensory characteristics of muffins gained higher scores (3.5) than average on a five-point scale, which indicates a good level of acceptance among the consumers.

The sensory assessment of baked goods enhanced with SCGs demonstrates a variety of results among cookies, bread, biscuits, cakes, and muffins [27,28,30,35,48,50]. Formulations with intermediate SCGs concentrations (about 10–20%) are frequently chosen in cookie studies due to their well-balanced flavor, color, and texture [27,32,33]. Nevertheless, higher SCGs levels may cause bitterness and strong coffee flavors, which would have a detrimental effect on acceptance [36]. Regarding bread, enhanced SCGs levels tend to diminish color acceptability, but the aroma remains largely unchanged [42]. Whole wheat bread containing 4% SCGs showed good acceptance in terms of taste and aroma [50]. Similarly to commercial biscuits, biscuits with SCGs ranging from 3.5% to 4.4% scored highest in sensory measures, suggesting that moderate SCGs additions have no discernible effect on texture or flavor [30]. Low amounts of SCGs (2–4%) improve the texture and flavor of cakes by imparting roasted and caramel-like characteristics. However, the sensory appeal of cakes with 6% SCGs decreases [31]. Finally, even though higher SCGs concentrations resulted in increased off-odor, muffins enhanced with 5.25% or 10.5%, SCGs demonstrated widespread acceptance [48]. Moderate amounts of SCGs can be added to bakery goods to improve sensory aspects without compromising taste or texture, making them acceptable to consumers.

4. Study Limitations

Analyses were conducted on the chemical, physical, microbiological, and sensory characteristics of baked goods (cookies, bread, biscuits, cakes, muffins, and pasta) that contained SCGs. The following problems were identified in various papers:

In some studies, the variety of coffee was not specified [7,27,31,35,36,46], and a few studies checked the microbiological properties of the baked goods [28,30,33,36]. Only one study was performed on pasta [19]. Some studies used small sample sizes or lacked diversity in participants or experimental conditions, potentially limiting the robustness and applicability of their results [28,38,47,50]. Some studies were relatively short and only provided a snapshot of the effects of SCGs incorporation [30,35,47]. Different storage conditions in studies affect the stability and sensory characteristics of cooked products containing SCGs [7,27,28,34,35,47,50]. Various processing methods used in different studies, such as differences in mixing, fermentation, and baking parameters, affected the final quality characteristics of baked products [7,29,33,34,50]. Differences in sensory evaluation methods, such as panel selection, training, and evaluation criteria, led to conflicting findings regarding consumer acceptance and preference [28,29,31,48,50]. The comparison of chemical, physical, microbiological, and sensory properties was also influenced in certain research reports by the standardization of the studies.

5. Conclusions

Incorporating spent coffee grounds (SCGs) into bakery products and pasta presents an opportunity to enhance their nutritional and sensory qualities. The chemical changes induced by SCGs' inclusion contribute to increased protein, fat, fiber, ash, and antioxidant content, thereby improving the overall nutritional value of baked goods. Furthermore, SCGs can positively impact the physical properties of bakery items, although these effects depend on factors such as SCGs' concentration, coffee variety, and processing methods.

Importantly, studies have shown that incorporating SCGs into bakery products does not compromise their microbial safety, ensuring the quality and safety of these goods for human consumption. To optimize the sensory properties of baked goods, careful consideration must be given to factors including SCGs' concentration and the type of baked product. Sensory properties encompass how food is perceived by the senses of taste, aroma, appearance, and mouthfeel, which are crucial for consumer acceptance. Optimizing these properties requires finding the right balance of SCGs in the recipe. Too much SCGs can overpower the flavor, while too little may not sufficiently enhance the taste or aroma. Different types of baked goods respond to SCGs uniquely due to their distinct structures and textures. For example, cookies, which have lower moisture and a higher density than bread or cakes, may benefit more from adding SCGs in terms of flavor and aroma enhancement. Moreover, storage conditions, such as temperature and humidity, are critical in maintaining sensory quality over time. SCGs can affect the shelf life, flavor retention, and freshness of baked goods, which is essential for ensuring continued consumer enjoyment even after extended storage.

From an economic perspective, upcycling SCGs supports a circular economy model by diverting coffee waste from landfills, thus contributing to significant waste reduction and minimizing the food industry's environmental footprint.

Compared to other food supply chain applications, such as animal feed, biofuel production, and composting, the use of SCGs in bakery products adds value through human nutrition while still contributing to sustainability goals. This highlights the versatility of SCGs as a functional ingredient, offering benefits beyond other common applications.

In conclusion, utilizing SCGs in bakery products holds substantial promise for creating innovative, nutritious, economically viable, and sustainable food formulations. Further research is needed to refine optimal formulations and processing techniques to fully harness the potential of SCGs incorporation into bakery applications, ensuring maximum benefits in terms of nutrition, sensory quality, and environmental impact.

Funding: This research received no external funding.

Institutional Review Board Statement: Not applicable.

Informed Consent Statement: Not applicable.

Data Availability Statement: No new data were created or analyzed in this study. Data sharing is not applicable to this article.

Acknowledgments: The authors would like to thank the School of Life and Environmental Sciences at the University of Sydney for their ongoing support.

Conflicts of Interest: The authors declare no conflicts of interest.

References

1. Hassan, S.M. Composition, Phyto-Chemical Properties, Recovery of Bio Active Components and Different Applications Food and Nonfood of the Spent Coffee: A Review. *Open J. Nutr. Food Sci.* **2023**, *5*, 1029.
2. Andrade, C.R.; Perestrelo, J.S.; Câmara, J.S. Bioactive Compounds and Antioxidant Activity from Spent Coffee Grounds as a Powerful Approach for its Valorization. *Molecules* **2022**, *27*, 7504. [CrossRef] [PubMed]
3. Lee, Y.-G.; Cho, E.-J.; Maskey, S.; Nguyen, D.-T.; Bae, H.-J. Value-Added Products from Coffee Waste: A Review. *Molecules* **2023**, *28*, 3562. [CrossRef]
4. International Coffee Organization—(ICO). Available online: https://www.ico.org/documents/cy2020-21/cmr-0221-e.pdf (accessed on 22 July 2021).

5. Zhao, N.; Liu, z.; Yu, T. Spent Coffee Grounds: Present and Future of an Environmentally Friendly Applications on Industries- A review. *Trends Food Sci. Technol.* **2023**, *143*, 104312. [CrossRef]
6. Cavanagh, Q.; Brooks, M.S.-L.; Rupasinghe, H.V. Innovative Technologies Used to Convert Spent Coffee Grounds into New Food Ingredients: Opportunities, Challenges, and Prospects. *Future Foods* **2023**, *8*, 100255. [CrossRef]
7. Koay, H.Y.; Azman, A.T.; Zin, Z.M.; Portman, K.L.; Hasmadi, M.; Rusli, N.D.; Zainol, M.K. Assessing the Impact of Spent Coffee Ground (SCG) Concentrations on Shortbread: A Study of Physicochemical Attributes and Sensory Acceptance. *Future Foods* **2023**, *8*, 100245. [CrossRef]
8. Salazar-López, N.J.; López-Rodríguez, C.V.; Hernández-Montoya, D.A.; Campos-Vega, R. Health Benefits of Spent Coffee Grounds. *Food Wastes and By-Products: Nutraceutical Health Potential* **2020**, *327*, 51.
9. Bevilacqua, E.; Cruzat, V.; Singh, I.; Rose'Meyer, R.B.; Panchal, S.K.; Brown, L. The Potential of Spent Coffee Grounds in Functional Food Development. *Nutrients* **2023**, *15*, 994. [CrossRef]
10. Díaz-Hernández, G.C.; Alvarez-Fitz, P.; Maldonado-Astudillo, Y.I.; Jiménez-Hernández, J.; Parra-Rojas, I.; Flores-Alfaro, E.; Salazar, R.; Ramírez, M. Antibacterial, Antiradical and Antiproliferative Potential of Green, Roasted, and Spent Coffee Extracts. *Appl. Sci.* **2022**, *12*, 1938. [CrossRef]
11. Rawangkan, A.; Siriphap, A.; Yosboonruang, A.; Kiddee, A.; Pook-In, G.; Saokaew, S.; Sutheinkul, O.; Duangjai, A. Potential Antimicrobial Properties of Coffee Beans and Coffee By-Products Against Drug-Resistant *Vibrio cholerae*. *Front. Nutr.* **2022**, *9*, 865684. [CrossRef]
12. Franca, A.S.; Oliveira, L.S. Potential Uses of Spent Coffee Grounds in the Food Industry. *Foods* **2022**, *11*, 2064. [CrossRef] [PubMed]
13. Rozhkova, I.V.; Yurova, E.A.; Leonova, V.A. Evaluation of the Amino Acid Composition and Content of Organic Acids of Complex Postbiotic Substances Obtained on the Basis of Metabolites of Probiotic Bacteria *Lacticaseibacillus paracasei* ABK and *Lactobacillus helveticus H9*. *Fermentation* **2023**, *9*, 460. [CrossRef]
14. Iriondo-DeHond, A.; Iriondo-DeHond, M.; Del Castillo, M.D. Applications of Compounds from Coffee Processing By-Products. *Biomolecules* **2020**, *10*, 1219. [CrossRef]
15. Klingel, T.; Kremer, J.I.; Gottstein, V.; Rajcic de Rezende, T.; Schwarz, S.; Lachenmeier, D.W. A Review of Coffee By-Products Including Leaf, Flower, Cherry, Husk, Silver Skin, and Spent Grounds as Novel Foods within the European Union. *Foods* **2020**, *9*, 665. [CrossRef]
16. Iriondo-DeHond, A.; Cornejo, F.S.; Fernandez-Gomez, B.; Vera, G.; Guisantes-Batan, E.; Alonso, S.G.; Andres, M.I.; Sanchez-Fortun, S.; Lopez-Gomez, L.; Uranga, J.A.; et al. Bioaccesibility, Metabolism, and Excretion of Lipids Composing Spent Coffee Grounds. *Nutrients* **2019**, *11*, 1411. [CrossRef]
17. Viegas, C.; Gomes, B.; Oliveira, F.; Dias, M.; Cervantes, R.; Pena, P.; Gomes, A.Q.; Caetano, L.A.; Carolino, E.; de Andrade, E.T.; et al. Microbial Contamination in the Coffee Industry: An Occupational Menace Besides a Food Safety Concern? *Int. J. Environ. Res. Public. Health* **2022**, *19*, 13488. [CrossRef]
18. Trà, T.T.T.; Yến, V.T.N.; Thu, N.T.A.; Nguyệt, T.N.M. Use of Wheat Flour and Spent Coffee Grounds in the Production of Cookies with High Fiber and Antioxidant Content: Effects of Spent Coffee Grounds Ratio on the Product Quality. *IOP Conf. Ser. Earth Environ. Sci.* **2021**, *947*, 012044. [CrossRef]
19. Sugianto, E.C. Incorporating Spent Coffee Grounds into Semolina Pasta: Effects on its Cooking Quality. Honours Thesis, School of Life and Environmental Sciences, Faculty of Science, The University of Sydney, Camperdown, Australia, 2023.
20. Komeroski, M.R.; Oliveira, V.R.d. Influence of the Amount and Type of Whey Protein on the Chemical, Technological, and Sensory Quality of Pasta and Bakery Products. *Foods* **2023**, *12*, 2801. [CrossRef]
21. Lin, S. Dietary Fiber in Bakery Products: Source, Processing, and Function. *Adv. Food Nutr. Res.* **2022**, *99*, 37–100.
22. Cappelli, A.; Cini, E. Challenges and Opportunities in Wheat Flour, Pasta, Bread, and Bakery Product Production Chains: A Systematic Review of Innovations and Improvement Strategies to Increase Sustainability, Productivity, and Product Quality. *Sustainability* **2021**, *13*, 2608. [CrossRef]
23. Melini, V.; Melini, F.; Luziatelli, F.; Ruzzi, M. Functional Ingredients from Agri-Food Waste: Effect of Inclusion Thereof on Phenolic Compound Content and Bioaccessibility in Bakery Products. *Antioxidants* **2020**, *9*, 1216. [CrossRef] [PubMed]
24. González-Montemayor, Á.M.; Flores-Gallegos, A.C.; Contreras-Esquivel, J.C.; Solanilla-Duque, J.F.; Rodríguez-Herrera, R. Functional Activities and its Applications in Bakery Products. *Trends Food Sci. Technol.* **2019**, *94*, 12–19. [CrossRef]
25. Ou, J. Incorporation of Polyphenols in Baked Products. *Adv. Food Nutr. Res.* **2021**, *98*, 207–252. [PubMed]
26. Mitelut, A.C.; Popa, E.E.; Popescu, P.A.; Popa, M.E. Trends of Innovation in Bread and Bakery Production. In *Trends in Wheat and Bread Making*; Galanakis, C.M., Ed.; Academic Press: Cambridge, MA, USA, 2021; pp. 199–226.
27. Aguilar-Raymundo, V.G.; Sánchez-Páez, R.; Gutiérrez-Salomón, A.L.; Barajas-Ramírez, J.A. Spent Coffee Grounds Cookies: Sensory and Texture Characteristics, Proximate Composition, Antioxidant Activity, and Total Phenolic Content. *J. Food Process. Preserv.* **2019**, *43*, 14223. [CrossRef]
28. Ahmed, A.R.; Alqahtani, N.K.; Ramadan, K.M.; Mohamed, H.I.; Mahmoud, M.A.; Elkatry, H.O. The Bioactive Substances in Spent Black Tea and Arabic Coffee Could Improve the Nutritional Value and Extend the Shelf Life of Sponge Cake after Fortification. *ACS Omega* **2023**, *8*, 33593–33609. [CrossRef]
29. Azuan, A.A.; Mohd, Z.Z.; Hasmadi, M.; Rusli, N.D.; Zainol, M.K. Physicochemical, Antioxidant and Sensory Characteristics of Cookies Supplemented with Different Levels of Spent Coffee Ground Extract. *Food Res.* **2020**, *4*, 1181–1190. [CrossRef]

30. Martinez-Saez, N.; García, A.T.; Pérez, I.D.; Rebollo-Hernanz, M.; Mesías, M.; Morales, F.J.; Del Castillo, M.D. Use of Spent Coffee Grounds as Food Ingredient in Bakery Products. *Food Chem.* **2017**, *216*, 114–122. [CrossRef]
31. Hussein, A.; Ali, H.; Bareh, G.; Farouk, A. Influence of Spent Coffee Ground as Fiber Source on Chemical, Rheological and Sensory Properties of Sponge Cake. *Pakistan J. Biol. Sci.* **2019**, *22*, 273–282. [CrossRef]
32. Oliveira Batista, J.; Car Cordeiro, C.; Klososki, S.J.; Mongruel Eleutério Dos Santos, C.; Leão, G.M.C.; Pimentel, T.C.; Rosset, M. Spent Coffee Grounds Improve the Nutritional Value and Technological Properties of Gluten-Free Cookies. *J. Culin. Sci. Technol.* **2023**, *21*, 994–1004. [CrossRef]
33. Desai, N.M.; Mallik, B.; Sakhare, S.D.; Murthy, P.S. Prebiotic Oligosaccharide Enriched Green Coffee Spent Cookies and their Nutritional, Physicochemical and Sensory Properties. *LWT Food Sci. Technol.* **2020**, *134*, 109924. [CrossRef]
34. Meerasri, J.; Sothornvit, R. Novel Development of Coffee oil Extracted from Spent Coffee Grounds as a Butter Substitute in Bakery Products. *J. Food Process. Preserv.* **2022**, *46*, e16687. [CrossRef]
35. Sérioa, A.; Passosa, C.P.; Kukurováb, K.; Ciesarováb, Z.; Nunesc, F.M.; Coimbraa, M.A. Supplementation of Cookies with Water Soluble Extracts Obtained by Microwave Assisted Extraction from Spent Coffee–Influence on Antioxidant and Sensory Properties. In Proceedings of the 11th Meeting on Food Chemistry, University of Averio, Bragança, Portugal, 16–19 September 2011.
36. Sharma, A.; Ray, A.; Singhal, R.S. A Biorefinery Approach Towards Valorization of Spent Coffee Ground: Extraction of the Oil by Supercritical Carbon Dioxide and Utilizing the Defatted Spent in Formulating Functional Cookies. *Future Foods* **2021**, *4*, 100090. [CrossRef]
37. Añón, A.; López, J.F.; Hernando, D.; Orriols, I.; Revilla, E.; Losada, M.M. Effect of Five Enological Practices and the General Phenolic Composition on Fermentation-Related Aroma Compounds in Mencia Young Red Wines. *Food Chem.* **2014**, *148*, 268–275. [CrossRef] [PubMed]
38. Ye, Z.; Arumugam, V.; Haugabrooks, E.; Williamson, P.; Hendrich, S. Soluble Dietary Fiber (Fibersol-2) Decreased Hunger and Increased Satiety Hormones in Humans When Ingested with a Meal. *Nutr. Res.* **2015**, *35*, 393–400. [CrossRef]
39. Castaldo, L.; Lombardi, S.; Gaspari, A.; Rubino, M.; Izzo, L.; Narváez, A.; Ritieni, A.; Grosso, M. In vitro Bioaccessibility and Antioxidant Activity of Polyphenolic Compounds from Spent Coffee Grounds-Enriched Cookies. *Foods* **2021**, *10*, 1837. [CrossRef]
40. Azmi, N.S.; Bhat, R.; Yeoh, T. Quality Evaluation of Novel Cookies Prepared by Supplementing with Fresh Turmeric Flower (*Curcuma longa* L.) Extracts as a Value Added Functional Ingredient. *Int. Food Res. J.* **2016**, *23*, 1514–1522.
41. Giuberti, G.; Rocchetti, G.; Sigolo, S.; Fortunati, P.; Lucini, L.; Gallo, A. Exploitation of Alfalfa Seed (*Medicago sativa* L.) Flour into Gluten-Free Rice Cookies: Nutritional, Antioxidant and Quality Characteristics. *Food Chem.* **2018**, *239*, 679–687. [CrossRef]
42. Daniel, T. Use of Spent Coffee Ground as Ingredient in Bread Formulation. Master's Thesis, College of Natural and Computational Sciences Center for Food Science and Nutrition, Addis Ababa University, Addis Ababa, Ethiopia, 2018.
43. Jenkins, R.W.; Stageman, N.E.; Fortune, C.M.; Chuck, C.J. Effect of the Type of Bean, Processing, and Geographical Location on the Biodiesel Produced from Waste Coffee Grounds. *Energy Fuels* **2014**, *28*, 1166–1174. [CrossRef]
44. Chau, A. Effect of Arabica and Robusta Spent Coffee Grounds on the Physical and Nutritional Quality of White Bread. Honours Thesis, School of Life and Environmental Sciences, Faculty of Science, The University of Sydney, Camperdown, Australia, 2023.
45. Khoddami, A.; Wilkes, M.A.; Roberts, T.H. Techniques for Analysis of Plant Phenolic Compounds. *Molecules* **2013**, *18*, 2328–2375. [CrossRef]
46. Ali, H.S.; Mansour, A.F.; Kamil, M.M.; Hussein, A.M.S. Formulation of Nutraceutical Biscuits Based on Dried Spent Coffee Grounds. *Int. J. Pharmacol.* **2018**, *14*, 584–594. [CrossRef]
47. Campos-Vega, R.; Arreguín-Campos, A.; Cruz-Medrano, M.A.; Del Castillo Bilbao, M.D. Spent Coffee (*Coffea arabica* L.) Grounds Promote Satiety and Attenuate Energy Intake: A Pilot Study. *J. Food Biochem.* **2020**, *44*, e13204. [CrossRef] [PubMed]
48. Severini, C.; Caporizzi, R.; Fiore, A.G.; Ricci, I.; Onur, O.M.; Derossi, A. Reuse of Spent Espresso Coffee as Sustainable Source of Fibre and Antioxidants. A Map on Functional, Microstructure and Sensory Effects of Novel Enriched Muffins. *LWT Food Sci. Technol.* **2020**, *119*, 108877. [CrossRef]
49. Benincá, D.B.; do Carmo, L.B.; Grancieri, M.; Aguiar, L.L.; Lima Filho, T.; Costa, A.G.V.; Silva, P.I. Incorporation of Spent Coffee Grounds in Muffins: A Promising Industrial Application. *Food Chem. Adv.* **2023**, *3*, 100329. [CrossRef]
50. Rwubatse, B.; Okoth, M.W.; Andago, A.A.; Ngala, S.; Kimonyo, A.; Bitwayiki, C. The Effect of Wheat Variety, Fermentation and Incorporation of Ingredients on the Texture Profile, Colour and Sensory Attributes of Whole Wheat Bread. *Croat. J. Food Sci. Technol.* **2021**, *13*, 227–235. [CrossRef]
51. Sulieman, A.A.; Zhu, K.-X.; Peng, W.; Hassan, H.A.; Obadi, M.; Siddeeg, A.; Zhou, H.M. Rheological and Quality Characteristics of Composite Gluten-Free Dough and Biscuits Supplemented with Fermented and Unfermented *Agaricus bisporus* Polysaccharide Flour. *Food Chem.* **2019**, *271*, 193–203. [CrossRef]
52. Paesani, C.; Bravo-Núñez, Á.; Gómez, M. Effect of Extrusion of Whole-Grain Maize Flour on the Characteristics of Gluten-Free Cookies. *LWT Food Sci. Technol.* **2020**, *132*, 109931. [CrossRef]
53. Chiş, M.S.; Păucean, A.; Man, S.M.; Mureşan, V.; Socaci, S.A.; Pop, A.; Muste, S. Textural and Sensory Features Changes of Gluten Free Muffins Based on Rice Sourdough Fermented with *Lactobacillus spicheri* DSM 15429. *Foods* **2020**, *9*, 363. [CrossRef]
54. American Association of Cereal Chemists. *AACC (2003): Guidelines for Measurement of Volume by Rapeseed Displacement. Method 10-05.01*, 11th ed.; American Association of Cereal Chemists: Saint Paul, MN, USA, 2003.
55. Boz, H.; Karaoğlu, M.M. Improving the Quality of Whole Wheat Bread by Using Various Plant Origin Materials. *Czech J. Food Sci.* **2014**, *31*, 457–466. [CrossRef]

56. Noort, M.W.; van Haaster, D.; Hemery, Y.; Schols, H.A.; Hamer, R.J. The Effect of Particle Size of Wheat Bran Fractions on Bread Quality–Evidence for Fibre–Protein Interactions. *J. Cereal Sci.* **2010**, *52*, 59–64. [CrossRef]
57. De Almeida, S.S.; Brito-Silva, L.; da Costa, G.B.M.; Barreto, M.S.; Freire, D.M.G.; Cadena, R.S.; Moura-Nunes, N. Whole- Wheat Bread Enzymatically Bioprocessed and Added with Green Coffee Infusion Had Improved Volume and Were Sensory Accepted When Consumers Were Informed of the Presence of Healthy Substances. *Int. J. Food Sci. Technol.* **2022**, *57*, 6112–6121. [CrossRef]
58. Xu, J.; Li, Y.; Zhao, Y.; Wang, D.; Wang, W. Influence of Antioxidant Dietary Fiber on Dough Properties and Bread Qualities: A Review. *J. Funct. Foods* **2021**, *80*, 104434. [CrossRef]
59. Mildner-Szkudlarz, S.; Zawirska-Wojtasiak, R.; Szwengiel, A.; Pacyński, M. Use of Grape By-Product as a Source of Dietary Fibre and Phenolic Compounds in Sourdough Mixed Rye Bread. *Int. J. Food Sci. Technol.* **2011**, *46*, 1485–1493. [CrossRef]
60. Siegwein, A.M.; Vodovotz, Y.; Fisher, E.L. Concentration of Soy Protein Isolate Affects Starch-Based Confections' Texture, Sensory, and Storage Properties. *J. Food Sci.* **2011**, *76*, E422–E428. [CrossRef] [PubMed]
61. Jagelaviciute, J.; Staniulyte, G.; Cizeikiene, D.; Basinskiene, L. Influence of Enzymatic Hydrolysis on Composition and Technological Properties of Apple Pomace and Its Application for Wheat Bread Making. *Plant Foods Hum. Nutr.* **2023**, *78*, 307–313. [CrossRef]
62. Villarino, C.B.J.; Jayasena, V.; Coorey, R.; Chakrabarti-Bell, S.; Johnson, S.K. The Effects of Australian Sweet Lupin (ASL) Variety on Physical Properties of Flours and Breads. *LWT Food Sci. Technol.* **2015**, *60*, 435–443. [CrossRef]
63. Mansour, A.F.; Pudil, F.; Janda, V.; Pokorný, J. Changes During the Extrusion of Semolina in Mixture with Sugars. *Czech J. Food Sci.* **2001**, *19*, 24–30. [CrossRef]
64. Wong, C.; Wijayanti, H.; Bhandari, B. Maillard Reaction in Limited Moisture and Low Water Activity Environment. In *Water Stress in Biological, Chemical, Pharmaceutical and Food Systems*; Gutiérrez-López, G., Alamilla-Beltrán, L., del Pilar Buera, M., Welti-Chanes, J., Parada-Arias, E., Barbosa-Cánovas, G., Eds.; Springer: New York, NY, USA, 2015; pp. 41–63.
65. American Association of Cereal Chemists. *Approved Methods of Analysis*, 10th ed.; American Association of Cereal Chemists: St. Paul, MN, USA, 2000; ISBN 978-1-891127-12-0.
66. Liu, T.; Hamid, N.; Kantono, K.; Pereira, L.; Farouk, M.M.; Knowles, S.O. Effects of Meat Addition on Pasta Structure, Nutrition and In vitro Digestibility. *Food Chem.* **2016**, *213*, 108–114. [CrossRef]
67. Symons, L.J.; Brennan, C. The Influence of $(1 \rightarrow 3)$ $(1 \rightarrow 4)$-β-D-Glucan-Rich Fractions from Barley on the Physicochemical Properties and In Vitro Reducing Sugar Release of White Wheat Breads. *J. Food Sci.* **2006**, *69*, C463–C467. [CrossRef]
68. Vignola, M.B.; Bustos, M.C.; Pérez, G.T. Comparison of Quality Attributes of Refined and Whole Wheat Extruded Pasta. *LWT Food Sci. Technol.* **2018**, *89*, 329–335. [CrossRef]
69. Nilusha, R.A.T.; Jayasinghe, J.M.J.K.; Perera, O.D.A.N.; Perera, P.I.P. Development of Pasta Products with Nonconventional Ingredients and Their Effect on Selected Quality Characteristics: A Brief Overview. *Int. J. Food Sci.* **2019**, *2019*, 6750726. [CrossRef]
70. Desai, A.; Brennan, M.A.; Brennan, C.S. The Effect of Semolina Replacement with Protein Powder from Fish (*Pseudophycis bachus*) on the physicochemical characteristics of pasta. *LWT Food Sci. Technol.* **2018**, *89*, 52–57. [CrossRef]

foods

MDPI

Article

Antioxidant Bio-Compounds from Chestnut Waste: A Value-Adding and Food Sustainability Strategy

Roberta Barletta [1,†], Alfonso Trezza [1,†], Andrea Bernini [1], Lia Millucci [1], Michela Geminiani [1,2] and Annalisa Santucci [1,2,*]

[1] Department of Biotechnology, Chemistry and Pharmacy, University of Siena, Via Aldo Moro, 53100 Siena, Italy; r.barletta@student.unisi.it (R.B.); alfonso.trezza2@unisi.it (A.T.); andrea.bernini@unisi.it (A.B.); lia.millucci@gmail.com (L.M.); geminiani2@unisi.it (M.G.)

[2] SienabioACTIVE, Università di Siena, Via Aldo Moro, 53100 Siena, Italy

[*] Correspondence: annalisa.santucci@unisi.it

[†] These authors contributed equally to this work.

Abstract: In an era of escalating environmental challenges, converting organic residues into high-value bioactive compounds provides a sustainable way to reduce waste and enhance resource efficiency. This study explores the potential of the circular bioeconomy through the valorization of agricultural byproducts, with a focus on the antioxidant properties of specific chestnut burr cultivars. Currently, over one-third of food production is wasted, contributing to both humanitarian and environmental crises. Through circular bioeconomy, we can transform biological waste into valuable products for use in fields like food innovation and sustainability. The antioxidant effects of three chestnut cultivars, *Bastarda Rossa*, *Cecio*, and *Marroni*, were assessed through in vitro assays, highlighting their potential to combat oxidative stress—an important factor for health-related applications. The characterization of the three cultivars showed the major presence of ellagic acid and gallic acid in the extract, renowned for their antioxidant activity. In vitro assays evaluated the phenolic and flavonoid content, as well as the antioxidant activity of the three extracts, confirming the cultivar *Cecio* as the richest in these bioactive compounds and the most performative in antioxidant assays. In vitro antioxidant and oxidative stress recovery assays on SaOS-2, fibroblast, and chondrocyte cell lines displayed a strong antioxidant activity. Furthermore, the cytotoxicity assay demonstrated the safety of all three extracts in the tested human cell lines. In silico docking simulations further validated the biological relevance of these compounds by predicting strong hydrophobic and polar interactions with oxidative stress-related protein targets. Overall, this study demonstrates the antioxidant properties of chestnut byproducts. The findings contribute to the development of functional foods, nutraceuticals, and other applications, underscoring the role of chestnut cultivars in advancing circular bioeconomy practices.

Keywords: circular bioeconomy; antioxidant activity; bioactive compounds; waste; oxidative stress; food innovation; sustainability; docking simulation; NMR

1. Introduction

The chestnut tree, considered for timber and fruit, has been seeing a notable increase in global production in the last years. Italy is the European Union's largest chestnut producer and contributes 38% of the European chestnut supply, with an annual yield of approximately 52,356 tons (FAOSTAT, accessed 18 October 2024). Chestnut cultivation has long been fundamental to Italian agriculture, especially in the mountainous regions where it once served as both a primary food source and an economic staple. In Tuscany, chestnut trees—easily recognizable by their serrated leaves and characteristic flowers—grow in specific areas, such as Piancastagnaio in Siena, which became synonymous with chestnut cultivation. These trees, which thrive in the distinctive microclimates of Mount Amiata, supported local livelihoods for generations, reinforcing the cultural and economic identity of the region. Numerous chestnut genotypes and cultivars were cultivated to meet both local and broader economic needs, increasing their significance to the communities surrounding Mount Amiata.

In addition to the historical significance, chestnut is long recognized for its health benefits, particularly in Chinese traditional medicine and more recent applications in cosmetics in Korea [1]. These uses mirror the chestnut's bioactive potential as chestnut-derived byproducts are known to contain bioactive compounds, including phenolics, that provide health benefits through their antioxidant and radical-scavenging properties [2–5].

Variability in the phytochemical concentration of chestnut fruits, influenced by the cultivar and growth conditions, underscores the importance of selective breeding to optimize yield and nutritional quality for functional food applications [6]. Such efforts aligned with the rising trend of utilizing natural sources—flavonoids, polyphenols, and other phytochemicals—as replacements for synthetic additives, particularly in the food and nutraceutical industries. These natural antioxidants are renowned for their broad spectrum of health benefits, which include anti-cancer, anti-inflammatory, antimicrobial, and cardioprotective effects [7–9].

In recent years, the growing environmental crisis caused by pollution and the waste of natural resources has become a major global concern. The increasing generation of biological waste, particularly agricultural residues, exacerbates this issue. However, these byproducts offer an untapped potential for sustainable applications. Specifically, plant-based waste, such as chestnut peels [10], is rich in bioactive compounds that can be harnessed for a variety of purposes, including antimicrobial and antioxidant uses. Despite the potential, there remains a gap in research regarding the effective repurposing of such waste materials into valuable, eco-friendly products that can address both environmental and health challenges. By investigating the antioxidant properties of *Castanea sativa* peel extracts, this study aims to contribute to closing this gap, providing an innovative and sustainable solution to the dual problem of waste management and public health.

The chestnut processing industry generates approximately 7000 tons of processed chestnut products annually, producing significant waste, with burrs and shells accounting for roughly 10% of the total chestnut weight. Historically, the management of chestnut burrs involved burning them to clear the land for optimal tree growth, a practice that, although effective in the short term, is costly, time-consuming, and environmentally detrimental due to CO_2 emissions [10].

The urgent need for more sustainable and economically viable approaches led to a growing interest in the valorization of these byproducts. Burrs are rich in tannins, ellagic acid, gallic acid, and other antioxidant biomolecules, which demonstrate significant antimicrobial, anti-carcinogenic, and cardioprotective properties [6,10–13].

This strategy followed the broader setting of the circular bioeconomy, which presents a valid alternative to the wasteful practices of the linear economy. While the linear economic model ultimately led to global resource depletion and contributed to environmental crises, the circular economy focuses on sustainable practices, encouraging resource reuse and recycling. In the case of chestnut burrs, rather than being discarded through dangerous practices like burning, these byproducts can be repurposed as valuable bioactive compounds for a variety of applications, from nutraceuticals to natural preservatives in food. This transition from a linear to a circular model is not only a matter of environmental sustainability but also offers the potential for economic growth, especially within the agri-food sector.

In fact, the circular bioeconomy has already made significant inroads in several industries. Food waste, once relegated to animal feed or discarded altogether, is now increasingly being seen as a valuable source of bioactive compounds [14,15]. Biorefineries, for example, are exploring ways to convert food waste into renewable energy, while the textile industry began utilizing natural dyes, such as those extracted from pomegranate seeds, to replace synthetic alternatives [16,17]. In the case of chestnut byproducts, including burrs, research revealed their potential to support the development of sustainable products such as natural antioxidants, antimicrobials, and even materials for use in textiles or biodegradable plastics. These bioactive compounds have the potential to be incorporated into food products, helping to combat oxidative damage, which is linked to chronic diseases like cancer, cardiovascular disease, and neurodegenerative disorders [5,9]. These extracts can also benefit the nutraceutical industry, particularly in supplements designed to address oxidative stress and inflammation. Moreover, their antimicrobial properties position chestnut extracts as valuable components for functional foods that promote gut health and food preservation, offering natural alternatives to synthetic preservatives [7,18–21]. Additionally, the bioactive compounds in chestnut extracts could be used in skincare products, offering protection from oxidative stress caused by pollutants and UV radiation [22,23].

Thus, this research addresses the critical knowledge gap on how to repurpose chestnut waste, particularly burrs, within circular bioeconomy, to yield value-added products with biological applications. By tackling the dual challenges of waste management and resource valorization, this study contributes to the development of sustainable solutions in the food and nutraceutical sectors.

As the world is facing dramatic environmental, economic, and social challenges, driven in part by the unsustainable practices of the linear economy, the circular economy offers a transformative solution [18,19]. By rethinking how natural resources, particularly agricultural byproducts, are used and valued, it is possible to create more sustainable, resilient, and resource-efficient systems. In this context, chestnut burrs exemplify the potential of agro-industrial waste to contribute to the circular bioeconomy, providing both environmental and economic benefits while offering innovative solutions for the food and health sectors [1,2].

This study aimed to dig into value-added applications for chestnut waste in the food and nutraceutical sectors. We analyzed and compared the chemical composition of hydroalcoholic extracts from three domestic cultivars certified under *Indicazione Geografica Protetta* (PGI) from Mount Amiata: *Bastarda Rossa* (RB), *Cecio* (RC), and *Marroni* (RM). These extracts showed significant phenolic contents, and ellagic and gallic acid were detected as the two most abundant compounds in all three extracts (mostly RC). Given the deep-sitting antioxidant properties of polyphenols, the total and cellular antioxidant capacities of these extracts were evaluated in vitro, along with their cytotoxicity on fibroblast, chondrocyte, and SaOS-2 cell lines. These were selected as highly relevant models: fibroblasts are critical in studying tissue repair [20,24], chondrocytes are involved in cartilage degeneration linked to oxidative damage [21,25], and SaOS-2 cells provide insights into oxidative stress in bone health [26].

All extracts demonstrated safety for cell lines and enhanced cellular recovery from oxidative stress, particularly RC and RB, which exhibited over a 50% recovery rate. The in vitro findings were supported by docking simulations, which indicated that the compounds could spontaneously bind to their predicted targets, initiating strong hydrophobic and polar interactions with key residues of the target.

Together, these findings underscore the potential of chestnut waste as a source of bioactive compounds for food innovation, and for nutraceutical development, advancing the chestnut sector's role in sustainable food production and supporting bioeconomy goals.

2. Materials and Methods

2.1. *C. sativa Shells*

The shells from C. *sativa* Mill were sourced from the PGI (Protected Geographical Indication) Castagna del Monte Amiata area, a renowned chestnut-producing region in Italy. These shells, from the *Bastarda Rossa*, *Cecio*, and *Marroni* cultivars, were directly peeled from the plant. Following peeling, the shells were collected, weighed (to determine fresh weight), dried, and weighed again to determine dry weight. After drying at room temperature, the samples were homogenized using a laboratory mill and stored in dark, sealed plastic bags at −80 °C until extraction.

2.2. *C. Sativa Shell Extract Preparation*

The extraction process was carried out as follows: 100 mL of a 70/30 ethanol/water solution (*v*/*v*) was combined with 10 g of chestnut burr powder (1:10 solid-to-solid ratio). This solvent mixture was chosen for its efficiency in extracting phenolic compounds from chestnut burrs. Ethanol (Sigma-Aldrich, St. Louis, MO, USA), being polar, dissolves a wide range of bioactive compounds, including polyphenols and tannins, while the addition of water enhances the solvent's penetration and increases the yield of these compounds. The 1/10 *w*/*v* ratio was selected to optimize the extraction of phenolics and tannins, ensuring a sufficiently concentrated extract without sacrificing practicality [27,28]. A thermostatic mantle stirred the mixture for 3 h at 80 °C. The extraction chamber was connected at the top to a net water-cooled bubble condenser, to allow the reflux of the evaporated solvent. Afterward, the mixture was centrifuged (30 min, 4000 rpm), and the supernatant was separated and filtered. The organic solvent was removed using a rotavapor, and the aqueous residue was freeze-dried to obtain the dry extract.

2.3. *Total Phenolic Content (TPC)*

The total phenolic content (TPC) was determined using the method introduced by Folin and Denis in 1912 and later modified by Folin and Ciocalteu in 1927 [29]. A calibration curve was established with gallic acid (GA) (Sigma-Aldrich, St. Louis, MO, USA) solutions. CSB samples were diluted from a stock solution (1 mg/mL) in milli-Q water. Standard and sample solutions were mixed with 1 mL of 1N FC reagent (Sigma-Aldrich, St. Louis, MO, USA) and incubated for 3 min. Subsequently, 1 mL of Na2CO3 (Sigma-Aldrich, St. Louis, MO, USA) and 7 mL of milli-Q water were added. The tubes were incubated for 90 min at room temperature, shielded from light. Absorbance was measured at 725 nm, with a blank solution serving as a control. The TPC was expressed as milligrams of GA equivalent (GAE) per 100 g of fresh material.

2.4. *Total Flavonoid Content (TFC)*

TFC was measured using the $NaNO_2$-$Al(NO_3)_3$ method with slight modifications [30]. In brief, 1 mL of a sample was combined with 0.3 mL of 5% $NaNO_2$ (Sigma-Aldrich, St. Louis, MO, USA) solution, and after 6 min, 0.3 mL of 10% $Al(NO_3)_3$ (Sigma-Aldrich, St. Louis, MO, USA) and 4 mL of 4% NaOH (Sigma-Aldrichm St. Louis, MO, USA) were added. The total volume was adjusted to 10 mL with distilled water, and the solution was gently shaken. Absorbance was read at 418 nm. The calibration curve was constructed using quercetin (Sigma-Aldrich, St. Louis, MO, USA), and the TFC was expressed as Quercetin Equivalents (QEs) per 100 g of fresh material.

2.5. Antioxidant Activity Evaluation

2.5.1. FRAP Assay

The antioxidant capacity was determined using the FRAP (Ferric Ion-Reducing Antioxidant Power) assay with minor modifications [31]. An aliquot of 1 mL containing different extract concentrations (25–400 μg) was mixed with 2.5 mL of 0.2 M phosphate buffer (pH = 6.6) and 1% potassium ferricyanide solution (Sigma-Aldrich, St. Louis, MO, USA). The mixture was incubated at 50 °C for 20 min. Afterward, 2.5 mL of 10% trichloroacetic acid (Sigma-Aldrich, St. Louis, MO, USA) and 2.5 mL of distilled water were added, followed by 0.5 mL of 0.1% ferric chloride solution (Sigma-Aldrich, St. Louis, MO, USA). Absorbance was recorded at 700 nm. Results were expressed as Ascorbic Acid Equivalent (AAE) per 100 g of fresh material.

2.5.2. DPPH Free Radical-Scavenging Activity

DPPH free radical-scavenging activity was assessed by mixing the extract solution with a freshly prepared DPPH (Sigma-Aldrich, St. Louis, MO, USA) solution [32]. The mixture was incubated for 20 min in the dark at room temperature, and absorbance was measured at 517 nm. Ascorbic acid (Sigma-Aldrich, St. Louis, MO, USA) was used for calibration. Radical-scavenging activity (RSA) was calculated using the formula $\%RSA = 100 \times (A0 - As)/A0$, where As is the absorbance of the sample and A0 is the absorbance of the control solution.

2.6. NMR Analysis

NMR samples were prepared by dissolving 50.0 mg of a dry extract in 0.5 mL of DMSO-d6 (Cambridge Isotope Laboratories, Inc., Tewksbury, MA, USA). TSP (3-(trimethylsilyl) propionic-2,2,3,3-d4 acid, Cambridge Isotope Laboratories, Inc., Tewksbury, MA, USA) was incorporated as an internal standard, reaching a final concentration of 0.3 mM. The spectra were recorded using a Bruker Advance III 600 spectrometer operating at 14.1 T. Metabolite quantification was conducted with the deconvolution feature of Chenomx 10 (Edmonton, AB, Canada). For 1H NMR, an NOESY 1D pulse sequence was employed with a 100 ms mixing time and a spectral width of 9600 Hz, digitized over 16K points. A total of 256 transients were collected with a 9 s repetition delay. Peak assignments for gallic and ellagic acids were confirmed by spiking in pure substances (Merck KGaA, Darmstadt, Germany). For 13C NMR, a 1D pulse sequence with WALTZ16 decoupling was used, covering a spectral width of 40,000 Hz and digitized over 64K points. A total of 4096 transients were accumulated with a repetition delay of 9 s, with the DMSO peak as the reference.

2.7. Cell Line Isolation and Culture

Primary human fibroblasts (available in our laboratory) were sourced from femoral cartilage fragments of arthropathy patients undergoing hip replacement surgery. Informed consent was obtained from the patients, and the study was approved by the Local Ethics Committee, following the guidelines of the Declaration of Helsinki (64th, 2013). Chondrocytes were isolated promptly post-surgery as previously outlined [33]. Cartilage was minced under sterile conditions and subjected to three rounds of enzymatic digestion: 1 mg/mL hyaluronidase (Sigma-Aldrich, St. Louis, MO, USA) (30 min), 5 mg/mL pronase (Sigma-Aldrich, St. Louis, MO, USA) (1 h), and 2 mg/mL collagenase (Type I, Sigma-Aldrich (St. Louis, MO, USA, 1 h) at 37 °C. The resulting cell suspension was filtered twice through 70 μm nylon meshes, washed, and centrifuged at 1500 rpm for 10 min. The pellet was resuspended in Dulbecco's Modified Eagle's Medium (DMEM) (Sigma-Aldrich, St. Louis, MO, USA), supplemented with 2% penicillin/streptomycin (P/S) (Sigma-Aldrich, St. Louis, MO, USA) and 10% fetal bovine serum (FBS) (Sigma-Aldrich, St. Louis, MO,

USA), and seeded into 10 cm Petri dishes. Cells were cultured at 37 °C in a humidified atmosphere with 95% air and 5% CO_2.

Primary human chondrocytes (from our laboratory) were collected from four male individuals (ages 42–50) undergoing abdominoplasty. The study was approved by the Local Ethics Committee and was conducted according to the principles of the Declaration of Helsinki (64th, 2013). Immediately after surgery, skin samples were excised and cut into 1–2 mm pieces (explants), which were placed in washing DMEM containing 10 times the usual antibiotic/antimycotic dose, plus gentamicin. These explants were minced using iris scissors, centrifuged at 200 rpm for 10 min to remove the supernatant, and then incubated with a collagenase solution (DMEM + 10% collagenase type II + 10× antibiotics/antimycotics) in a 60 mm dish at 37 °C with 5% CO_2 for 21 h. Afterward, the explants were washed twice with PBS. For the next 1–3 days, the explants were cultured in a dish with an explant medium containing 20% heat-inactivated FBS and the usual antibiotic dose. Once collagenase was removed, the cultures were observed daily under a microscope to monitor cell migration around the explants. The explants were then transferred to a culture flask.

Once the fibroblast cultures reached confluence, they were expanded further. The medium was discarded, and the cultures were washed three times with PBS. In the second and third passages, keratinocytes were removed by brief trypsinization (0.1% trypsin/EDTA, Sigma-Aldrich St. Louis, MO, USA). The volume was adjusted for the vessel size (e.g., 1.0 mL of 0.1% trypsin/EDTA for T75 flasks), and the cultures were incubated at 37 °C. After 2–3 min, fibroblasts detached from the flask surface, as keratinocytes and fibroblasts exhibit different attachment properties, allowing for their separation.

Primary human sarcoma osteogenic cells (SaOS-2) (ATCC-HTB-85) were obtained from the American Type Culture Collection (ATCC, Manassas, VA, USA) and cultured in DMEM (Dulbecco's Modified Eagle Medium) enriched with L-glutamine (Sigma-Aldrich, St. Louis, MO, USA) and 10% FBS and 1% antibiotics (penicillin/streptomycin). Growth occurred as mentioned above.

2.8. Cytotoxicity Assay

Shell extract cytotoxicity was determined on human primary fibroblasts, chondrocytes, and SaOS-2 cells. Cells were cultured with the supplementation of different concentrations (range: 31.5–500 μg/mL) of shell extracts. Cell viability was evaluated at sub-confluence in 96-well plates (Thermo Fisher Scientific, Seoul, Republic of Korea) by the MTT (3-(4,5-dimethylthiazol-2-yl)-2,5-diphenyl tetrazolium bromide) method, as previously described [34].

2.9. Protective Effect Under Oxidative Stress Conditions

Shell extracts were evaluated for their potential to prevent oxidative stress in 0.2 mM H_2O_2-treated cells. After cells reached total confluence in 96-well plates, previously filtered (0.2 μm) shell extracts (range: 3.125–100 μg/mL) were added for 24 h. Subsequently, 0.2 mM H_2O_2 (Sigma-Aldrich, St. Louis, MO, USA) was added, and incubation occurred for 2 h. The protective effect of shell extracts was evaluated by the MTT assay as SaOS-2 cell viability compared with untreated control.

2.10. Statistical Analysis

Data are presented as the means ± standard deviation (SD) of at least three independent experiments performed in triplicate. The statistical analysis was performed using Student's *t*-test and one-way ANOVA, and differences were considered significant at $p < 0.05$ and $p < 0.01$.

2.11. Structural Optimization and Resources

To further investigate the antioxidant activity observed in the *C. sativa* extract, in silico studies were conducted to identify the potential biological target for ellagic and gallic acids, the extract's two primary compounds.

First, "PASS protein prediction" was used to retrieve the list of potential targets, sorted by confidence [35]. For both ellagic and gallic acid, the target with the highest confidence was selected, and its involvement with oxidative stress was checked from the DrugBank "target section" (typing in the words "oxidative stress") [36], which consists of 212 targets associated with cellular oxidative stress, either directly or indirectly.

The 3D structures and FASTA sequences of the validated targets were generated with AlphaFold and retrieved from the UniProt database [37], respectively.

The potential missing side chains and steric clashes in the 3D structures reported in the PDB files were addressed and resolved through molecular modeling using MODELLER v.9.3, implemented in PyMOD3.0 (PyMOL2.5 plugin) [38]. The 3D structures were analyzed and validated with PROCHECK v.3.5.4 [39,40]. To minimize high-energy intramolecular interactions before docking simulations, GROMACS 2019.3 [41] with the CHARMM36 force field [42] was employed. All parameters for the biological targets and ligands were assigned using CHARMM-GUI v.3.8 [43].

In detail, prior to conducting further simulations, the starting conformation sequence was aligned with its primary structure, allowing for the addition of any missing side chains. Loop modeling in MODELLER v.9.3 was then used to optimize the initial orientation of each loop. The structures were subsequently validated using PROCHECK, where a Ramachandran plot (assessing the backbone ϕ and ψ angles) and Chi1–Chi2 plots for side chains confirmed the validity of the starting conformations. Energy minimization was then performed on each structure using GROMACS 2019.3 with the CHARMM36 force field, aiming to eliminate steric clashes and optimize energy values.

The optimized structures were then placed in a cubic box filled with TIP3P water molecules, and the system was neutralized by adding counterions. Simulations were conducted under periodic boundary conditions. Energy minimization was carried out for 5000 steps using the steepest-descent algorithm, reaching a minimum energy with forces less than 10 kJ/mol/nm. A short 25 ns classic molecular dynamics (cMD) simulation was performed to further relax the system. During the cMD simulations, each time step was integrated with a 2 fs step size. The temperature was maintained at 300 K using a Nose–Hoover thermostat, while the system pressure was regulated at 1 atm with a Parrinello–Rahman barostat, using a damping coefficient of 1 ps−1. Bond lengths involving hydrogen atoms were constrained with the LINCS algorithm [41,44].

2.12. Docking Simulations and Interaction Network

Docking simulations were then performed to further unravel the interactions between ellagic and gallic acid with their respective targets as described in our previous works [5,7]. For greater consistency in docking results, the exhaustiveness was adjusted from 8 to 32 and only binding poses with an RMSD within 2 Å of the optimal docked pose were accepted; all other parameters were kept as default. The 3D structures of gallic acid and ellagic acid were obtained from the PubChem database [44,45], with compound IDs 370 and 5281855, respectively, and downloaded in SDF format. A box was configured around the residues previously shown to be involved in interactions with their inhibitors using DockingPie2.0 [46] integrated into PyMOL 3.0, with AutoDock Tools v.4.2 [36,47].

The AutoDock/VinaXB v.1.1.2 tool performed docking simulations [48]. Protein and ligand files were converted using MGLTOOLS v.1.5.7 scripts, while Gasteiger partial charges were added with OpenBabel v.3.1.0 [49]. The 3D structures of gallic acid and ellagic acid were obtained from the PubChem database (CID: 370 and 5281855, respectively) [44].

The interaction networks were analyzed using the P.L.I.P. v. 2.3.0 tool [50], and sequence alignments to identify key target residues were conducted with ClustalW v.2.1 [51,52].

3. Results

3.1. *C. sativa* Extracts' Total Phenolic Contents

The extraction yields for the three chestnut cultivars and their total phenolic contents were evaluated and shown in Table 1. The total phenolic contents of the cultivars varied from 7.041 (RC) to 1.893 (RM) GAE/100 g fresh material (GAE, gallic acid equivalent). The cultivar *Cecio* (RC) showed the highest total phenolic content of 7 GAE/100 g fresh material, followed by the cultivar *Bastarda Rossa* (RB) with 4.8 GAE/100 g fresh material of total phenolics. The lowest phenolic content was reported for the cultivar *Marroni* (RM), with 1.8 GAE/100 g fresh material.

Table 1. Total Phenolic Content of the three analyzed cultivars: *Bastarda Rossa* (RB); *Cecio* (RC); *Marroni* (RM).

Samples	Extraction Yield (%)	GAE/mg of Extract	GAE/100 g of Fresh Material
RB	16.64	0.292961609	4.866
RC	20.84	0.338665448	7.041
RM	7.6	0.257312614	1.893

These values confirmed that the mixture EtOH:H_2O at 70:30 is optimal for the extraction of phenolic compounds, indicating the potential of *Cecio* and *Bastarda Rossa* chestnut shell extracts to formulate nutraceuticals or cosmetics as natural antioxidants.

3.2. *C. sativa* Extracts' Flavonoid Contents

The flavonoid content of the extracts of the three cultivars was evaluated and reported in Table 2. The total flavonoid content varied from 319.2 (RM) to 1000 (RC) quercetin/100 g fresh material.

Table 2. Flavonoid Content of the three analyzed cultivars: *Bastarda Rossa* (RB); *Cecio* (RC); *Marroni* (RM).

Samples	Extraction Yield (%)	QE/mg of Extract	QE/100 g of Fresh Material
RB	16.64	0.0483	798.7
RC	20.84	0.0508	1000
RM	7.6	0.0425	319.2

Cecio (RC) showed the highest value, followed by the cultivar *Bastarda Rossa* (RB) and then *Marroni* (RM). These findings would suggest that the total flavonoid content in the three cultivars follows the line of TPC and contributes to the antioxidant power of the extracts.

3.3. *C. sativa* Extracts' Antioxidant Activity

The total antioxidant activities of three different chestnut cultivar shell extracts were evaluated by the FRAP assay. As shown in Table 3, the highest total antioxidant activity was found in cultivar *Cecio* (RC) with the value of 146.5 AAE/100 g fresh material, followed by *Bastarda Rossa* (RB) and *Marroni* (RM), with the values of 91.5 and 34.9 AAE/100 g fresh material, respectively. These data align with previous findings that *Cecio* and *Bastarda Rossa* are the best cultivars for their antioxidant power.

Table 3. Antioxidant activity of three cultivars analyzed by FRAP assay: *Bastarda Rossa* (RB); *Cecio* (RC); *Marroni* (RM).

Samples	Extraction Yield (%)	mmol AAE/mg Extract	AAE/100 g of Fresh Material
RB	16.64	5.530995	91.520
RC	20.84	7.037723	146.5
RM	7.6	4.564747	34.96

3.4. *C. sativa* Extracts' Antiradical Activity

The DPPH colorimetric assay is based on the ability of reducing molecules to scavenge the 2,2-diphenyl-1-picrylhydrazyl synthetic radical dissolved in an alcoholic solution, thus obtaining its neutral form. The process is monitored following the conversion of the violet-colored radical into a pale-yellow solution at 517 nm, and the antiradical activity is reported at a percentage decrease in such absorbance. The antiradical activity was evaluated by the DPPH assay and the results are shown in Figure 1 and Table 4. Even in this case, the cultivar *Cecio* was the most effective in terms of antiradical activity with a 76.6% Abs decrease.

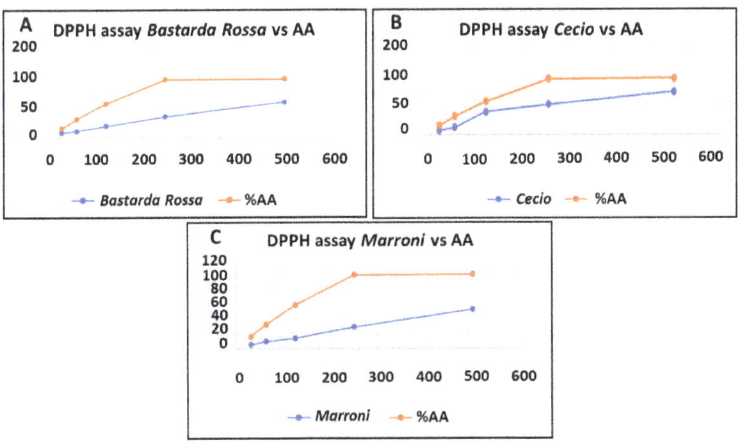

Figure 1. DPPH assay. Absorbance decrease in three cultivar extracts analyzed compared to control, ascorbic acid (AA). Antioxidant activity (expressed as %AA) of chestnut varieties was measured at varying extract concentrations (0–500 µg/mL). (**A**–**C**) represent cultivars *Bastarda Rossa*, *Cecio*, and *Marroni*, respectively.

Table 4. Antiradical activity of three cultivar extracts analyzed by DPPH assay. Results are expressed as Abs decrease (%) at 517 nm. *Bastarda Rossa* (RB); *Cecio* (RC); *Marroni* (RM).

Samples	Extraction Yield (%)	% Abs Decrease, 517 nm
RB	16.64	61.11
RC	20.84	76.6
RM	7.6	52.8

Combining the results obtained from the DPPH colorimetric assay and FRAP assay, information was gathered about the redox properties and radical-scavenging activity induced by electron and hydrogen transfer, which are key processes related to tannins' antioxidant capacity.

The antioxidant activity of the three cultivar extracts and their TPC values were positively correlated (antioxidant capacity vs. TPC: $p = 0.009159382$; antiradical power vs. TPC: $p = 0.012930563$), implying that the measured antioxidant capacity is reasonably attributable to the phenolic compounds present in the extracts. Additionally, *Cecio* and *Bastarda Rossa* cultivars represented better sources of bioactive antioxidant molecules than the *Marroni* cultivar.

3.5. Phenolic Compound Determination by NMR Analysis

Bioactive compounds of *Bastarda Rossa*, *Cecio*, and *Marroni* shells were elucidated by the NMR metabolomics approach. Indeed, the metabolite profiling associated with the antioxidant potential of chestnut shell extracts of the three main cultivars of *Monte Amiata* PGI represents an important step in the bioactive compound's characterization due to the chestnut biodiversity in this region.

Through the analysis of 1H NMR spectra, gallic and ellagic acid were identified as the main components in the aqueous extracts of the shells from the three cultivars (see Figure 2). As shown in Table 5, NMR quantification indicates that *Bastarda Rossa* (RB) had the highest concentration of ellagic acid, followed by *Cecio* (RC) and *Marroni* (RM). In contrast, RC had a greater gallic acid concentration than both RB and RM. Overall, *Marroni* was found to have the lowest levels of these compounds.

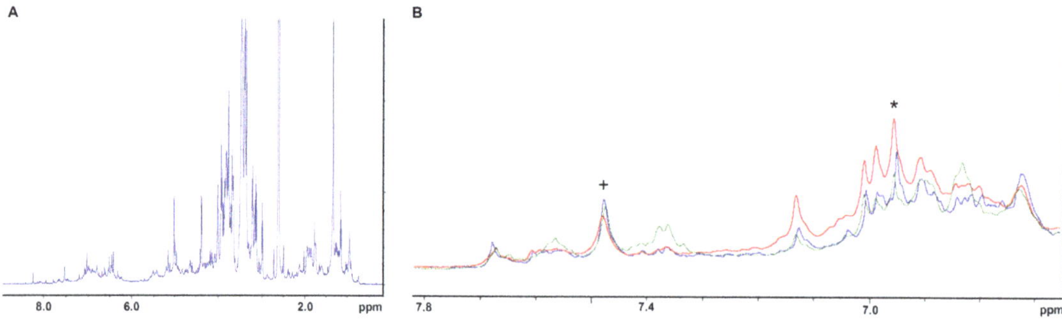

Figure 2. (**A**) The 1H NMR spectrum of an aqueous extract. (**B**) The aromatic region of the spectrum shows signals of gallic acid (*) and ellagic acid (+) for the cultivar *Bastarda Rossa* (RB, blue), *Cecio* (RC, red), and *Marroni* (RM, green). Signal integration allowed for polyphenol quantification, reported in Table 5.

Table 5. The concentration of (+) ellagic and (*) gallic acid reported as mg/100 mg extract in the three cultivars determined by 1H NMR spectroscopy. *Bastarda Rossa* (RB); *Cecio* (RC); *Marroni* (RM).

Sample	RB	RC	RM
(+) ellagic acid	0.61 ± 0.12	0.56 ± 0.11	0.49 ± 0.10
(*) gallic acid	0.41 ± 0.08	0.56 ± 0.11	0.22 ± 0.04

The screening conducted using 13C NMR spectroscopy provided a quantitative profile of the phytochemicals present in each material, as shown in Figure 3 and Table 6. This study demonstrated that 1D 13C data improve the identification of metabolites and facilitate the better separation of groups for comparison. The 13C NMR spectrum of the hydroalcoholic shell extracts from the three cultivars revealed signals in the 140–160 ppm region, indicating the presence of carbons derived from phenolic compounds. Additionally, alkoxy carbons were identified in the region from 50 to 110 ppm, likely associated with sugars such as glucose. In the *Marroni* (RM) cultivar, these sugars appear to be more abundant than in the other cultivars (refer to Table 6). This could be attributed to the fact that *Marroni* is a cultivated variety selected for specific nut qualities and consumer uses, such as candying, roasting, boiling, and drying, which require a sweeter flavor. Furthermore, the *Marroni* tree has more specific growing requirements and is generally less productive in terms of the number of chestnuts produced per tree compared to ordinary chestnuts. The findings regarding phenolic compounds are consistent with previous data and indicate that the *Cecio* (RC) cultivar is the richest in phenolic content.

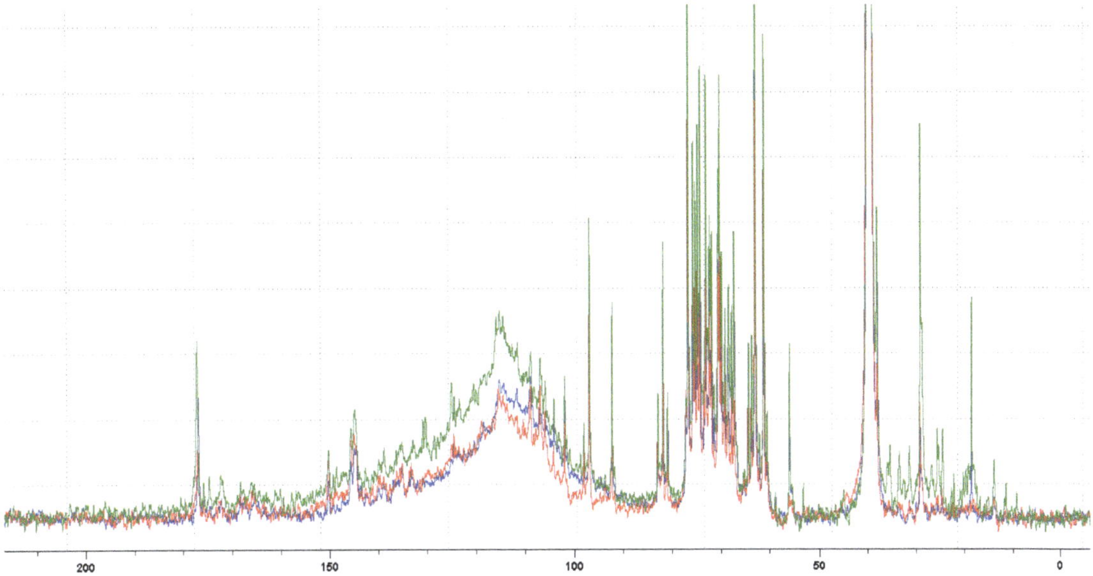

Figure 3. 13C NMR normalized spectra of *Bastarda Rossa* (RB, blue), *Cecio* (RC, red), and *Marroni* (RM, green).

Table 6. Phytochemicals present in the three cultivars obtained by 13C NMR spectroscopy. (Quantities were normalized to 100 in relation to the internal calibrant peak.) *Bastarda Rossa* (RB); *Cecio* (RC); *Marroni* (RM).

	C_{alkyl} (0–35 ppm)	C_{alkoxy} (50–100 ppm)	$C_{aromatic}$ (100–140 ppm)	$C_{phenolic}$ (140–160 ppm)	$C_{carboxyl/carbonyl}$ (160–215 ppm)
RB	2.5	47.0	63.2	5.5	4.1
RC	3.4	38.5	54.13	7.9	1.9
RM	14.5	60.7	81.2	7.0	6.3

3.6. Citotoxicity of *C. sativa* Shell Extracts

The cytotoxic effect of the three extracts was assessed in the SaOS-2 cell line, primary chondrocytes, and human primary fibroblasts.

Cells were treated with increasing concentrations of the three different extracts (31.5–500 μg/mL) for 24 h and 48 h and the cytotoxic effect on cell viability was assessed by the MTT assay (Figure 4).

Results showed that for all cultivars, no cytotoxic effect was observed in SaOS-2 or primary chondrocytes and human primary fibroblasts for concentrations ranging from 31.5 to 250 μg/mL, whilst extracts reduced the viability at 500 μg/mL. A cytotoxic effect on the viability of SaOS-2 cells was observed at 24 h at 500 μg/mL only for the cultivar *Marroni*, but the extract did not affect the viability until the concentration of 250 μg/mL also at 48 h.

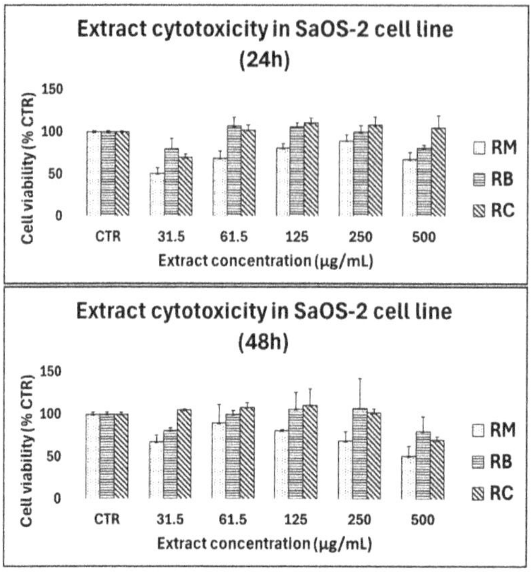

Figure 4. Cytotoxicity of chestnut extracts (RM: *Marroni*, RB: *Bastarda Rossa*, RC: *Cecio*) on SaOS-2 cells at varying concentrations (31.5–500 µg/mL) after 24 h (**top panel**) and 48 h (**bottom panel**) of treatment. Cell viability was measured using the MTT assay, and the results are expressed as a percentage of the untreated control group (CTR). Extract concentrations ranged from 31.5 to 500 µg/mL, with bars representing the mean ± standard deviation from three independent experiments.

The three extracts were safe even on primary human chondrocytes, in concentrations ranging from 31.5 to 250 µg/mL (Figure 5).

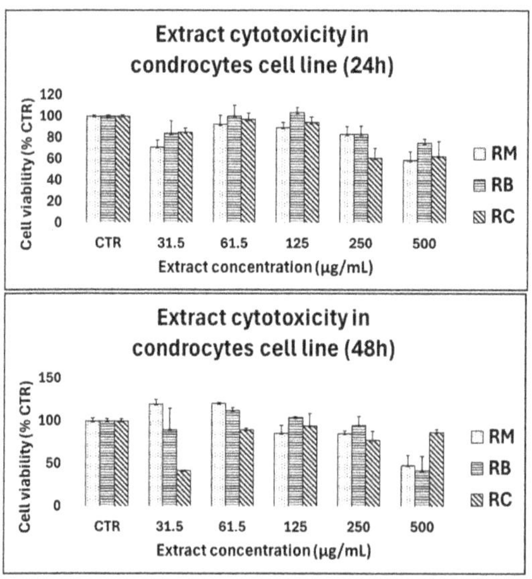

Figure 5. Cytotoxicity of chestnut extracts (RM: *Marroni*, RB: *Bastarda Rossa*, RC: *Cecio*) on chondrocyte varying concentrations (31.5–500 µg/mL) after 24 h (**top panel**) and 48 h (**bottom panel**) of treatment. Cell viability was measured using the MTT assay, and the results are expressed as a percentage of the untreated control group (CTR). Extract concentrations ranged from 31.5 to 500 µg/mL, with bars representing the mean ± standard deviation from three independent experiments.

In human primary fibroblasts (Figure 6), the three extracts showed a similar trend, being the three cultivars cytotoxic at the concentrations 250–500 μg/mL, but safe at minor concentrations. According to the obtained results, the extract did not decrease the cell viability in the range of 31.5–250 μg/mL in both cell types, supporting their potential safety in human cells at appropriate concentrations.

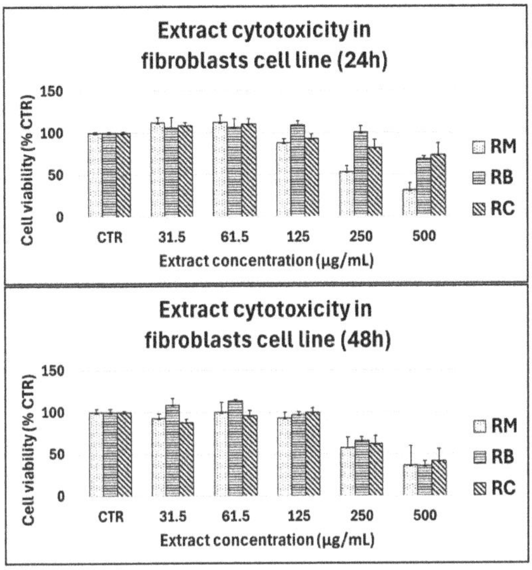

Figure 6. Cytotoxicity of chestnut extracts (RM: *Marroni*, RB: *Bastarda Rossa*, RC: *Cecio*) on fibroblasts at varying concentrations (31.5–500 μg/mL) after 24 h (**top panel**) and 48 h (**bottom panel**) of treatment. Cell viability was measured using the MTT assay, and the results are expressed as a percentage of the untreated control group (CTR). Extract concentrations ranged from 31.5 to 500 μg/mL, with bars representing the mean ± standard deviation from three independent experiments.

3.7. *Evaluation of Antioxidant Power of Three Cultivar Extracts*

Strong evidence states that oxidative stress is responsible for the pathophysiology of the aging process and may also be involved in the pathogenesis of atherosclerosis, neurodegenerative diseases, cancer, and diabetes. Epidemiological evidence indicated a link between dietary intake of antioxidants and bone health and metabolism, while numerous studies showed the health-promoting properties of polyphenols, providing additional mechanisms through which they promote skeletal health by reducing resorption caused by high oxidative stress [8,53].

In this study, chestnut shells were used to obtain bioactive compounds given their richness in phenolic compounds with antioxidant power, responsible for biological effects that contribute to benefits in human health and potential food applications in the development of functional foods. The three cultivar extracts were also assessed for their ability to counteract the ROS effects on cultured human cells, thereby legitimizing their use in the alimentary field. Briefly, SaOS-2 cells, human primary chondrocytes, and human primary fibroblasts were exposed to H_2O_2 (0.2 mM), which, in preliminary experiments, reduced cell viability by 50% (Figure 7).

For SaOS-2 cells, all three extracts showed a viability increase, compared to treated controls, in the range of 3125 to 25 μg/mL, reaching a recovery of over 50% with *Bastarda Rossa* (RB) and *Cecio* (RC). Results on primary chondrocytes were particularly interesting, with a viability recovery over 50% for RC at the concentration of 2.5 μg/mL. The same recovery was found for RM in the range of 6.25–50 μg/mL. RB was a little less efficient, but, at 12.5 μg/mL, it allowed for recovery of approximately 25% of the viability compared to the stressed control. On human primary fibroblasts, all the tested extracts showed a noticeable ability to recover from oxidative stress. The RC extract was the best for antioxidant ability on cells, showing a higher percentage of viability recovery (Figure 7). Overall, for every extract, a recovery of cell viability was assessed from 30 to 50% for RC on chondrocytes and SaOS-2, and 25% for human primary fibroblasts. In total, 6.25 and 12.5–25 μg/mL are the ranges of better concentrations presenting the highest protective effect in all the tested extracts in the SaOS-2 cell line, and in chondrocytes and fibroblasts, respectively (Figure 7).

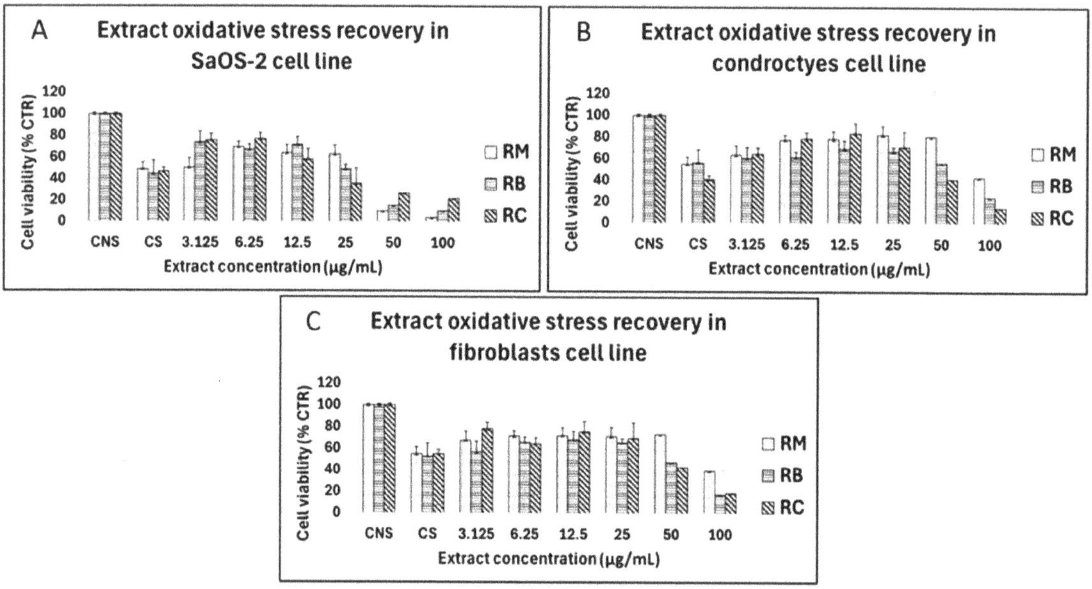

Figure 7. Oxidative stress recovery in various cell lines treated with chestnut extracts. RM = *Marroni*; RB = *Bastarda Rossa*; RC = *Cecio*. (**A**) SaOS-2; (**B**) chondrocytes; and (**C**) fibroblasts, at different concentrations (3.125–100 μg/mL). Oxidative stress was induced in cells using hydrogen peroxide (H_2O_2), with untreated stressed cells (CNS) and untreated non-stressed cells (CS) serving as controls. Cell viability was measured using MTT, and results are expressed as a percentage of the untreated non-stressed control (CS). Data represent the mean ± standard deviation from three independent experiments.

These results further contributed to evaluating the high potential of this chestnut byproduct as a new source of antioxidant compounds to be used in various fields, such as developing functional foods and nutraceuticals.

3.8. In Silico Results

Target Identification and Docking Simulations

In silico studies were meant to confirm the in vitro-observed antioxidant activity by predicting the biological targets for the two most abundant compounds within the extracts (ellagic acid and gallic acid) and evaluating their interactions at the molecular level.

To identify potential targets of the two most abundant compounds within the extract, namely ellagic and gallic acid, a ligand-based virtual screening was conducted through the "PASS Protein Prediction" [35], which, for each compound, sorted the predicted targets by confidence. The most confident target was then checked for its presence in the complete oxidative stress target network, identified using the DrugBank database "target section" [36], which consists of 212 targets associated with cellular oxidative stress, either directly or indirectly. The individuated targets were serine/threonine-protein kinase NEK6 for ellagic acid and carbonic anhydrase III for gallic acid, respectively, displaying 77% and 87% confidence. For each target, its primary sequence was retrieved from the UniProt database (serine/threonine-protein kinase NEK6, UniprotKB entry: Q9HC98; carbonic anhydrase III, UniprotKB entry: P07451). AlphaFold generated the targets' 3D structure, further optimized through molecular modeling to solve structure gaps and steric hindrances.

Gallic acid and ellagic acid were then evaluated against their targets using docking simulations (Figure 8). Two strategies were employed to select the top complexes: (i) selecting compounds with a binding free energy (docking score) lower than −6 kcal/mol and (ii) assessing the compound's ability to form strong interactions with consensus binding residues of the target. The results indicated a docking score of −6.9 kcal/mol for the complex ellagic acid/NEK6, and a docking score of −5.7 kcal/mol for the complex gallic acid/carbonic anhydrase III.

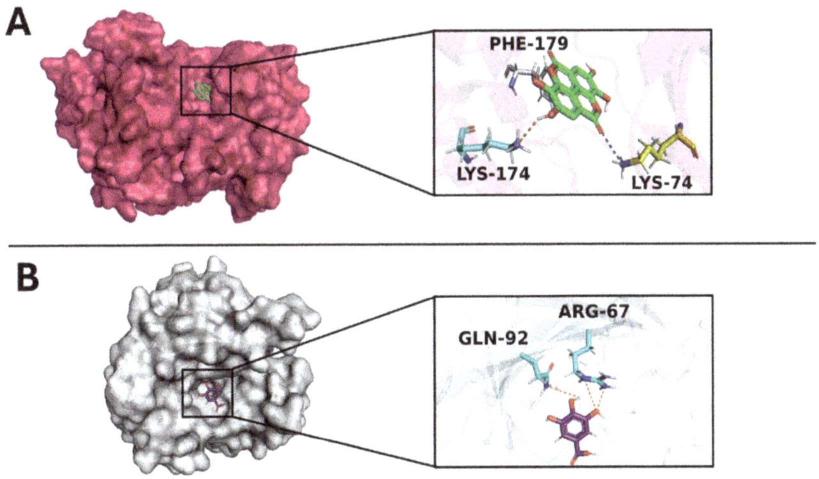

Figure 8. An overview of the target/ligand complexes. (**A**) is ellagic acid in complex with the NEK6 kinase 3D structure; (**B**) is gallic acid in complex with the carbonic anhydrase III 3D structure. The enlarged pictures display the interaction network established among the target-binding residues and the ligand. The binding residues involved in hydrogen bonds, salt bridges, and hydrophobic interactions are labeled cyan, yellow, and gray, respectively. Hydrogen bonds and salt bridges are pictured as orange and blue dashed lines, respectively.

The potential inhibitory effects of ellagic and gallic acid on their targets were further explored by analyzing the interaction networks, showing wide polar and non-polar interactions between each compound and its target that were identified. Specifically, ellagic acid formed a broad interaction network with key NEK6 residues, such as Phe-179 (hydrophobic interaction), Lys-174 (hydrogen bond), and Lys-74 (salt bridge) [54]. Likewise, gallic acid established hydrophobic interactions with carbonic anhydrase III critical residues, that is, Arg-67 and Gln-92 [55].

Our findings suggest that these compounds exhibit binding capabilities with their targets in key residues, which may result in the potential inhibitory activity displayed in vitro. Overall, computational studies confirmed the in vitro results and provided additional insights on the mechanism of action of the extracts' major compounds, through which they exert their activity.

4. Discussion

Natural sources like plants have long presented promising alternatives for the development of bioproducts with healthy properties. This is a field of growing importance, especially within the food and nutraceutical industries, where demand for natural, sustainable ingredients has risen. For centuries, natural compounds derived from plants have been foundational in traditional medicine systems, including traditional Chinese medicine. These compounds offer therapeutic effects by influencing crucial biological processes such as cell proliferation, angiogenesis, survival, and apoptosis. Additionally, plant-derived bioactive molecules are valued for their strong antioxidant and anti-inflammatory properties, which make them integral to therapeutic practices aimed at combating oxidative stress and chronic inflammation [56]. This knowledge of plant bioactive molecules laid the foundation for a transition toward the use of sustainably derived natural compounds to develop health-focused products. This shift is particularly evident in the food and nutraceutical sectors, where sourcing effective bioactive molecules from natural, plant-based materials became a priority. As consumers are becoming more healthy and environmentally aware, there is an increasing demand for sustainable products. This growing interest might open new routes for innovation, especially in the creation of functional foods and natural health supplements that promote wellness through diets [57–59].

One of the most reliable sources of such bioactive compounds is chestnut (*Castanea sativa*), a tree species whose cultivation is deeply rooted in Italian agricultural traditions. However, in recent years, chestnut cultivation has lowered, particularly due to the diminishing economic viability of traditional chestnut-based products, and, while the decline in chestnut cultivation poses challenges, adopting a holistic approach that combines economic, environmental, and technological strategies can ensure sustainability. By leveraging chestnut byproducts and promoting eco-friendly practices, it is possible to sustainably obtain antioxidant-rich extracts while supporting the revival of this valuable crop [60]. This decline urged the need to explore new applications for chestnut byproducts, which account for approximately 10–15% of the total chestnut weight. These byproducts, often discarded as waste, in fact contain a wealth of valuable bioactive compounds that could contribute significantly to food sustainability and nutraceutical innovation [61,62]. This approach would also reduce pollution and CO_2 emissions from burning the chestnut burrs to ensure optimal tree growth [17].

This study investigates the underexplored potential of chestnut shells as a rich source of bioactive compounds like tannins, phenolic acids, and flavonoids, which are known for their antioxidant, anti-inflammatory, and antimicrobial properties. The two most abundant compounds found by NMR from this study were ellagic acid and gallic acid, both presenting a sustainable alternative to synthetic antioxidants in food systems. Specifically, chestnut shell-derived antioxidants can be incorporated into food products to combat oxidative damage, a contributing factor to chronic diseases such as cancer, cardiovascular disease, and neurodegenerative disorders [5,7,63]. Additionally, chestnut extracts could be used in the nutraceutical industry for supplements targeting oxidative stress and inflammation [19,20]. Beyond health, their antimicrobial properties make them valuable for developing functional foods that promote gut health and food preservation, potentially replacing synthetic preservatives with natural alternatives by combating both food-borne and health-threatening pathogens [7,21]. Furthermore, these bioactive compounds could be harnessed in skincare products, offering natural protection from oxidative stress due to environmental pollutants and UV radiation [22,23].

Importantly, the use of chestnut burrs aligns with the principles of the circular bioeconomy by reducing food waste while creating high-value products [64].

As part of this study, the hydroalcoholic extracts from three *Monte Amiata* PGI chestnut cultivars—*Bastarda Rossa* (RB), *Cecio* (RC), and *Marroni* (RM)—were biochemically characterized for phenolic, flavonoid, and antioxidant content.

The FRAP and DPPH assays were selected due to their ease of use and proven reliability for assessing antioxidant activity. These methods are widely recognized for their straightforward protocols, adaptability to a high-throughput analysis, and consistent reproducibility. They effectively provide foundational antioxidant profiles, enabling the quick evaluation of the potential of various plant-based compounds. Furthermore, their robust nature makes them suitable for initial screening and comparative studies, offering valuable insights into the efficacy and bioactivity of the extracts under investigation [65–67]. These assays provide a straightforward method to compare the radical-scavenging capacities of various extracts, and they are particularly useful for establishing preliminary antioxidant profiles, guiding the selection of extracts for more comprehensive studies.

RC showed the highest concentration of bioactive compounds such as gallic acid, ellagic acid, and flavonoids, and consequently, its potential application in the nutraceutical field. RM, with a milder profile of these molecules due to selective breeding, appeared more suited for food consumption, aligning well with evolving consumer behavior toward functional foods [22–24]. Then, the three cultivars' extracts were assessed for cytotoxicity and protection from oxidative stress in three different cell lines: SaOS-2, fibroblasts, and chondrocytes. Although a slight variation in the cell model viability response used in this study was noted upon treatment with the three cultivar extracts, none of the considered extracts yielded a toxic effect in the cultured cells in a wide range of concentrations. The variations in cellular responses to the extracts could be attributed to the distinct biological characteristics of each cell type. SaOS-2 cells, an osteosarcoma cell line, are typically more sensitive to oxidative stress and reactive oxygen species (ROS) due to their heightened metabolic activity. As shown in several studies, osteosarcoma cells can exhibit significant reductions in viability when exposed to high concentrations of plant extracts that induce oxidative damage [68]. Fibroblasts, on the other hand, tend to be more resilient to oxidative stress. They have evolved efficient antioxidant defense mechanisms, allowing them to better tolerate prolonged exposure to potentially harmful compounds. Studies have demonstrated that fibroblasts can endure higher concentrations of bioactive compounds without a drastic decrease in cell viability [69]. Chondrocytes exhibit slower metabolic activity and lower antioxidant capacities, making them more vulnerable to oxidative stress. This characteristic makes chondrocytes more sensitive to prolonged exposure to high concentrations of plant extracts, as evidenced by similar findings in other studies on cartilage cells [70]. These inherent differences in cellular resilience to oxidative stress and metabolic activity explain the observed differential responses in cell viability across the different cell types used in this study upon treatment with the chestnut extract. In regards to the cytoprotective activity, the *Cecio* and *Bastarda Rossa* extracts demonstrated strong cytoprotective effects against H_2O_2-induced oxidative stress in all cell lines, likely due to their elevated levels of gallic acid, ellagic acid, and flavonoids, which may modulate oxidative stress pathways or stimulate cellular repair mechanisms (as shown by cell viability percentages compared to the untreated controls). According to the literature body, all these bioactive compounds display well-documented protective activities against oxidative stress [19], underlining the prospects of chestnut byproducts to support food waste, adding value through innovative applications in food sustainability and nutraceuticals.

To elucidate more the possible mechanism of action of the extracts' bioactive compounds, in silico studies were conducted. The potential targets for ellagic acid and gallic acid, the two most abundant compounds within the extracts, were pinpointed as the serine/threonine-protein kinase NEK6 and carbonic anhydrase III. The following docking simulation provided useful insights into the potential interaction between each ligand with its target. As a result, ellagic acid and gallic acid both exhibited a high binding free energy score, triggering hydrogen bonds, salt bridges, and hydrophobic interactions with key target residues [54,55]. Altogether, these simulations strongly strengthened the experimental evidence provided in our study and demonstrated the potential ability of the extracts' bioactive compounds to interfere with the target biological function by interacting with critical target residues.

This study found that the three analyzed chestnut shell extracts from *Cecio*, *Bastarda Rossa*, and *Marroni* cultivars could significantly contribute to food system innovation as naturally innovative sources of antioxidant compounds.

Given the promising bioactivity of the chestnut burr extracts, particularly the presence of phenolic compounds such as ellagic acid and gallic acid, there is potential for translating these findings into practical applications within the food and nutraceutical industries [8–13]. These compounds are well known for their antioxidant and antimicrobial properties, making them ideal candidates for incorporation into functional foods, natural preservatives, or supplements. For instance, ellagic acid demonstrated potential in reducing oxidative stress and inflammation, which could be leveraged in formulating health supplements aimed at supporting cardiovascular or anti-cancer benefits [19,20,22,23]. Additionally, the antimicrobial properties observed in these extracts could lead to their application as natural preservatives in food products, offering an eco-friendly alternative to synthetic preservatives [7,21].

Harnessing these byproducts from chestnut processing not only supports circular bioeconomic goals but also aligns with global efforts to enhance food sustainability, reduce food loss, and potentially improve food security by creating value-added health-promoting solutions from otherwise discarded resources.

However, while we value the importance of investigating the stability, bioavailability, dosages, and practical applications of the three extracts within food systems, this study primarily focuses on the antioxidant potential of chestnut burr extracts at the in vitro level. These aspects might be explored in future studies, as they represent an essential next step in understanding how such bioactive compounds can be incorporated into food or nutraceutical formulations. This work serves as a foundation for future research aimed at harnessing the benefits of chestnut-derived compounds in sustainable food production and health-related products [19].

5. Conclusions

Our study was conducted alongside a circular bioeconomy framework, aiming to repurpose chestnut shells for their valuable biological properties. By focusing on three cultivars of *Castanea sativa* shells—*Cecio*, *Bastarda Rossa*, and *Marroni*—the antioxidant potential and phenolic content of their hydroalcoholic extracts were investigated. Findings from this study demonstrated that the shell extracts from the *Cecio* and *Bastarda Rossa* cultivars showed notably higher antioxidant activity and concentrations of key bioactive compounds, including ellagic acid, gallic acid, and flavonoids. Fundamentally, ellagic and gallic acid were the two most abundant metabolites to be found. The biochemical characterization further confirmed that *Cecio* and *Bastarda Rossa* are particularly promising cultivars as their extracts exhibit strong cytoprotective and antioxidant effects in cellular models under oxidative stress. H_2O_2-induced cytotoxicity and oxidative damage were significantly reduced in human cells, underscoring the protective potential of their phenolic compounds against aging-related oxidative stress, representing valuable sources of bioactive compounds with potential applications in nutraceuticals, food innovation, and food sustainability.

To further investigate the potential antioxidant activity of the cultivars' extracts, in silico studies were performed to identify the target for the two majorly represented compounds in the extracts, namely ellagic acid and gallic acid. Their interaction with the target was explored through docking simulations and an interaction network analysis, displaying a range of interactions, which would potentially justify the inhibitory effect of the ligands on their respective targets.

Ellagic and gallic acid were selected as the most abundant compounds in the extracts. In silico findings indicated their potential bioactivity against their predicted targets, warranting further investigation. Importantly, this study was not focused on proposing potential target inhibitors, rather to highlight the importance and the innovation to align the circular bioeconomy principles by identifying natural wastes as antioxidant sources, reducing time, and costs, and minimizing environmental pollutants. Exploring the extracts' potential within a sustainable, integrated bioprocessing approach offers an eco-friendly pathway to assess the biological activity of plant byproduct compounds. However, further investigations are needed to explore their stability, bioavailability, dosages, and potential uses in nutraceuticals or functional foods.

Author Contributions: Conceptualization, R.B. and A.T.; methodology, R.B., A.T. and A.B.; software, A.B.; validation, R.B., A.T. and A.B.; formal analysis, R.B., A.T. and A.B.; investigation, R.B., A.T. and A.B.; resources, A.B. and A.S.; data curation, R.B., A.T. and A.B.; writing—original draft preparation, R.B. and A.B.; writing—review and editing, R.B., A.T., A.B. and A.S.; visualization, R.B., A.T., A.B., L.M., M.G. and A.S.; supervision, A.S.; project administration, A.S.; funding acquisition, A.S. All authors have read and agreed to the published version of the manuscript.

Funding: This research received no external funding.

Institutional Review Board Statement: Comitato Etico Policlinico Universitario di Siena, number: GGP10058, 21 July 2010.

Informed Consent Statement: Informed consent was obtained from all subjects involved in the study.

Data Availability Statement: The original contributions presented in the study are included in the article; further inquiries can be directed to the corresponding author.

Acknowledgments: The authors thank PRIN 2022-Prot. 2022LW54KC "Bio-waste products as a source of nutraceuticals to target the common trait of inflammation in cancer and neurodegeneration".

Conflicts of Interest: The authors declare no conflicts of interest.

References

1. Chang, X.; Liu, F.; Lin, Z.; Qiu, J.; Peng, C.; Lu, Y.; Guo, X. Phytochemical Profiles and Cellular Antioxidant Activities in Chestnut (*Castanea mollissima* BL.) Kernels of Five Different Cultivars. *Molecules* **2020**, *25*, 178. [CrossRef] [PubMed]
2. Cravotto, C.; Grillo, G.; Binello, A.; Gallina, L.; Olivares-Vicente, M.; Herranz-López, M.; Micol, V.; Barrajón-Catalán, E.; Cravotto, G. Bioactive Antioxidant Compounds from Chestnut Peels through Semi-Industrial Subcritical Water Extraction. *Antioxidants* **2022**, *11*, 988. [CrossRef] [PubMed]
3. Echegaray, N.; Gómez, B.; Barba, F.J.; Franco, D.; Estévez, M.; Carballo, J.; Marszałek, K.; Lorenzo, J.M. Chestnuts and By-products as Source of Natural Antioxidants in Meat and Meat Products: A Review. *Trends Food Sci. Technol.* **2018**, *82*, 110–121. [CrossRef]
4. Comandini, P.; Lerma-García, M.J.; Simó-Alfonso, E.F.; Toschi, T.G. Tannin Analysis of Chestnut Bark Samples (*Castanea sativa* Mill.) by HPLC-DAD-MS. *Food Chem.* **2014**, *157*, 290–295. [CrossRef]
5. Trezza, A.; Geminiani, M.; Cutrera, G.; Dreassi, E.; Frusciante, L.; Lamponi, S.; Spiga, O.; Santucci, A. A Drug Discovery Approach to Reveal a Novel Antioxidant Natural Source: The Case of Chestnut Burr Biomass. *Int. J. Mol. Sci.* **2024**, *25*, 2517. [CrossRef]
6. De Vasconcelos, M.C.; Bennett, R.N.; Rosa, E.A.; Ferreira-Cardoso, J.V. Composition of European Chestnut (*Castanea sativa* Mill.) and Association with Health Effects: Fresh and Processed Products. *J. Sci. Food Agric.* **2010**, *90*, 1578–1589. [CrossRef]
7. Trezza, A.; Barletta, R.; Geminiani, M.; Frusciante, L.; Olmastroni, T.; Sannio, F.; Docquier, J.-D.; Santucci, A. Chestnut Burrs as Natural Source of Antimicrobial Bioactive Compounds: A Valorization of Agri-Food Waste. *Appl. Sci.* **2024**, *14*, 6552. [CrossRef]

8. Sorice, A.; Siano, F.; Capone, F.; Guerriero, E.; Picariello, G.; Budillon, A.; Ciliberto, G.; Paolucci, M.; Costantini, S.; Volpe, M.G. Potential Anticancer Effects of Polyphenols from Chestnut Shell Extracts: Modulation of Cell Growth, and Cytokinomic and Metabolomic Profiles. *Molecules* **2016**, *21*, 1411. [CrossRef]
9. Pinto, D.; Braga, N.; Silva, A.M.; Costa, P.; Delerue-Matos, C.; Rodrigues, F. *Valorization of Fruit Processing By-Products*; Academic Press: Cambridge, MA, USA, 2019; pp. 128–141.
10. Liberti, A.; Goretti, G.; Russo, M.V. PCDD and PCDF formation in the combustion of vegetable wastes. *Chemosphere* **1983**, *12*, 661–663. [CrossRef]
11. Silva, V.; Falco, V.; Dias, M.I.; Barros, L.; Silva, A.; Capita, R.; Alonso-Calleja, C.; Amaral, J.S.; Igrejas, G.; Ferreira, C.F.R.; et al. Evaluation of the Phenolic Profile of *Castanea sativa* Mill. By-Products and Their Antioxidant and Antimicrobial Activity against Multiresistant Bacteria. *Antioxidants* **2020**, *9*, 87. [CrossRef]
12. Neves, J.M.; Matos, C.; Moutinho, C.; Queiroz, G.; Gomes, L.R. Ethnopharmacological Notes about Ancient Uses of Medicinal Plants in Trás-os-Montes (Northern Portugal). *J. Ethnopharmacol.* **2009**, *124*, 270–283. [CrossRef] [PubMed]
13. Corregidor, V.; Antonio, A.L.; Alves, L.C.; Cabo Verde, S. Castanea sativa Shells and Fruits: Compositional Analysis by Proton Induced X-Ray Emission. *Nucl. Instrum. Methods Phys. Res. Sect. B* **2020**, *477*, 98–103. [CrossRef]
14. Bongaarts, J. Human Population Growth and the Demographic Transition. *Philos. Trans. R. Soc. Lond. B Biol. Sci.* **2009**, *364*, 2985–2990. [CrossRef] [PubMed]
15. Di Vaio, A.; Hasan, S.; Palladino, R.; Hassan, R. The Transition Towards Circular Economy and Waste within Accounting and Accountability Models: A Systematic Literature Review and Conceptual Framework. *Environ. Dev. Sustain.* **2023**, *25*, 734–810. [CrossRef] [PubMed]
16. Venkatesh, G. Circular Bio-economy—Paradigm for the Future: Systematic Review of Scientific Journal Publications from 2015 to 2021. *Circ. Econ. Sustain.* **2022**, *2*, 231–279. [CrossRef]
17. Ghangrekar, M.M.; Das, S.; Das, S. 15—Microbial Electrochemical Technologies for CO_2 Sequestration. In *Biomass, Biofuels, Biochemicals*; Pandey, A., Tyagi, R.D., Varjani, S., Eds.; Elsevier: Amsterdam, The Netherlands, 2021; pp. 413–443.
18. Faraoni, P.; Laschi, S. Bioactive Compounds from Agrifood Byproducts: Their Use in Medicine and Biology. *Int. J. Mol. Sci.* **2024**, *25*, 5776. [CrossRef]
19. Hadidi, M.; Liñán-Atero, R.; Tarahi, M.; Christodoulou, C.M.; Aghababaei, F. The Potential Health Benefits of Gallic Acid: Therapeutic and Food Applications. *Antioxidants* **2024**, *13*, 1001. [CrossRef]
20. Golmei, P.; Kasna, S.; Roy, K.P.; Kumar, S. A Review on Pharmacological Advancement of Ellagic Acid. *J. Pharmacol. Pharmacother.* **2024**, *15*, 93–104. [CrossRef]
21. De, R.; Sarkar, A.; Ghosh, P.; Ganguly, M.; Karmakar, B.C.; Saha, D.R.; Halder, A.; Chowdhury, A.; Mukhopadhyay, A.K. Antimicrobial Activity of Ellagic Acid against *Helicobacter pylori* Isolates from India and during Infections in Mice. *J. Antimicrob. Chemother.* **2018**, *73*, 1595–1603. [CrossRef]
22. Khan, B.A.; Mahmood, T.; Menaa, F.; Shahzad, Y.; Yousaf, A.M.; Hussain, T.; Ray, S.D. New Perspectives on the Efficacy of Gallic Acid in Cosmetics & Nanocosmeceuticals. *Curr. Pharm. Des.* **2018**, *24*, 5181–5187.
23. Evtyugin, D.D.; Magina, S.; Evtuguin, D.V. Recent Advances in the Production and Applications of Ellagic Acid and Its Derivatives: A Review. *Molecules* **2020**, *25*, 2745. [CrossRef] [PubMed]
24. Younesi, F.S.; Miller, A.E.; Barker, T.H.; Rossi, F.M.V.; Hinz, B. Fibroblast and Myofibroblast Activation in Normal Tissue Repair and Fibrosis. *Nat. Rev. Mol. Cell Biol.* **2024**, *25*, 617–638. [CrossRef] [PubMed]
25. Goldring, M.B. Chondrogenesis, Chondrocyte Differentiation, and Articular Cartilage Metabolism in Health and Osteoarthritis. *Ther. Adv. Musculoskelet. Dis.* **2012**, *4*, 269–285. [CrossRef] [PubMed]
26. Marcucci, G.; Domazetovic, V.; Nediani, C.; Ruzzolini, J.; Favre, C.; Brandi, M.L. Oxidative Stress and Natural Antioxidants in Osteoporosis: Novel Preventive and Therapeutic Approaches. *Antioxidants* **2023**, *12*, 373. [CrossRef]
27. Dai, J.; Mumper, R.J. Plant Phenolics: Extraction, Analysis and Their Antioxidant and Anticancer Properties. *Molecules* **2010**, *15*, 7313–7352. [CrossRef]
28. Liyana-Pathirana, C.; Shahidi, F. Optimization of Extraction of Phenolic Compounds from Wheat Using Response Surface Methodology. *Food Chem.* **2005**, *93*, 47–56. [CrossRef]
29. Singleton, V.L.; Orthofer, R.; Lamuela-Raventós, R.M. Analysis of Total Phenols and Other Oxidation Substrates and Antioxidants by Means of Folin-Ciocalteu Reagent. *Methods Enzymol.* **1999**, *299*, 152–178.
30. Chang, C.F.; Yang, M.-H.; Wen, H.-M.; Chern, J.C. Estimation of Total Flavonoid Content in Propolis by Two Complementary Colorimetric Methods. *J. Food Drug Anal.* **2020**, *28*, 413–418.
31. Jayaprakasha, G.K.; Singh, R.P.; Sakariah, K.K. Antioxidant Activity of Grape Seed (*Vitis vinifera*) Extracts on Peroxidation Models in Vitro. *Food Chem.* **2001**, *73*, 285–290. [CrossRef]
32. Yen, G.-C.; Chen, H.-Y. Antioxidant Activity of Various Tea Extracts in Relation to Their Antimutagenicity. *J. Agric. Food Chem.* **1995**, *43*, 27–32. [CrossRef]

33. Braconi, D.; Bernardini, G.; Bianchini, C.; Laschi, M.; Millucci, L.; Amato, L.; Tinti, L.; Serchi, T.; Chellini, F.; Spreafico, A.; et al. Biochemical and Proteomic Characterization of Alkaptonuric Chondrocytes. *J. Cell. Physiol.* **2012**, *227*, 3333–3343. [CrossRef] [PubMed]
34. Mosmann, T. Rapid Colorimetric Assay for Cellular Growth and Survival: Application to Proliferation and Cytotoxicity Assays. *J. Immunol. Methods* **1983**, *65*, 55–63. [CrossRef] [PubMed]
35. Filimonov, D.A.; Lagunin, A.A.; Gloriozova, T.A.; Rudik, A.V.; Druzhilovskii, D.S.; Pogodin, P.V.; Poroikov, V.V. Prediction of the Biological Activity Spectra of Organic Compounds Using the Pass Online Web Resource. *Chem. Heterocycl. Compd.* **2014**, *50*, 444–457. [CrossRef]
36. Wishart, D.S.; Feunang, Y.D.; Guo, A.C.; Lo, E.J.; Marcu, A.; Grant, J.R.; Sajed, T.; Johnson, D.; Li, C.; Sayeeda, Z.; et al. DrugBank 5.0: A Major Update to the DrugBank Database for 2018. *Nucleic Acids Res.* **2018**, *46*, D1074–D1082. [CrossRef]
37. Bateman, A.; Martin, M.-J.; Orchard, S.; Magrane, M.; Ahmad, S.; Alpi, E.; Bowler-Barnett, E.; Britto, R.; Bye-A-Jee, H.; Cukura, A.; et al. UniProt: The Universal Protein Knowledgebase in 2023. *Nucleic Acids Res.* **2022**, *51*, D523–D531. [CrossRef]
38. Janson, G.; Paiardini, A. PyMod 3: A complete suite for structural bioinformatics in PyMOL. *Bioinformatics* **2021**, *37*, 1471–1472. [CrossRef]
39. Laskowski, R.A.; Rullmann, J.A.; MacArthur, M.W.; Kaptein, R.; Thornton, J.M. AQUA and PROCHECK-NMR: Programs for Checking the Quality of Protein Structures Solved by NMR. *J. Biomol. NMR* **1996**, *8*, 477–486. [CrossRef]
40. Laskowski, R.A.; MacArthur, M.W.; Moss, D.S.; Thornton, J.M. PROCHECK: A Program to Check the Stereochemical Quality of Protein Structures. *J. Appl. Cryst.* **1993**, *26*, 283–291. [CrossRef]
41. Berendsen, H.J.C.; van der Spoel, D.; van Drunen, R. GROMACS: A Message-Passing Parallel Molecular Dynamics Implementation. *Comput. Phys. Commun.* **1995**, *91*, 43–56. [CrossRef]
42. Vanommeslaeghe, K.; Hatcher, E.; Acharya, C.; Kundu, S.; Zhong, S.; Shim, J.; Darian, E.; Guvench, O.; Lopes, P.; Vorobyov, I.; et al. CHARMM General Force Field: A Force Field for Drug-Like Molecules Compatible with the CHARMM All-Atom Additive Biological Force Fields. *J. Comput. Chem.* **2010**, *31*, 671–690. [CrossRef]
43. Jo, S.; Kim, T.; Iyer, V.G.; Im, W. CHARMM-GUI: A Web-Based Graphical User Interface for CHARMM. *J. Comput. Chem.* **2008**, *29*, 1859–1865. [CrossRef] [PubMed]
44. Kim, S.; Chen, J.; Cheng, T.; Gindulyte, A.; He, J.; He, S.; Li, Q.; Shoemaker, B.A.; Thiessen, P.A.; Yu, B.; et al. PubChem 2019 Update: Improved Access to Chemical Data. *Nucleic Acids Res.* **2019**, *47*, D1102–D1109. [CrossRef] [PubMed]
45. Garullo, G.; Saponara, S.; Ahmed, A.; Gorelli, B.; Mazzotta, S.; Trezza, A.; Gianibbi, B.; Campiani, G.; Fusi, F.; Aiello, F. Novel Labdane Diterpenes-Based Synthetic Derivatives: Identification of a Bifunctional Vasodilator That Inhibits CaV1.2 and Stimulates KCa1.1 Channels. *Mar. Drugs* **2022**, *20*, 515. [CrossRef] [PubMed]
46. Rosignoli, S.; Paiardini, A. DockingPie: A Consensus Docking Plugin for PyMOL. *Bioinformatics* **2022**, *38*, 4233–4234. [CrossRef]
47. Morris, G.M.; Huey, R.; Lindstrom, W.; Sanner, M.F.; Belew, R.K.; Goodsell, D.S.; Olson, A.J. AutoDock4 and AutoDockTools4: Automated Docking with Selective Receptor Flexibility. *J. Comput. Chem.* **2009**, *30*, 2785–2791. [CrossRef]
48. Trott, O.; Olson, A.J. AutoDock Vina: Improving the Speed and Accuracy of Docking with a New Scoring Function, Efficient Optimization, and Multithreading. *J. Comput. Chem.* **2010**, *31*, 455–461. [CrossRef]
49. O'Boyle, N.M.; Banck, M.; James, C.A.; Morley, C.; Vandermeersch, T.; Hutchison, G.R. Open Babel: An Open Chemical Toolbox. *J. Cheminform.* **2011**, *3*, 33. [CrossRef]
50. Salentin, S.; Schreiber, S.; Haupt, V.J.; Adasme, M.F.; Schroeder, M. PLIP: Fully Automated Protein-Ligand Interaction Profiler. *Nucleic Acids Res.* **2015**, *43*, W443–W447. [CrossRef]
51. Thompson, J.D.; Higgins, D.G.; Gibson, T.J. CLUSTAL W: Improving the Sensitivity of Progressive Multiple Sequence Alignment through Sequence Weighting, Position-Specific Gap Penalties and Weight Matrix Choice. *Nucleic Acids Res.* **1994**, *22*, 4673–4680. [CrossRef]
52. Sievers, F.; Wilm, A.; Dineen, D.; Gibson, T.J.; Karplus, K.; Li, W.; Lopez, R.; McWilliam, H.; Remmert, M.; Söding, J.; et al. Fast, Scalable Generation of High-Quality Protein Multiple Sequence Alignments Using Clustal Omega. *Mol. Syst. Biol.* **2011**, *7*, 539. [CrossRef]
53. Rudrapal, M.; Khairnar, S.J.; Khan, J.; Dukhyil, A.B.; Ansari, M.A.; Alomary, M.N.; Alshabrmi, F.M.; Palai, S.; Deb, P.K.; Devi, R. Dietary Polyphenols and Their Role in Oxidative Stress-Induced Human Diseases: Insights into Protective Effects, Antioxidant Potentials and Mechanism(s) of Action. *Front. Pharmacol.* **2022**, *13*, 806470. [CrossRef] [PubMed]
54. Gassel, M.; Breitenlechner, C.B.; Rüger, P.; Jucknischke, U.; Schneider, T.; Huber, R.; Bossemeyer, D.; Engh, R.A. Mutants of Protein Kinase A That Mimic the ATP-Binding Site of Protein Kinase B (AKT). *J. Mol. Biol.* **2003**, *329*, 1021–1034. [CrossRef] [PubMed]
55. Fares, M.; Eldehna, W.M.; Bua, S.; Lanzi, C.; Lucarini, L.; Masini, E.; Peat, T.S.; Abdel-Aziz, H.A.; Nocentini, A.; Keller, P.A.; et al. Discovery of Potent Dual-Tailed Benzenesulfonamide Inhibitors of Human Carbonic Anhydrases Implicated in Glaucoma and in Vivo Profiling of Their Intraocular Pressure-Lowering Action. *J. Med. Chem.* **2020**, *63*, 3317–3326. [CrossRef]
56. Song, F.-L.; Gan, R.-Y.; Zhang, Y.; Xiao, Q.; Kuang, L.; Li, H.-B. Total Phenolic Contents and Antioxidant Capacities of Selected Chinese Medicinal Plants. *Int. J. Mol. Sci.* **2010**, *11*, 2362–2372. [CrossRef]

57. Eliopoulos, C.; Markou, G.; Langousi, I.; Arapoglou, D. Reintegration of Food Industry By-Products: Potential Applications. *Foods* **2022**, *11*, 3743. [CrossRef]
58. Guine, R.; Costa, C.; Florença, S.; Correia, P. A Review of the Use of Chestnut in Traditional and Innovative Food Products. *J. Nuts* **2023**, *1*, 1–18.
59. Raţu, R.N.; Veleşcu, I.D.; Stoica, F.; Usturoi, A.; Arsenoaia, V.N.; Crivei, I.C.; Postolache, A.N.; Lipşa, F.D.; Filipov, F.; Florea, A.M.; et al. Application of Agri-Food By-Products in the Food Industry. *Agriculture* **2023**, *13*, 1559. [CrossRef]
60. Salvador, R.; Barros, M.V.; Donner, M.; Brito, P.; Halog, A.; De Francisco, A.C. How to Advance Regional Circular Bioeconomy Systems? Identifying Barriers, Challenges, Drivers, and Opportunities. *Sustain. Prod. Consum.* **2022**, *32*, 248–269. [CrossRef]
61. Vannacci, G.; Bellini, E. Chestnut (*Castanea sativa*): A Multipurpose European Tree. In Proceedings of the Workshop, Bruxelles, Belgium, 30 September 2010.
62. Conedera, M.; Krebs, P.; Tinner, W.; Pradella, M.; Torriani, D. The Cultivation of *Castanea sativa* (Mill.) in Europe, from Its Origin to Its Diffusion on a Continental Scale. *Veg. Hist. Archaeobot.* **2004**, *13*, 161–179. [CrossRef]
63. Frusciante, L.; Geminiani, M.; Olmastroni, T.; Mastroeni, P.; Trezza, A.; Salvini, L.; Lamponi, S.; Spiga, O.; Santucci, A. Repurposing *Castanea sativa* Spiny Burr By-Products Extract as a Potentially Effective Anti-Inflammatory Agent for Novel Future Biotechnological Applications. *Life* **2024**, *14*, 763. [CrossRef]
64. Pal, P.; Singh, A.K.; Srivastava, R.K.; Rathore, S.S.; Sahoo, U.K.; Subudhi, S.; Sarangi, P.K.; Prus, P. Circular Bioeconomy in Action: Transforming Food Wastes into Renewable Food Resources. *Foods* **2024**, *13*, 3007. [CrossRef] [PubMed]
65. Munteanu, I.G.; Apetrei, C. Analytical Methods Used in Determining Antioxidant Activity: A Review. *Int. J. Mol. Sci.* **2021**, *22*, 3380. [CrossRef] [PubMed]
66. Rumpf, J.; Burger, R.; Schulze, M. Statistical Evaluation of DPPH, ABTS, FRAP, and Folin-Ciocalteu Assays to Assess the Antioxidant Capacity of Lignins. *Int. J. Biol. Macromol.* **2023**, *233*, 123470. [CrossRef] [PubMed]
67. Sethi, S.; Joshi, A.; Arora, B.; Sharma, R. Significance of FRAP, DPPH, and CUPRAC Assays for Antioxidant Activity Determination in Apple Fruit Extracts. *Eur. Food Res. Technol.* **2020**, *246*, 591–598. [CrossRef]
68. Sunjic, S.B.; Gasparovic, A.C.; Jaganjac, M.; Rechberger, G.; Meinitzer, A.; Grune, T.; Kohlwein, S.D.; Mihaljevic, B.; Zarkovic, N. Sensitivity of Osteosarcoma Cells to Concentration-Dependent Bioactivities of Lipid Peroxidation Product 4-Hydroxynonenal Depend on Their Level of Differentiation. *Cells* **2021**, *10*, 269. [CrossRef]
69. Sadowska-Bartosz, I.; Bartosz, G. Effect of Antioxidants on the Fibroblast Replicative Lifespan In Vitro. *Oxid. Med. Cell. Longev.* **2020**, *2020*, 6423783. [CrossRef]
70. Yudoh, K.; van Trieu, N.; Nakamura, H.; Hongo-Masuko, K.; Kato, T.; Nishioka, K. Potential Involvement of Oxidative Stress in Cartilage Senescence and Development of Osteoarthritis: Oxidative Stress Induces Chondrocyte Telomere Instability and Downregulation of Chondrocyte Function. *Arthritis Res. Ther.* **2005**, *7*, R380. [CrossRef]

foods

MDPI

Article

Tackling Food Waste: An Exploratory Case Study on Consumer Behavior in Romania

Cristina-Anca Danciu [1] Alin Croitoru [2], Iuliana Antonie [1], Anca Tulbure [1], Agatha Popescu [3], Cristian Stanciu [4], Camelia Sava [1] and Mirela Stanciu [1,*]

[1] Faculty of Agricultural Sciences, Food Industry and Environmental Protection, Lucian Blaga University of Sibiu, 7-9 Dr. Ion Ratiu Str., 550012 Sibiu, Romania; cristina.danciu@ulbsibiu.ro (C.-A.D.); iuliana.antonie@ulbsibiu.ro (I.A.); anca.tulbure@ulbsibiu.ro (A.T.); camelia.sava@ulbsibiu.ro (C.S.)

[2] Faculty of Social Sciences and Humanities, Lucian Blaga University of Sibiu, 5-7 Victoriei Blvd., 550324 Sibiu, Romania; alin.croitoru@ulbsibiu.ro

[3] Faculty of Management and Rural Development, University of Agronomic Sciences and Veterinary Medicine, 59 Marasti Blvd., District 1, 011464 Bucharest, Romania; agatha_popescu@yahoo.com

[4] Faculty of Sciences, Lucian Blaga University of Sibiu, 5-7 Dr. Ion Rațiu Str., 550012 Sibiu, Romania; cristiandumitru.stanciu@ulbsibiu.ro

* Correspondence: mirela.stanciu@ulbsibiu.ro; Tel.: +40-269211338

Abstract: The scourge of food waste (FW) is a significant global challenge, impacting climate change, food security, and the sustainability of agrifood systems. The objective of this paper is to identify, analyze, and understand the factors influencing household consumer behaviors in Romania regarding the reduction of FW. Three primary research objectives were established to assess food consumption behaviors within households, to explore attitudes toward FW, and to understand the motivations for reducing FW along with the measures implemented by households to address this issue. Methodology: Data were collected through an online self-administered questionnaire, designed to investigate consumer behaviors related to the avoidance of FW. A descriptive statistical analysis was performed, and a linear regression model was developed to evaluate a composite index measuring Romanian consumers' behavior towards FW reduction. Results: The resulting model identifies key predictors that drive concrete actions to minimize FW, including the desire to mitigate the environmental impact, household conversations about FW and strategies to reduce it, established food routines, the influence of one's social circle, individual ecological and social responsibility, and the effectiveness of awareness campaigns addressing the consequences of FW. Practical and social implications: The findings highlight the necessity of education and awareness initiatives to shift attitudes and behaviors concerning FW. Future research is warranted to deepen understanding and enhance interventions. Originality: This study represents a pioneering and innovative inquiry into FW behavior in Romania, filling a gap in the existing literature and contributing to the broader discourse on this pressing environmental issue.

Keywords: food waste; consumer behavior; food waste avoidance; composite index for measuring food waste avoidance; Romania

1. Introduction

Food waste (FW) refers to food discarded by retailers, consumers, or other sectors of the food supply chain. The Food and Agriculture Organization (FAO) estimates that around 14 percent of the world's food (valued at USD 400 billion per year) continues to be wasted after it is harvested and before it reaches the shops. At the same time, UNEP's Food Waste Index Report shows that a further 17% of our food is wasted in retail and by consumers, particularly in households [1] (UNEP Food Waste Index Report 2021 | UNEP—UN Environment Programme). The FAO estimates that the food that is wasted could feed 1.26 billion hungry people every year [2].

Statistical data published by Eurostat conclude that the level of food waste in the EU in 2021 was 58 million tons of fresh food, of which approximately 54% was generated by the household consumer sector. For the same year, the level of FW per capita reached 131 kg. Misniakiewicz et al., (2024) show that around 65% of the food waste generated in Europe comes from five countries [3]. These include Germany (10,922,321 tons), France (8,764,999 tons), Italy (8,291,265 tons), Poland (4,281,212 tons), and Spain (4,260,845 tons). At the opposite pole are Malta (79,589 tons) and Luxembourg (83,622 tons) [4]. A report published by EUFIC states that losses and waste are more carbon-intensive for processed products that encompass many additional resources accumulated along the supply chain. The carbon footprint of FW differs according to the category of food that generates food waste, with a direct impact on the environment [5].

According to the FUSIONS (Food Use for Social Innovation by Optimizing Waste Prevention Strategies), a project with 21 partners from 13 EU countries, an estimated 100 Mtons of food waste is produced each year in the EU and the FW which can be avoided represents an average economic cost of EUR 595 per household per year [6]..

FW waste can occur at various stages of the supply chain due to factors such as inefficiencies in production, transportation, storage, distribution, consumer behavior, cosmetic standards, and expiration dates [7–11]. FW contributes to environmental degradation through greenhouse gas emissions from decomposing organic matter in landfills, deforestation, water wastage, and energy consumption associated with production and transportation [12,13]. From a social and economic point of view, FW represents a missed opportunity to alleviate hunger and food insecurity, especially in regions where food access is limited. It also translates into economic losses for producers, processors, transporters, retailers, and consumers [14–16].

The forecasts related to the numerical growth of the population throughout the world show that humanity will reach approximately 9.8 billion people in 2050. Numerous authors claim that halving FW in the agri-food chain by the year 2030 will meet the current and future food needs of humanity, by the United Nations' Sustainable Development Goal 12.3 [17,18]. This reduction can only be achieved under the joint efforts of all interested stakeholders [19]. The methods used to estimate the level of food waste must be relevant, representative, and reliable [20].

At a time of rising global hunger and surging food prices, various countries have developed national strategies and action plans to address the FW issue [2,21]. Also, organizations and businesses are implementing multiple strategies to tackle FW [22,23]. These include improving storage and transportation infrastructure, enhancing supply chain efficiency through planning, promoting consumer awareness and behavior change, redistributing surplus food to those in need, preventing FW through feeding animals, and developing innovative technologies and digital apps [24] and tools [25] for food preservation and waste reduction [26–31].

In households, FW is closely related to dietary behavior acquired throughout life [17] and depends on motivational factors, skills, and opportunities as well as current personal food management practices [32,33]. FW in households is produced in three stages, starting from purchasing food, storage, and consumption [34]. All these are closely related to the category of food [35], the purchasing place [36], the production system of food, and eating habits [37]. Household food consumption habits, frequency of cooking and consumption of home-prepared meals, preferences for different places of food purchase, type of diet, and preference for food produced in different farming systems (conventional, traditional, or organic), are an integral part of a responsible food behavior with direct effects on the protection of resources, the avoidance of FW, and the minimization of the negative impact on the environment.

Consumers play a crucial role in reducing FW by making mindful purchasing decisions by buying only the necessities, properly storing and using food items, minimizing overeating, and supporting initiatives that redistribute surplus food to vulnerable populations [38–41].

Key aspects of the research on FW in Romania

Removing the scarcity of studies regarding FW in Romania is an important objective, considering the FW's negative impact on the economy, the environment, and food security. In Romania, FW research is mostly coordinated by governmental institutions, non-governmental organizations, and academic research institutions. According to the report published in 2024 by UNEP regarding the Food Waste Index, Romania is the only member state that did not report data in this regard to Eurostat [42].

National-wide studies: These are carried out by the National Institute of Statistics (INS) or the Ministry of Agriculture and Rural Development. They aim to assess the extent and characteristics of food waste along the entire agri-food chain [43].

Sector analyses: Research often focuses on specific sectors, such as agricultural production, food processing, distribution, and retail, to identify vulnerabilities and opportunities to reduce FW [44].

Multiple data sources: Researchers use a variety of data sources, including official statistics, sector organization reports, market studies, and field surveys to gain a comprehensive picture of the FW phenomenon [45–47].

Economic and social impact assessment: Studies analyze the impact of food waste on the national economy, including financial costs for producers, processors, and consumers [48,49], as well as the social implications for food security and social inclusion [50,51].

Identifying effective solutions and interventions: Research proposes strategies and policies to reduce food waste, including legislative measures, technological innovations [52], education and awareness programs [53], and collaborative initiatives between different stakeholders [54,55]. Some authors show that to reduce food waste, the understanding of the social value of religion and environmental behavior [56] must also be taken into account [57]. Other authors emphasize the need to develop digital technologies that encourage sustainable eating habits, efficiency, and sustainability of the food system [58], and contribute to a national culture of avoiding FW [31].

Monitoring and evaluation: After interventions and policies are implemented, research continues to monitor and evaluate progress in reducing food waste and identify new challenges and opportunities [59].

The analysis of the Romanian consumer's behavior, regarding FW, highlights the main causes: There has been insufficient attention related to the shelf life of food, improper storage [60], behavioral differences due to gender [61], residence [62], and income [63]. Other causes that have been reported are a lack of sustainable food strategies [64], responsible food behavior, different moral attitudes towards FW, the perception of self-control over one's behavior [65], and the purchase of excessive quantities because of retail discounts and promotions [66]. A recently published study identified five clusters of food consumption patterns in Romanians closely related to FW [67].

This study complements the specialized literature on Romanian consumer behavior about avoiding FW. The main aims are to identify, analyze, and understand the factors that influence consumer behavior regarding the avoidance of FW, to inform policymakers, guide investment, and mobilize action to reduce FW in Romania.

The article is structured as follows: Section 1: *Introduction*, where the problem addressed is presented and reference is made to the state of knowledge on this subject internationally and also in the Romanian context; Section 2: *Materials and Methods*, in which the study design, research objectives, data about the study sample, and the methods used for statistical analysis are presented; and Section 3. *Results and discussions*, which presents the results of the descriptive statistics for both the dependent and independent variables concerning FW avoidance behavior. A composite index has been developed to facilitate an objective measurement of FW avoidance behavior among consumers. The validity of this composite index was evaluated through a stepwise linear regression analysis, which aims to elucidate the impact of various determinants on the propensity to avoid FW in Romania. Discussions also include a comparative analysis with results documented by researchers in other countries. This comparison serves to highlight the nuances of FW avoidance behavior across different cultural and socio-economic settings, thereby enriching the comprehension of the multifaceted elements influencing consumer attitudes toward FW; Section 4: *Conclusions*.

2. Materials and Methods

2.1. Study Design

The article is based on survey-type research in which an online self-administered questionnaire was used, employing a non-probability sample from Romania. The data were collected in 2024, and the total sample size was 369 consumers. They were informed about the purpose of the study and data protection, and by completing the questionnaire they agreed to participate. The questionnaire was distributed online to the authors' knowledge network and students, and through them, to people responsible for purchasing and cooking food in their families. The study participants were 18 to over 65 years old. The thematic questionnaire was specifically constructed for this research and was based both on the specialized literature and on original research ideas designed to explore these topics in a national context where there is not yet a tradition of FW research. Most of the questions included a Lickert-type scale for registering responses and the questionnaire collected standard socio-demographic information. For data processing, we used the IBM SPSS 26 software package.

The general and specific objectives of the study are represented in Figure 1.

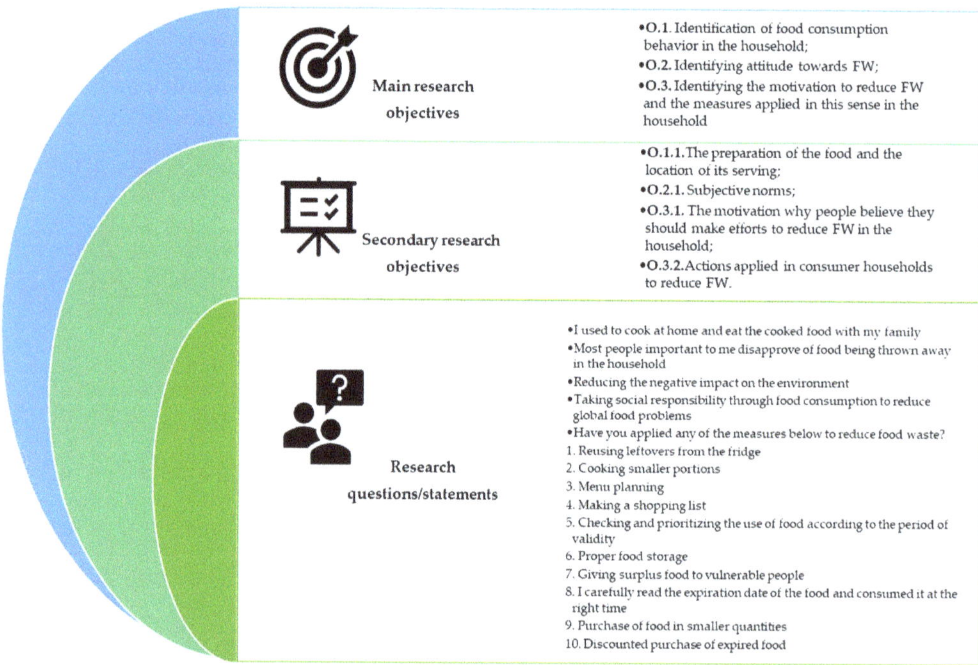

Figure 1. Study design. Source: Developed by the authors.

2.2. Participants

The questionnaire was completed by 376 people residing in 27 counties, out of the 41 counties and the municipality of Bucharest existing in Romania. Of these, only 369 people sent complete answers. Out of the total number of respondents, 221 people (59.89%) lived in Sibiu County. Other socio-demographic data of the sample are shown in Tables 1 and 2.

Table 1. Data of respondents.

Specification		Frequency	%	Specification		Frequency	%
Gender	Male	121	32.79	Age groups (years)	18–20	99	26.8
	Female	248	67.21		21–30	140	37.9
	Total	369	100		31–40	42	11.4
Domicile	Urban	253	68.56		41–50	56	15.2
	Rural	116	31.44		Above 50	32	8.7
	Total	369	100		Total	369	100

Source: Authors 'own calculation.

Table 2. Level of education, income, and number of people in the respondents' household.

Specification		Frequency	%	Specification		Frequency	%
The highest level of education completed	Education level below tertiary education	199	53.9%	Monthly expenditure of the family for the purchase of food *	Under 1000 lei	86	23.30
	Tertiary education level completed	170	46.1%		1001–2000 lei	182	49.32
					2001–3000 lei	75	20.33
					Above 3001 lei	26	7.05
					Total	369	100
	Total	369	100	Persons in Household	1	15	4.06
Family income *	Under 2500 lei	15	4.06		2	85	23.04
	2501–5000 lei	91	24.67		3	93	25.20
	5001–7500 lei	100	27.10		4	111	30.08
	7501–10,000 lei	85	23.04		5	25	6.78
	Above 10,001 lei	78	21.13		Above 6	40	10.84
	Total	369	100		Total	369	100

* 1 euro = 4.98 lei on 1 June 2024. Source: Authors' own calculation.

From the above table analysis, it is shown that approximately 67% of the respondents were female, approximately 69% lived in urban areas, and approximately 76% of them were young, aged between 18 and 40. Regarding the level of education of the respondents, more than half of them were high school graduates (53.9%), and 46.01% had completed university studies (bachelor's, master's, and PhD). In general, as expected, the household structure included 3–4 people (55.28%). More than half of the respondents' families (50.14%) had an average monthly income between 5001–10,000 lei (1000–2000 EUR). For 49.32% of the respondents, the monthly expenses for purchasing food necessary for the household were between 1001 and 2000 lei (200–400 EUR).

2.3. Statistical Analysis

The dataset allowed us to build a composite index for assessing self-declared behaviors for food waste avoidance at the individual level. This index type obtains an overall representation of a concept or construct that cannot be directly measured by a single item. The procedure of the summative index is a composite measure created by summing the individual scores of multiple items further described in this paper's section of analyses. The composite index was also tested by employing a factor analysis, a statistical technique to uncover latent structures underlying the relationships among a set of observed variables. The primary goal of factor analysis is to reduce the dimensionality of data and identify clusters of correlated variables that can be represented by a smaller number of underlying factors.

While we employed descriptive analyses to describe and summarize the characteristics of the variables included in the analyses (mean and standard deviation), we further designed a stepwise linear regression model to test the relationships between the variance of the summative index (dependent variable) and a set of predictors explained in the section of analyses. Even though the fact that the sample is not representative of the entire population of Romania and does not allow us to generalize the conclusions, the stepwise linear regression model enabled us to test the type and the significance of the relationships between certain variables at the sample level and to draw meaningful conclusions about FW waste avoidance.

3. Results and Discussions

3.1. Descriptive Analyses

3.1.1. Dependent Variable

The dependent variable was a composite index reflecting individual self-assessment regarding ten FW avoidance behaviors (Table 3). For each of these, study participants were asked to rate on a scale from 1 to 7 the frequency with which they engaged in these behaviors (where 1 represents never, and 7 always). To maintain the relevance and interpretability of the index, we opted to calculate the individual-level mean with a formula as follows: sum of individual values from i1 to i10 divided by 10 [Index = Σ (i1: i10) ÷ 10]. At the sample level, concerning the 10 indicators introduced in the analysis (Table 3), we can say that the most frequent FW avoidance behaviors were "Proper food storage" (M = 5.58) and "Reading food expiration dates carefully and consuming them on time" (M = 5.18). In contrast, the least frequent behaviors were related to "Donating excess food to vulnerable individuals" (M = 3.89). All other measured behaviors fell between these extremes of the scale and registered mean values in the range between 4 and 5.

Table 3. List of indicators included in the composite index for measuring FW avoidance (ranked by the mean value recorded at the sample level).

Behaviors	Mean	Std. Dev.	Variance	Coefficient of Variation (%)
i1 Proper food storage	5.58	1.49	2.21	26.71
i2 Reading food expiration dates carefully and consuming them on time	5.18	1.61	2.60	31.05
i3 Checking and prioritizing the use of food based on expiration dates	4.96	1.65	2.72	33.24
i4 Making a shopping list	4.72	1.76	3.08	37.34
i5 Cooking smaller portions	4.69	1.47	2.15	31.34
i6 Purchasing food in smaller quantities	4.66	1.54	2.35	32.98
i7 Reusing leftovers from the fridge	4.20	1.82	3.30	43.33
i8 Meal planning	4.18	1.68	2.81	40.18
i9 Purchasing discounted food nearing its expiration date	4.15	1.85	3.42	44.58
i10 Donating excess food to vulnerable individuals	3.89	1.72	2.97	44.19

Source: Authors' own elaboration.

The calculated mean scores, the standard deviation, the variance, and the coefficient of variation resulting from individual self-reports on ten food waste avoidance behaviors are shown below.

The coefficient of variation had values between 26.71% and 44.58%, below 35% in the case of 5 self-declared behaviors regarding the avoidance of FW. To verify if the indicators could be grouped into a one-dimensional index, we conducted a factor analysis with principal component extraction (Figure 2). This showed factors clustering on one dimension, and the Kaiser–Meyer–Olkin Measure of Sampling Adequacy (KMO) coefficient was 0.905 (the detailed results are presented in Supplementary Materials File S1). At the same time, the statistical analysis indicated excellent reliability of the instrument used, suggesting that the items were consistent and that the constructed assessment had an adequate foundation (Cronbach's Alpha coefficient of 0.88).

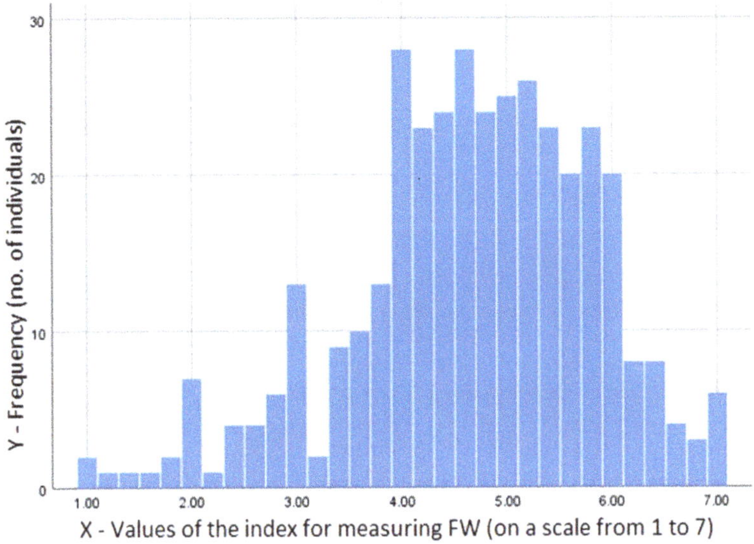

Figure 2. Histogram—Frequency (DV: The composite index for measuring food waste avoidance). Mean = 4.62; Std. Dev. 1.154; N = 369; (Authors' own representation).

3.1.2. Independent Variables

To understand the variation in FW avoidance behaviors, we considered it useful to analyze the influence that a set of social values, consumption behaviors, and individual characteristics may have. In this regard, we selected the variables in Table 4 and some descriptive statistics.

Table 4. List of independent variables.

Indicators	Mean (M)	Std. Dev. (SD)	Min	Max
The individual usually cooks at home (scale 1 to 5)	4.30	0.86	1	5
Reducing the negative environmental impact is an important motivation for making an effort to reduce food waste (scale 1 to 5)	3.93	0.97	1	5
Taking social responsibility through food consumption to alleviate global food issues is an important motivation for making an effort to reduce food waste (scale 1 to 5)	3.97	1.01	1	5
Most of the important people in individual's life disapprove of food being wasted (scale 1 to 5)	2.82	1.30	1	5
The individuals had discussions in their household about food waste and ways to reduce it (yes = 1)	0.83	0.37	0	1
The individuals heard about information campaigns or news that raise awareness about food waste and offer tips for reducing it (yes = 1)	0.62	0.49	0	1
Gender (male)	0.33	0.47	0	1
Age (number of years)	30.22	12.91	18	73

Source: Authors' own elaboration.

The descriptive statistics revealed participants' behaviors and attitudes towards FW reduction. The results showed a strong inclination towards cooking at home, as indicated by a high mean score of 4.30 with an SD of 0.859, signifying consistency. Motivations such as environmental impact (M = 3.93, SD = 0.969) and social responsibility (M = 3.97, SD = 1.013) were somewhat lower but still significant, exhibiting varying opinions among respondents. Participants noted perceived disapproval of FW by important individuals (M = 2.82, SD = 1.304), reflecting diverse views. Household discussions on FW (M = 0.83, SD = 0.374) and awareness of information campaigns (M = 0.62, SD = 0.487) were reported, with varying levels of exposure among respondents. The sample included more female respondents and the mean for males was 0.33 and SD of 0.470. The average age was 30.22 years (SD = 12.909), indicating diversity in age ranges. These findings provide insights into the sample specificity although the sample does not represent the entire population.

3.2. Linear Regression Model

Firstly, the adjusted R square values (Table 5) increased from 0.184 to 0.270 with the development of the model and this indicates that the final model (M6) explains approximately 27% of the variance in the behavior of avoiding food waste. This means that there are other factors not accounted for in the model that may also influence the behavior, but our results provide meaningful explanations regarding individuals' behaviors in this register. From the total of IV included in the model, six factors were in statistically significant relationships with the dependent variable (DV). At the same time, gender, age, education, number of persons in the household, and household earnings were excluded from the list of variables and had no statistically significant influence within this model (their coefficients are presented in Supplementary Materials File S2). Looking at the coefficients of each IV maintained in the analysis, we can see how each variable relates to the (DV), the composite index for measuring food waste avoidance. A recently published study shows that the number of household members influences the amount of waste produced in the household, while other variables such as age, education level, or family income do not significantly influence the level of FW [68]. Other studies show that demographic features such as age and gender do not considerably influence food waste avoidance within the studied sample. The specialized literature reports the influence of consumers' socio-demographic characteristics on sustainable food behavior [69], while food waste avoidance behavior does not depend on these characteristics [70].

Reducing the negative environmental impact (IV1) was the first and, in relative terms, the most important predictor for the index of FW avoidance within the model. It had a strong statistically significant relationship with the DV ($p < 0.001$) and a standardized coefficient 0.304. In a regression model that included multiple predictors, the fact that reducing the negative environmental impact was the most important factor indicates a very strong connection between individuals' concerns for environmental conservation and the adjustment of their own consumption behaviors in this direction.

The second predictor included in the statistical model was related to household-level discussions about food waste and ways to reduce it (IV2). The predictor is in a positive and strong statistically significant relationship with the DV (** $p < 0.01$) and its standardized coefficient of 0.157 indicates that people who approached this issue in discussions within their household have higher values in terms of behaviors of FW waste avoidance.

This indicates that FW avoidance behaviors emerge as a result of a family environment in which the individual becomes aware of this issue and feels the need to take a stance by engaging in conversations. Encouraging discussions at the household level seems to be a direction worth considering, and in the longterm, it can have a significant spillover effect through the socialization of new generations in this spirit.

Table 5. Results of the stepwise linear regression model: DV—The composite index for measuring food waste avoidance.

Models \ Predictors	Standardized Coefficients (Beta) M1	M2	M3	M4	M5	M6
IV1. Reducing the negative environmental impact is an important motivation for making an effort to reduce food waste	0.432 ***	0.416 ***	0.414 ***	0.425 ***	0.302 ***	0.304 ***
IV2. The individuals had discussions in their household about food waste and ways to reduce it		0.198 ***	0.184 ***	0.182 ***	0.171 ***	0.157 **
IV3. The individual usually cooks at home			0.138 **	0.138 **	0.129 **	0.122 **
IV4. Most of the important people in an individual's life disapprove of food being wasted				−0.123 **	-0.125 **	-0.126 **
IV5. Taking social responsibility through food consumption to alleviate global food issues is an important motivation for making an effort to reduce food waste					0.169 *	0.170 *
IV6. The individuals heard about information campaigns or news that raise awareness about food waste and offer tips for reducing it						0.102 *
Adjusted R Square	0.184	0.221	0.238	0.251	0.262	0.270

Notes: * $p < 0.05$; ** $p < 0.01$; *** $p < 0.001$ (please see Supplementary Materials File S2 for the complete results of the stepwise linear regression analysis). Source: Authors' own elaboration.

The individuals who usually cook at home (IV3) had a standardized coefficient of 0.122, indicating that this behavior correlated to food waste avoidance (p-value < 0.01). This result is important because it indicates how cooking at home empowers individuals and contributes to food waste reduction in Romanian society. The obtained result agrees with the one highlighted by Bobeica et al., (2024), which shows that routine and individual behavior in the household is important for reducing FW [71].

The fourth predictor included in the model was linked to the individual's social circle. Within our sample, those who have important people in their lives disapproving of food waste (IV4) registered a negative standardized coefficient of −0.126 and the relationship was statistically significant ($p < 0.01$). Food waste avoidance behaviors have a negative relationship with the social circle in which individuals live. This can indicate that those who exhibit such behaviors perceive themselves more as exceptions about important people in their lives. This result may seem counterintuitive, but it shows that in Romania, there is not yet a culture of food waste avoidance. This may be the result, as other authors show, that most people expect actual measures to avoid food waste to be taken on different levels [71].

In a similar fashion to reducing the negative environmental impact, people motivated to shape their behaviors by taking social responsibility through food consumption to alleviate global food issues (IV5) registered higher values of the FW index. It was a positive coefficient of 0.170 and the relationship was statistically significant at p-value < 0.05. This suggests that individuals who are environmentally and socially responsible are more inclined to exhibit behaviors of avoiding food waste.

Last but not least, individuals' awareness of information campaigns about food waste (IV6) had a standardized coefficient of 0.102, which means that individuals who were exposed to information about food waste were more likely to exhibit behaviors of avoiding food waste. This relationship was statistically significant with a p-value < 0.05. This shows that such communication programs are effective, and it is important for public authorities to make efforts to communicate on this issue or to support social initiatives in this direction.

For a significant impact on the environment and in the area of social responsibility related to FW reduction, a comprehensive approach is needed, which integrates all stages of the food supply chain, consumer education, effective collaboration between various civil society organizations and the private sector [72], implementation of innovative and sustainable production and consumption practices [35,60,73–75], sustainable energy production [76], expansion of national food bank branches [77], fueling innovations (technological, organizational, or marketing) [78], application of blockchain technology [79], the conversion of agri-food waste into innovative bioproducts [80,81], and the use of food waste in animal feed [82]. As Biggi et al. (2024) show, corporate social responsibility and legislative measures can have a synergistic effect on environmental management and FW reduction [83]. However, the need to create a societal culture to avoid and reduce FW was also reported internationally [84].

Environmental and social responsibility exhibit stronger positive associations with food waste avoidance, emphasizing the impact of individuals' values on their waste reduction behaviors. Household discussions about food waste and exposure to information campaigns enhance food waste avoidance tendencies.

An overview on the variables that do not significantly influence the DV can be also informative. At the sample level, there were no statistically significant differences between men and women regarding the reporting of certain behaviors associated with FW. Similarly, with respect to age groups, the findings were not markedly different, using the reference category of the youngest individuals in the sample (20 years or below). Perhaps more surprisingly, the completion of tertiary education programs did not significantly influence FW behaviors. This may be attributed to the fact that such subjects are not covered in the university curriculum in Romania, and extracurricular programs of this nature aimed at students are lacking. The number of individuals in the household did not statistically impact the value of the FW index. Furthermore, income levels did not have a statistically significant relationship with the FW index. The lack of statistical significance in some relationships within this context may be influenced by the nuances already captured by the previously presented attitudinal-value model. The absence of a clear demographic and socio-economic profile regarding FW may also be relevant because, in Romania, there has not yet been sufficient time for such behaviors to crystallize solely within certain segments of the population. It is more likely that multiple directions of influence exist at both generational and socio-economic levels. For instance, some behaviors may still bear the imprint of the significant food shortages experienced during the Communist period of the country or the consumerism developed during the transition to a market economy since 1989. Nevertheless, in the absence of representative national studies, our research opens important niches for further investigation.

The issue of FW is complex and requires multiple and timely interventions along the food supply chain [85,86]. In the final stage of the chain, information campaigns and the need to educate household consumers, especially young people regarding the need to avoid and reduce FW is highlighted in numerous research papers [3,46,87]. Food literacy is an important determinant of dietary behavior [86]. Education in this direction must be carried out in the context of the three pillars of sustainability (social, economic, and environmental), within each generation and between generations, emphasizing an understanding of the resources involved in food production [87].

4. Conclusions

The present paper is an exploratory study on the behavior of domestic consumers in Romania regarding FW. The results have determined a composite index that measures concrete actions to avoid FW. The determined composite index assesses self-reported food waste avoidance behaviors at the individual level. A concept or construct that cannot be directly measured was thus represented. The model was tested using the stepwise linear regression model. According to the resulting model, the main predictors that determine concrete actions to avoid FW are the desire to reduce the negative impact on the environment; discussions about FW held at the household level and ways to reduce it; household food routines, especially regarding cooking at home; social circle; ecological and social responsibility of individuals; and awareness campaigns regarding the environmental, social, and economic impact of FW, carried out in all stages of the agri-food chain.

The social values and personal beliefs formed through an appropriate education for each generation can be used to promote a culture of avoiding food waste, which must be created and consolidated in Romania. A national culture of avoiding FW and sustainable FW management is essential in sustainable development, the current need for action is to mitigate the effects of climate change and ensure food safety and security.

Informing and educating conscious and responsible generations is the key to tackling this problem. Education and awareness efforts with the support of all stakeholders, followed by FW monitoring, are essential to changing attitudes and behaviors regarding this issue. Public awareness campaigns, school programs, and community initiatives can help promote a culture of less wasteful food consumption. In this context, the recommendations for politicians and government institutions refer to the need for education and awareness efforts, which are essential to change attitudes and behaviors regarding FW in Romania.

The research has some limitations related to the number of respondents. The findings offer a comprehensive image of the current FW behavior based on an online survey and a self-administered questionnaire, which does not necessarily represent the entire population of Romania.

Further studies are necessary regarding the Romanian consumer behavior related to FW, to obtain a general picture of this behavior at the national level. The research will be extended to the other links of the food supply chain (producers, processors, traders, and HoReCa).

Supplementary Materials: The following supporting information can be downloaded at: https://www.mdpi.com/article/10.3390/foods13203313/s1, File S1: Factor analysis-FW index, File S2: Stepwise regression model.

Author Contributions: Conceptualization, M.S. and C.-A.D.; methodology, M.S., C.-A.D. and I.A.; software, A.C.; validation, A.P., C.S. (Camelia Sava) and M.S.; formal analysis, A.C.; investigation, A.T. and C.S. (Cristian Stanciu); resources, M.S. and I.A.; data curation, A.P.; writing—original draft preparation, C.-A.D. and M.S.; writing—review and editing, M.S. and C.S. (Cristian Stanciu); visualization, C.S. (Cristian Stanciu); supervision, A.P. and C.S. (Camelia Sava); project administration, M.S.; funding acquisition, M.S. All authors have read and agreed to the published version of the manuscript.

Funding: The project was financed by Lucian Blaga University of Sibiu through the research grant LBUS-IRG-2022-08.

Institutional Review Board Statement: Not applicable.

Informed Consent Statement: Not applicable.

Data Availability Statement: The original contributions presented in the study are included in the article/supplementary material, further inquiries can be directed to the corresponding author.

Acknowledgments: The authors express their gratitude to all individuals who took the time to respond to the questionnaire developed in this research study. This research was conducted as part of a national project ADER 17.1.1 "Research on the evolution of the phenomenon of food waste at the national level, from the perspective of Romania's commitment to achieving the objective of sustainable development", funded by the Ministry of Agriculture and Rural Development. The authors thank the "Lucian Blaga" University in Sibiu for support and funding.

Conflicts of Interest: The authors declare no conflicts of interest. The funders had no role in the study's design; in the collection, analyses, or interpretation of data; in the writing of the manuscript; or in the decision to publish the results.

References

1. UNEP Food Waste Index Report 2021 | UNEP—UN Environment Programme. Available online: https://www.unep.org/resources/report/unep-food-waste-index-report-2021 (accessed on 25 March 2024).
2. FAO; IFAD; UNICEF; WFP; WHO. The State of Food Security and Nutrition in the World 2022. Repurposing Food and Agricultural Policies to Make Healthy Diets More Affordable. Available online: https://openknowledge.fao.org/handle/20.500.14283/cc0639en (accessed on 25 March 2024).
3. Miśniakiewicz, M.; Amicarelli, V.; Chrobak, G.; Górka-Chowaniec, A.; Bux, C. Do Living Arrangements and Eating Habits Influence University Students' Food Waste Perception in Italy and Poland? *Sustainability* **2024**, *16*, 2102. [CrossRef]
4. *Food Waste and Food Waste Prevention-Estimates*; Eurostat—Statistics Explained; European Commission: Brussels, Belgium, 2023.
5. *Food Waste in Europe: Statistics and Facts about the Problem*; Eufic: Brussels, Belgium, 2021.
6. FUSIONS, EU. "Estimates of European Food Waste Levels." Διαθέσιμο Στο: 2016. Available online: http://www.eufusions.org/phocadownload/publications/estimates%20of%20european%20food%20waste2 (accessed on 25 March 2024).
7. Rösler, F.; Kreyenschmidt, J.; Ritter, G. Recommendation of Good Practice in the Food-Processing Industry for Preventing and Handling Food Loss and Waste. *Sustainability* **2021**, *13*, 9569. [CrossRef]
8. Parfitt, J.; Croker, T.; Brockhaus, A. Global Food Loss and Waste in Primary Production: A Reassessment of Its Scale and Significance. *Sustainability* **2021**, *13*, 12087. [CrossRef]
9. Karimi, N. Assessing Global Waste Management: Alternatives to Landfilling in Different Waste Streams—A Scoping Review. *Sustainability* **2023**, *15*, 13290. [CrossRef]
10. Caldeira, C.; De Laurentiis, V.; Corrado, S.; van Holsteijn, F.; Sala, S. Quantification of Food Waste per Product Group along the Food Supply Chain in the European Union: A Mass Flow Analysis. *Resour. Conserv. Recycl.* **2019**, *149*, 479–488. [CrossRef]
11. Corrado, S.; Sala, S. Food Waste Accounting along Global and European Food Supply Chains: State of the Art and Outlook. *Waste Manag.* **2018**, *79*, 120–131. [CrossRef]
12. Liu, Z.; Rousta, K.; Bolton, K.; Brancoli, P. Waste Quantification and Carbon Footprint of Street Market Food Waste in Brazil. *Cogent Food Agric.* **2024**, *10*, 2305520. [CrossRef]
13. Zhao, P.; Zhang, S.; Wang, X.; Sun, H.; Guo, Y.; Wang, Q.; Sun, X. Research Progress on Anaerobic Digestion of Cellulose Waste Based on Bibliometric Analysis. *Sustainability* **2023**, *15*, 16060. [CrossRef]
14. Herzberg, R.; Schneider, F.; Banse, M. Policy Instruments to Reduce Food Loss before Retail—Perspectives of Fruit and Vegetable Supply Chain Actors in Europe. *Waste Manag.* **2023**, *170*, 354–365. [CrossRef] [PubMed]
15. Garcia-Herrero, I.; Hoehn, D.; Margallo, M.; Laso, J.; Bala, A.; Batlle-Bayer, L.; Fullana, P.; Vazquez-Rowe, I.; Gonzalez, M.J.; Durá, M.J.; et al. On Estimating Potential Food Waste Reduction to Support Sustainable Production and Consumption Policies. *Food Policy* **2018**, *80*, 24–38. [CrossRef]
16. Gruber, V.; Holweg, C.; Teller, C. What a Waste! Exploring the Human Reality of Food Waste from the Store Manager's Perspective. *J. Public Policy Mark.* **2016**, *35*, 3–25. [CrossRef]
17. Deliberador, L.R.; Batalha, M.O.; da Silva César, A.; Masood Azeem, M.; Lane, J.L.; Rodrigues Silva Carrijo, P. Why Do We Waste so Much Food? Understanding Household Food Waste through a Theoretical Framework. *J. Clean. Prod.* **2023**, *419*, 137974. [CrossRef]
18. Pocol, C.B.; Amuza, A.; Moldovan, M.G.; Stanca, L.; Dabija, D.C. Consequences of Recent Crises on the FLW Consumer Behaviour: A National Wide Representative Research—The Case of Romania. *Foods* **2023**, *12*, 1973. [CrossRef] [PubMed]
19. Amicarelli, V.; Aluculesei, A.C.; Lagioia, G.; Pamfilie, R.; Bux, C. How to Manage and Minimize Food Waste in the Hotel Industry: An Exploratory Research. *Int. J. Cult. Tour. Hosp. Res.* **2022**, *16*, 152–167. [CrossRef]
20. Vittuari, M.; García, H.L.; Caldeira, C.; Bruns, H.; Sala, S. *Literature Review on Drivers of Consumer Food Waste and Levers for Behavioural Change*; The Evolution of the Literature on Consumer Food Waste Prevention: Luxembourg, 2023.
21. Garske, B.; Heyl, K.; Ekardt, F.; Weber, L.M.; Gradzka, W. Challenges of Food Waste Governance: An Assessment of European Legislation on Food Waste and Recommendations for Improvement by Economic Instruments. *Land* **2020**, *9*, 231. [CrossRef]
22. Zaharia, A.; Diaconeasa, M.C.; Maehle, N.; Szolnoki, G.; Capitello, R. Developing Sustainable Food Systems in Europe: National Policies and Stakeholder Perspectives in a Four-Country Analysis. *Int. J. Environ. Res. Public Health* **2021**, *18*, 7701. [CrossRef]
23. Mastorakis, G.; Kopanakis, I.; Makridis, J.; Chroni, C.; Synani, K.; Lasaridi, K.; Abeliotis, K.; Louloudakis, I.; Daliakopoulos, I.N.; Manios, T. Managing Household Food Waste with the FoodSaveShare Mobile Application. *Sustainability* **2024**, *16*, 2800. [CrossRef]
24. Economou, F.; Chatziparaskeva, G.; Papamichael, I.; Loizia, P.; Voukkali, I.; Navarro-Pedreño, J.; Klontza, E.; Lekkas, D.F.; Naddeo, V.; Zorpas, A.A. The Concept of Food Waste and Food Loss Prevention and Measuring Tools. *Waste Manag. Res.* **2024**, *42*, 651–669. [CrossRef]
25. Dumitru, O.M.; Iorga, C.S.; Mustatea, G. Food Waste along the Food Chain in Romania: An Impact Analysis. *Foods* **2021**, *10*, 2280. [CrossRef]
26. Dumitru, O.M.; Iorga, S.C.; Sanmartin, Á.M. Food Waste Impact on Romanian Households. *Rom. Biotechnol. Lett.* **2021**, *26*, 2207–2213. [CrossRef]
27. Campoy-Muñoz, P.; Cardenete, M.A.; Delgado, M.C. Economic Impact Assessment of Food Waste Reduction on European Countries through Social Accounting Matrices. *Resour. Conserv. Recycl.* **2017**, *122*, 202–209. [CrossRef]

28. Grosu, R.M. "Green" Practices in the Food Retail Sector: Evidence from the Romanian Market. *Br. Food J.* **2024**, *126*, 173–190. [CrossRef]
29. van Geffen, L.; van Herpen, E.; van Trijp, H.; Quested, T.; Diaz-Ruiz, R.; Bos-Brouwers, H.; Bajzelj, B.; Ujhelyi, K. *Causes & Determinants of Consumers Food Waste. A Theoretical Framework*; European Commission: Brussels, Belgium, 2016.
30. Moldovan, M.-G.; Dabija, D.-C.; Pocol, C.B. Innovative Strategies for Food Waste Reduction and the Use of Mobile Applications in the Agri-Food Sector. *Sci. Pap. Ser. Manag. Econ. Eng. Agric. Rural Dev.* **2024**, *24*, 675–688.
31. van Geffen, L.; van Herpen, E.; Sijtsema, S.; van Trijp, H. Food Waste as the Consequence of Competing Motivations, Lack of Opportunities, and Insufficient Abilities. *Resour. Conserv. Recycl. X* **2020**, *5*, 100026. [CrossRef]
32. van Rooijen, M.A.; Gerdessen, J.C.; Claassen, G.D.H.; de Leeuw, S.L.J.M. Optimizing Household Food Waste: The Impact of Meal Planning, Package Sizes, and Performance Indicators. *Resour. Conserv. Recycl.* **2024**, *205*, 107559. [CrossRef]
33. Szymkowiak, A.; Borusiak, B.; Pierański, B.; Kotyza, P.; Smutka, L. Household Food Waste: The Meaning of Product's Attributes and Food-Related Lifestyle. *Front. Environ. Sci.* **2022**, *10*, 918485. [CrossRef]
34. Gorgan, C.; Chersan, I.C.; Dragomir, V.D.; Dumitru, M. Food Waste Prevention Solutions in the Annual Reports of European Companies. *Amfiteatru Econ.* **2022**, *24*, 309. [CrossRef]
35. Vermeir, I.; Petrescu, D.C.; Petrescu-Mag, R.M. What Are the 'Shape Friendly' Locations to Sell Misshapen Tomatoes? The Effect of Point of Purchase on Consumers' Abnormality Perception and Probability to Buy. *Food Qual. Prefer.* **2023**, *106*, 104809. [CrossRef]
36. Loebnitz, N.; Grunert, K.G. The Impact of Abnormally Shaped Vegetables on Consumers' Risk Perception. *Food Qual. Prefer.* **2018**, *63*, 80–87. [CrossRef]
37. Chinie, C.; Biclesanu, I.; Bellini, F. The Impact of Awareness Campaigns on Combating the Food Wasting Behavior of Consumers. *Sustainability* **2021**, *13*, 11423. [CrossRef]
38. Ribeiro, A.; Madureira, L.; Carvalho, R. Citizens' Deliberation on Solutions to Fight Urban Household Food Waste and Nexus with Growing Urban Gardens: The Case of Porto Metropolitan Area in Portugal. *Clean. Responsible Consum.* **2024**, *13*, 100188. [CrossRef]
39. Pocol, C.B.; Pinoteau, M.; Amuza, A.; Burlea-Schiopoiu, A.; Glogovețan, A.I. Food Waste Behavior among Romanian Consumers: A Cluster Analysis. *Sustainability* **2020**, *12*, 9708. [CrossRef]
40. Gaiani, S.; Caldeira, S.; Adorno, V.; Segrè, A.; Vittuari, M. Food Wasters: Profiling Consumers' Attitude to Waste Food in Italy. *Waste Manag.* **2018**, *72*, 17–24. [CrossRef]
41. UNEP. *Food Waste Index Report 2024 Think Eat Save Tracking Progress to Halve Global Food Waste*; UNEP: Nairobi, Kenya, 2024.
42. Firoiu, D.; Ionescu, G.H.; Bandoi, A.; Florea, N.M.; Jianu, E. Achieving Sustainable Development Goals (SDG): Implementation of the 2030 Agenda in Romania. *Sustainability* **2019**, *11*, 2156. [CrossRef]
43. Danciu, C.A.; Tulbure, A.; Stanciu, M.A.; Antonie, I.; Capatana, C.; Zerbeș, M.V.; Giurea, R.; Rada, E.C. Overview of the Sustainable Valorization of Using Waste and By-Products in Grain Processing. *Foods* **2023**, *12*, 3770. [CrossRef] [PubMed]
44. Bucea-Manea-Țoniș, R.; Martins, O.M.D.; Ilic, D.; Belous, M.; Bucea-Manea-țoniș, R.; Braicu, C.; Simion, V.E. Green and Sustainable Public Procurement—An Instrument for Nudging Consumer Behavior. A Case Study on Romanian Green Public Agriculture across Different Sectors of Activity. *Sustainability* **2021**, *13*, 12. [CrossRef]
45. Nijloveanu, D.; Tița, V.; Bold, N.; Fîntîneru, G.; Smedescu, D.; Chiurciu, I.-A.; Smedescu, C.; Gheorghe Pătrăchioiu, N. Dynamics of Food Loss and Waste Causes along the Agri-Food Chain in Romania. *Sci. Pap. Ser. Manag. Econ. Eng. Agric. Rural. Dev.* **2023**, *23*, 569–579.
46. Pocol, C.B.; Marinescu, V.; Amuza, A.; Cadar, R.L.; Rodideal, A.A. Sustainable vs. Unsustainable Food Consumption Behaviour: A Study among Students from Romania, Bulgaria, and Moldova. *Sustainability* **2020**, *12*, 4699. [CrossRef]
47. Voinea, L.; Vrânceanu, D.M.; Filip, A.; Popescu, D.V.; Negrea, T.M.; Dina, R. Research on Food Behavior in Romania from the Perspective of Supporting Healthy Eating Habits. *Sustainability* **2019**, *11*, 5255. [CrossRef]
48. Voinea, L.; Popescu, D.V.; Bucur, M.; Negrea, T.M.; Dina, R.; Enache, C. Reshaping the Traditional Pattern of Food Consumption in Romania through the Integration of Sustainable Diet Principles. A Qualitative Study. *Sustainability* **2020**, *12*, 5826. [CrossRef]
49. Muresan, I.C.; Harun, R.; Andreica, I.; Chiciudean, G.O.; Kovacs, E.; Oroian, C.F.; Brata, A.M.; Dumitras, D.E. Household Attitudes and Behavior towards the Food Waste Generation before and during the COVID-19 Pandemic in Romania. *Agronomy* **2022**, *12*, 746. [CrossRef]
50. Oroian, C.; Mureșan, C.; Oroian, I.; Burduhos, P. Romanian Consumers Behavior towards Domestic Food Waste. *Sci. Pap. Ser. A Agron.* **2021**, *64*, 724–731.
51. Ferdeș, M.; Zăbavă, B.Ș.; Paraschiv, G.; Ionescu, M.; Dincă, M.N.; Moiceanu, G. Food Waste Management for Biogas Production in the Context of Sustainable Development. *Energies* **2022**, *15*, 6268. [CrossRef]
52. Chereji, A.I.; Chiurciu, I.A.; Popa, A.; Chereji, I.; Iorga, A.M. Consumer Behaviour Regarding Food Waste in Romania, Rural versus Urban. *Agronomy* **2023**, *13*, 571. [CrossRef]
53. Archip, B.C.; Banatean-Dunea, I.; Petrescu, D.C.; Petrescu-Mag, R.M. Determinants of Food Waste in Cluj-Napoca (Romania): A Community-Based System Dynamics Approach. *Int. J. Environ. Res. Public Health* **2023**, *20*, 2140. [CrossRef]
54. Stanciu, M.; Popescu, A.; Antonie, I.; Sava, C.; Nistoreanu, B.G. Good Practices on Reducing Food Waste Throughout the Food Supply Chain. *Amfiteatru Econ.* **2022**, *24*, 566–582. [CrossRef]
55. Trasca, T.I.; Ocnean, M.; Gherman, R.; Lile, R.A.; Balan, I.M.; Brad, I.; Tulcan, C.; Firu Negoescu, G.A. Synergy between the Waste of Natural Resources and Food Waste Related to Meat Consumption in Romania. *Agriculture* **2024**, *14*, 644. [CrossRef]

56. Baran, T.; Lupu, C.; Privitera, D. Faith and Sustainability: Exploring Religiosity's Impact on Intentions to Reduce Food Waste. *Sustainability* **2024**, *16*, 4852. [CrossRef]
57. Bocean, C.G. A Longitudinal Analysis of the Impact of Digital Technologies on Sustainable Food Production and Consumption in the European Union. *Foods* **2024**, *13*, 1281. [CrossRef]
58. Petrariu, R.; Constantin, M.; Dinu, M.; Pătărlăgeanu, S.R.; Deaconu, M.E. Water, Energy, Food, Waste Nexus: Between Synergy and Trade-Offs in Romania Based on Entrepreneurship and Economic Performance. *Energies* **2021**, *14*, 5172. [CrossRef]
59. Ghinea, C.; Ghiuta, O.A. Household Food Waste Generation: Young Consumers Behaviour, Habits and Attitudes. *Int. J. Environ. Sci. Technol.* **2019**, *16*, 2185–2200. [CrossRef]
60. Cantaragiu, R. The Impact of Gender on Food Waste at the Consumer Level. *Stud. Univ. Vasile Goldis Arad Econ. Ser.* **2019**, *29*, 41–57. [CrossRef]
61. Costuleanu, C.L.; Brezuleanu, S.; Ignat, G.; Boldureanu, G.; Cristian Toma, M.; Gladiola Andruseac, G. Relationships of Food Packaging Waste Processing and Costs with Environmental Protection in Iasi County. *Rev. Chim.* **2016**, *67*, 1990–1993.
62. Costuleanu, C.L.; Vintu, R.; Robu, A.; Ignat, G.; Brezuleanu, S. Food Waste Processing Relationships with Environment Protection. *Rev. Chim.* **2015**, *66*, 743–745.
63. Moldovan, M.G.; Dabija, D.C.; Stanca, L.; Pocol, C.B. A Qualitative Study on the Consumer Behaviour Related to Food Waste: Romanian Perspectives through Word Cloud and Sentiment Analysis. *Sustainability* **2024**, *16*, 4193. [CrossRef]
64. Stefan, V.; van Herpen, E.; Tudoran, A.A.; Lähteenmäki, L. Avoiding Food Waste by Romanian Consumers: The Importance of Planning and Shopping Routines. *Food Qual. Prefer.* **2013**, *28*, 375–381. [CrossRef]
65. Chivu, R.G.; Stoica, I.; Orzan, G.; Radu, A.V. Discounts or Happy Money for Efficient Inventory Management? A New Trend on Romanian Markets in Changing the Consumer Behaviour. In *New Trends and Issues Proceedings on Humanities and Social Sciences, Proceedings of the 2th International Congress on Nursing, Rotterdam, The Netherlands, 26–29 August 2018*; Brooklyn Research and Publishing Institute: Brooklyn, NY, USA, 2018; Volume 5, pp. 82–88.
66. Nijloveanu, D.; Tița, V.; Bold, N.; Smedescu, D.; Fîntîneru, A.; Tudor, V.; Smedescu, C.; Jerca, E. A Statistical Study Analysis on Exploring Consumption Patterns Regarding Food Loss and Waste. *Sci. Pap. Ser. Manag. Econ. Eng. Agric. Rural Dev.* **2024**, *24*, 685–694.
67. Peronti, B.; Di Veroli, J.N.; Scognamiglio, U.; Baiamonte, I.; Stefanovic, L.; Bügel, S.G.; Matthiessen, L.E.; Aboussaleh, Y.; Belfakira, C.; Średnicka-Tober, D.; et al. Household Food Waste in Five Territories in Europe and Northern Africa: Evaluation of Differences and Similarities as Implication for Actions. *J. Clean. Prod.* **2024**, *452*, 142086. [CrossRef]
68. Muresan, I.C.; Harun, R.; Arion, F.H.; Brata, A.M.; Chereches, I.A.; Chiciudean, G.O.; Dumitras, D.E.; Oroian, C.F.; Tirpe, O.P. Consumers' Attitude towards Sustainable Food Consumption during the COVID-19 Pandemic in Romania. *Agriculture* **2021**, *11*, 1050. [CrossRef]
69. Aschemann-Witzel, J.; de Hooge, I.; Amani, P.; Bech-Larsen, T.; Oostindjer, M. Consumer-Related Food Waste: Causes and Potential for Action. *Sustainability* **2015**, *7*, 6457–6477. [CrossRef]
70. Bobeică, M.; Arghiroiu, G.A.; Beciu, S. Exploratory Analysis of Food Waste Causes in Romanian Households. *Cienc. Rural* **2024**, *54*, 20230247. [CrossRef]
71. Petrescu-Mag, R.M.; Petrescu, D.C.; Robinson, G.M. Adopting Temperance-Oriented Behavior? New Possibilities for Consumers and Their Food Waste. *J. Agric. Environ. Ethics* **2019**, *32*, 5–26. [CrossRef]
72. Cantaragiu, R. Corporate Social Entrepreneurship Initiatives against Food Waste—The Case of Lidl in Romania. In Proceedings of the International Conference on Business Excellence; Walter de Gruyter GmbH: Berlin, Germany, 2019; Volume 13, pp. 505–514.
73. Mihai, F.C.; Grozavu, A. Role of Waste Collection Efficiency in Providing a Cleaner Rural Environment. *Sustainability* **2019**, *11*, 6855. [CrossRef]
74. Afriyie, E.; Gatzweiler, F.; Zurek, M.; Asem, F.E.; Ahiakpa, J.K.; Okpattah, B.; Aidoo, E.K.; Zhu, Y.G. Determinants of Household-Level Food Storage Practices and Outcomes on Food Safety and Security in Accra, Ghana. *Foods* **2022**, *11*, 3266. [CrossRef]
75. Cioabla, A.E.; Popescu, F.; Trif-Tordai, G.; Dumitrel, A.G.; Oprisa-Stanescu, P.D.; Lelea, D. Sustainable Development of Romanian Cities through Biogas Production from Municipal Wastes and Application in Co-Combustion Processes. *Therm. Sci.* **2018**, *22*, 1071–1076. [CrossRef]
76. Dumitru, I.; Burghiu, A.G. Romanian Food Waste Analysis. In *New Trends in Sustainable Business and Consumption, Proceedings of the BASIQ International Conference, Bari, Italy, 30 May–1 June 2019*; Pamfilie, R., Dinu, V., Tachiciu, L., Plesea, D., Vasiliu, C., Eds.; ASE Publishing: Bucharest, Romania, 2019; pp. 441–447.
77. Aramyan, L.; Grainger, M.; Logatcheva, K.; Piras, S.; Setti, M.; Stewart, G.; Vittuari, M. Food Waste Reduction in Supply Chains through Innovations: A Review. *Meas. Bus. Excell.* **2021**, *25*, 475–492. [CrossRef]
78. Monica, M.; Marin, I.; Vidu, L.; Marin, M.-P. Learning about the Reduction of Food Waste Using Blockchain Technology. *arXiv* **2021**, arXiv:2101.02026.
79. Vodnar, D.C.; Călinoiu, L.F.; Dulf, F.V.; Ştefănescu, B.E.; Crişan, G.; Socaciu, C. Identification of the Bioactive Compounds and Antioxidant, Antimutagenic and Antimicrobial Activities of Thermally Processed Agro-Industrial Waste. *Food Chem.* **2017**, *231*, 131–140. [CrossRef]
80. Rashwan, A.K.; Bai, H.; Osman, A.I.; Eltohamy, K.M.; Chen, Z.; Younis, H.A.; Al-Fatesh, A.; Rooney, D.W.; Yap, P.S. Recycling Food and Agriculture By-Products to Mitigate Climate Change: A Review. *Environ. Chem. Lett.* **2023**, *21*, 3351–3375. [CrossRef]

81. Wang, Y.; Rassler, S.; Stefanovski, D.; Bender, J.; Deutsch, J.; Chen, T.; Cui, Z.; Dou, Z. Evidence of Animal Productivity Outcomes When Fed Diets Including Food Waste: A Systematic Review of Global Primary Data. *Resour. Conserv. Recycl.* **2024**, *203*, 107411. [CrossRef]
82. Biggi, G.; Principato, L.; Castellacci, F. Food Waste Reduction, Corporate Responsibility and National Policies: Evidence from Europe. *Br. Food J.* **2024**, *126*, 470–485. [CrossRef]
83. Griffin, M.; Sobal, J.; Lyson, T.A. An Analysis of a Community Food Waste Stream. *Agric. Hum. Values* **2009**, *26*, 67–81. [CrossRef]
84. Canali, M.; Amani, P.; Aramyan, L.; Gheoldus, M.; Moates, G.; Östergren, K.; Silvennoinen, K.; Waldron, K.; Vittuari, M. Food Waste Drivers in Europe, from Identification to Possible Interventions. *Sustainability* **2017**, *9*, 37. [CrossRef]
85. Petrescu-Mag, R.M.; Petrescu, D.C.; Ajtai, I.; Roba, C.A.; Gica, O.A.; Cuibus, L.; Tenter, A.; Toader, V.; Negrusa, A.L.; Bican-Brișan, N. Causes and Solutions for Fruit and Vegetable Waste: A Participatory Approach with Romanian Farmers for Sustainable Agriculture. *Int. J. Agric. Sustain.* **2024**, *22*, 2329391. [CrossRef]
86. Amicarelli, V.; Tricase, C.; Spada, A.; Bux, C. Households' Food Waste Behavior at Local Scale: A Cluster Analysis after the COVID-19 Lockdown. *Sustainability* **2021**, *13*, 3283. [CrossRef]
87. Silva, S.S.; Pereira, A.C.; Marques, B.; Liz Martins, M. Inadequacy of Meals Served and Food Waste in a Portuguese University Canteen. *Sustainability* **2024**, *16*, 4317. [CrossRef]

foods

MDPI

Article

Solid-State Fermentation of Cereal Waste Improves the Bioavailability and Yield of Bacterial Cellulose Production by a *Novacetimonas* sp. Isolate

Shriya Henry [1], Sushil Dhital [2], Huseyin Sumer [1] and Vito Butardo, Jr. [1,*]

[1] Department of Chemistry and Biotechnology, Swinburne University of Technology, Hawthorn, Melbourne, VIC 3122, Australia; shriyahenry@swin.edu.au (S.H.); hsumer@swin.edu.au (H.S.)
[2] Department of Chemical and Biological Engineering, Faculty of Engineering, Monash University, Clayton, VIC 3800, Australia; sushil.dhital@monash.edu
* Correspondence: vbutardo@swin.edu.au

Abstract: Cereal wastes such as rice bran and cereal dust are valuable yet underutilised by-products of grain processing. This study aimed to bio-convert these wastes into bacterial cellulose (BC), an emerging sustainable and renewable biomaterial, via an inexpensive solid-state fermentation (SSF) pre-treatment using three mould isolates. Medium substitution by directly using untreated rice bran or cereal dust did not significantly increase the yield of bacterial cellulose produced by *Novacetimonas* sp. (NCBI accession number PP421219) compared to the standard Hestrin–Schramm (HS) medium. In contrast, rice bran fermented with *Rhizopus oligosporus* yielded the highest bacterial cellulose (1.55 ± 0.6 g/L dry weight) compared to the untreated control (0.45 ± 0.1 g/L dry weight), demonstrating an up to 22% increase in yield. Using the SSF process, the media production costs were reduced by up to 90% compared to the standard HS medium. Physicochemical characterisation using SEM, EDS, FTIR, XPS, XRD, and TGA was performed to gain insights into the internal structure, morphology, and chemical bonding of differently produced BC, which revealed comparable biopolymer properties between BC produced in standard and waste-based media. Hence, our findings demonstrate the effectiveness of fungal SSF for transforming abundant cereal waste into BC, providing a circular economy solution to reduce waste and convert it into by-products to enhance the sustainability of the cereal industry.

Keywords: bacterial cellulose; biopolymer; solid-state fermentation; rice bran; cereal dust; cereal waste; *Novacetimonas*

1. Introduction

Bacterial cellulose (BC) is an emerging biopolymer with unique properties such as a high purity, crystallinity, and mechanical strength, making it superior to plant-derived cellulose for various applications [1,2]. Various genera within acetic acid bacteria (AAB), including *Gluconacetobacter*, *Acetobacter*, and *Komagateibacter*, are known for their ability to produce cellulose. *Acetobacter*, commonly found in vinegar, is a heterotrophic bacteria capable of converting glucose, glycerol, sugars, and other organic compounds into cellulose [3]. Although several bacteria that generate bacterial cellulose have been identified, much research has concentrated on model organisms like *Gluconacetobacter xylinus*. This species is particularly notable for its ability to utilise a wide variety of carbohydrates, making it the most important cellulose-producing bacteria [4].

BC and its derivatives are in high demand due to their extensive commercial applications across various industries, such as biomedical, pharmaceutical, cosmetics, packaging, electronics, and food. The global market for BC is projected to expand at a compound annual growth rate (CAGR) of 12.6% from 2019 to 2027, potentially reaching USD 771.76 million. With an estimated annual production of 1012 tons of this biopolymer, the strong market demand ensures a consistent supply of renewable and biodegradable BC raw materials [5]. However, the scaling of industrial BC manufacturing remains restricted by the high costs of conventional culture media [2]. Solid-state fermentation (SSF) has emerged as a promising strategy to enhance the nutritional potential of carbohydrate-rich waste streams, offering advantages such as a reduced reactor volume, lower energy requirements, and simplified downstream processing [6].

Rice bran and cereal dust, abundant waste of grain processing, represent promising yet underutilised substrates for BC production. Annually, rice bran production exceeds 29 million tonnes globally, while cereal dust constitutes approximately 2–3% of grain weight during milling [7,8]. These materials are rich in complex carbohydrates, proteins, and essential nutrients, potentially supporting microbial growth and metabolism [9–11]. Previous studies have demonstrated the feasibility of using rice bran as an alternative nitrogen source for BC production, with yields increasing up to 2.4-fold compared to the standard medium [8]. Substrates such as cereal grains have been explored, but there is a noticeable gap in the existing literature regarding the use of cereal dust specifically to produce bacterial cellulose. Limited research has been conducted in this area, making this study particularly novel and impactful. However, the fibrous architecture of these residues can limit nutrient accessibility without preprocessing, and to address that, SSF was used as an approach.

SSF has emerged as an optimal bioprocessing strategy to enhance the nutritional and functional potential of carbohydrate-rich waste streams from cereal milling and rice production [12]. SSF offers advantages such as being suitable for tiny effluent amounts and simple substrates, being cost-effective, and having no extra considerations for standard procedure control. This technique is usually applied to agro-industrial waste resources [13]. SSF utilises filamentous fungi perfectly adapted to solid substrates in nature to break down complex-matrix polysaccharides through secreted enzyme cocktails into readily assimilable sugars, proteins, and metabolites. *Aspergillus niger*, *A. oryzae*, *A. parasiticus*, *Acremonium crysogenum*, and *Rhizopus oryzae* are some of the microorganisms which have been explored for the SSF of rice bran, wheat bran, and other agricultural wastes and by-products [13–18]. The literature confirms that SSF significantly increased the protein accessibility of rice bran by unlocking nutrients entrapped within fibrillar networks using natural enzymatic biocatalysts [19].

SSF is particularly advantageous for bioprocessing agro-industrial waste streams, as it operates with a low moisture content (typically < 70%), making it well-suited for materials with low water activity, such as lignocellulosic biomass. The technique offers several economic and sustainability benefits, including a reduced reactor volume, lower energy requirements for aeration and mixing, and simplified downstream processing compared to submerged fermentation. SSF also allows for the direct use of unfractionated solid substrates, minimising the need for costly pre-treatment steps. Moreover, the low water activity in SSF helps prevent bacterial contamination, allowing for a semi-sterile process without the need for strict aseptic conditions. These advantages have made SSF an attractive bioprocessing strategy for the valorisation of agro-industrial wastes into value-added products like enzymes, organic acids, and biofuels [12,13,17,20].

To address the challenge of producing BC sustainably and at a reduced cost, this study aimed to develop an SSF-based bioprocess for valorising rice bran and cereal dust into BC using three fungal strains: *Rhizopus oryzae*, *Pleurotus ostreatus*, and *Rhizopus oligosporus*. The specific objectives were to (a) assess the impact of SSF pre-treatment on nutrient bioavailability and BC production potential, (b) compare the yield and physicochemical properties of BC produced from untreated and SSF-treated waste streams, and (c) characterise the resulting BC materials using a suite of analytical techniques such as SEM, EDS, FTIR, XRD, and TGA. Additionally, we conducted a preliminary cost analysis to assess the economic feasibility of our approach. By addressing these objectives, this study provides a proof-of-concept for the sustainable valorisation of agro-food wastes into high-value BC biomaterials, potentially paving the way for cost-effective and eco-friendly BC production processes (Figure 1).

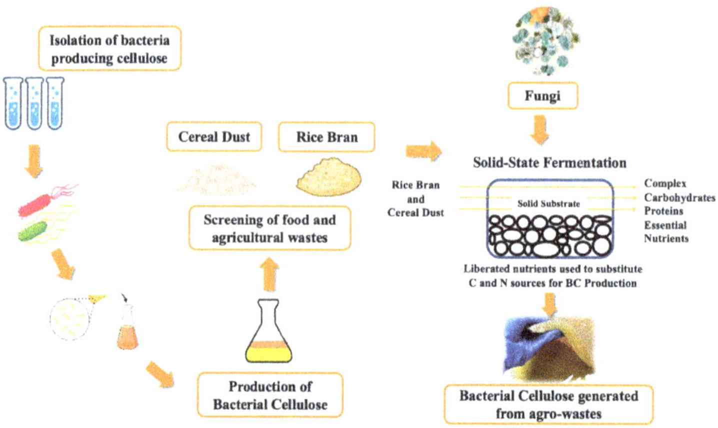

Figure 1. Summarised representation of bacterial cellulose production from solid-state fermented cereal wastes [21].

2. Materials and Methods

2.1. Microorganisms and Preparation of Inoculum

A novel *Novacetimonas* sp., an acetic acid bacterium isolated from apple cider vinegar, was used in this study [21]. The isolation process involved screening multiple fermented liquid samples by subculturing them in Hestrin–Schramm (HS) medium to test for consistent bacterial cellulose pellicle formation. This isolate from apple cider vinegar demonstrated stable BC production in repeated subcultures. An initial characterisation through biochemical testing revealed that this isolate belongs to the acetic acid bacteria group. The genomic DNA extraction was carried out using an in-house buffer solution containing Tris-HCl, EDTA, and SDS. Genomic DNA samples were amplified using 27F: 5′-AGAGTTTGATCCTGGCTCAG-3′ and 1492R: 5′-TACGGTTACCTTGTTACGACTT-3′ 16S rRNA universal primers [22]. Further identification was performed using 16S rRNA gene sequencing, which identified that the isolate belongs to the *Novacetimonas* genus with a 99% similarity to known sequences. The 16S rRNA sequence was submitted to NCBI with the accession number PP421219. Gram staining and electron microscopy illustrated the rod-shaped morphology characteristic of this species [21]. The growth profile analysis exhibited an extended exponential phase, faster cell growth in agitated cultures, and higher BC production in static cultures [21]. This isolate represents a promising candidate for BC production due to its robust growth and cellulose-synthesising capabilities. Comparative genomic analyses are currently ongoing to identify this novel isolate down to the species level and to characterise its BC biosynthetic and secretion pathways.

The isolate was grown in HS culture medium for two weeks at 28 °C in a Thermoline Refrigerated Incubator (Model Number TRISLH-495-1-SD, Sydney, Australia) by the static cultivation method to prepare the inoculum. After the incubation period, the flask used to prepare the inoculum was shaken vigorously each time to let the bacterial cells detach from the BC pellicle and mix with the HS inoculate culture medium. This cell suspension, adjusted to 0.1 OD, served as the inoculum. A 10% (*v*/*v*) inoculum concentration was used for all experiments.

2.2. Waste Materials Used for the Reformulation of Bacterial Cellulose Media

Stabilised rice bran was sourced from SunRice, Yanco, New South Wales, Australia. The bran stabilisation process, which involves heat treatment to deactivate lipase enzymes while preserving nutritional value, uses a proprietary and confidential commercial method. Cereal dust was obtained from Rex James Stockfeed Ltd., Nathalia, Victoria, Australia. Proximate analyses of both the rice bran and cereal dust substrates were conducted by certified commercial testing laboratories to determine their carbohydrate and protein content for media reformulation purposes. The results of these analyses are presented in Supplementary Table S1. This approach ensures the accuracy and reliability of the compositional data used in our study.

2.3. Reformulation of HS Culture Medium Using Agro Wastes for BC Production

Based on a proximate analysis, the standard HS bacterial cellulose production medium was reformulated by substituting key carbon and nitrogen sources. The reformulated media composition is presented in Supplementary Table S2.

2.4. Solid-State Fermentation of Rice Bran (RB) and Cereal Dust (CD)

The solid-state fermentation of rice bran and cereal dust was carried out using three fungal species: *Rhizopus oryzae* (RO) from NATURITAS, Barcelona, Spain; *Pleurotus osteratus* (PO) from Urban Spores, Queensland, Australia; and *Rhizopus oligosporus* (T) from Nourish Me Organics, Victoria, Australia. The pre-treatment of RB and CD using SSF was carried out in 5 L Erlenmeyer flasks for a better surface area and controlled aeration during the treatment. For each substrate, 200 g of material was evenly distributed at the bottom of the flask. The samples were moistened with distilled water prior to sterilisation. Each flask was inoculated with a fungal spore suspension containing 4.0×10^6 spores medium, which was prepared using a haemocytometer [23]. The flasks were covered with sterile gauze to allow ventilation and incubated at 25 °C for 14 days in a Thermoline Refrigerated Incubator (Model Number—TRISLH-495-1-SD). After incubation, the fermented samples were sterilised at 121 °C for 16 min to decontaminate the fungal spores. The treated substrates were then oven-dried (LABEC) at 60 °C for 48 h until moisture-free, ground into a fine powder, and used as culture media components for BC production [13].

2.5. BC Production from Treated and Untreated Waste Substrates

Small-scale and large-scale experiments compared BC production using treated and untreated rice bran (RB) and cereal dust (CD) against standard HS medium. For small-scale trials, 50 mL tubes contained waste media formulations substituting HS carbon and nitrogen sources. SSF-treated and untreated substrates were used in equal amounts. The media were sterilised, adjusted to pH 6.2 (Mettler Toledo SevenEasy S20 pH Meter, Australia), and inoculated with *Novacetomonas* sp. (0.1 OD). The cultures were incubated statically at 28 °C for 14 days. BC pellicles were harvested, purified, and weighed after oven-drying and lyophilisation. On the other hand, large-scale production used 1L Schott bottles to increase the surface area, following the same protocol. This approach was based on studies showing BC yield is proportional to vessel surface area [24,25]. Inoculum size, pH, temperature, incubation period, and processing methods remained consistent with the small-scale trials.

2.6. Characterisation of BC Produced from SSF-Treated Substrates

Analytical techniques such as Fourier-transform infrared spectroscopy (FTIR), scanning electron microscopy (SEM), energy-dispersive X-ray spectroscopy (EDS), X-ray diffraction (XRD), thermogravimetric analysis (TGA), and derivative thermogravimetry (DTG) were used to study and evaluate the comparable properties of BC produced from SSF-treated waste media. The physico-chemical characterisations of these BC films were compared to the BC films generated from HS medium taken as a control, nata de coco as a commercially available BC control, and untreated BC samples (RB and CD). The comparable properties of these BC films generated from different waste media formulations were also studied based on two drying techniques: oven-drying (OD) in a hot air oven at 60 °C for 8 h and freeze-drying (FD) by lyophilising the samples at −80 °C for 6 h and moving them to freeze dryer (LABCONCO FreeZone 4.5 Plus, Australia) at a condenser temperature of −120 °C and vacuum pressure of 0.5 mbar until the weight remained constant. Following are the methods used for the preparation and analysis of BC films for each characterisation performed.

2.7. Scanning Electron Microscopy (SEM) and Energy-Dispersive X-ray Spectroscopy (EDS)

A field-emission scanning electron microscope (Zeiss Supra 40 VP, Jena, Germany) was used to observe the structural morphology of the ultrafine fibre network. BC pellicles obtained after the oven and freeze-drying were cut into small pieces and coated with a thin layer of gold onto the sample by a sputter coating device for 15 s, followed by the attachment of double-sided conducting carbon tape to the slide for better conductivity. The coated samples were viewed at different magnifications. Energy-dispersive X-ray spectroscopy (EDS) (Oxford Inca Energy 200, Thornleigh, Australia) was also performed simultaneously on the samples coated for SEM analysis.

2.8. Fourier-Transformed Infrared Spectroscopy (FTIR)

The infrared spectra of the BC were recorded using FTIR-VERTEX 70, Burker, Australia. BC samples were prepared as films to analyse, performing over the range of 400–4000 cm^{-1} with a resolution of 16 cm^{-1} and averaged over 200 scans. The spectrum line was baseline-corrected and optimised on the generated IR spectra using Quasar 1.7.0 and Origin 2023 software to compare and study the spectra of BC film samples from different reformulated media and drying methods.

2.9. X-ray Diffraction (XRD)

The bacterial cellulose crystal structure was analysed using an X-ray diffractometer (Bruker D8 Advance, Bruker, Australia) 5–80, 0.02, 0.5 s, 0 rpm, 10 × 10 mm variable slit, automatic air scatter (32 min). The oven- and freeze-dried samples of BC were prepared as thin and flat layers using glass slides and mounted on the sample stage. The samples were scanned at the rate of 0.5 s per step. The X-ray diffractogram was analysed and processed using Origin 2023 Peak Analyser Software.

The crystallinity index (CI) was calculated from the obtained diffractogram using the following equation [26]:

$$\text{Crystallinity Index (CI)} = (I_{002} - I_{AM}/I_{002}) \times 100$$

The I_{002} represents the highest intensity of the 2θ peak, which is to say around 22.5° corresponds to the crystalline region, and I_{AM} represents the intensity of the peak situated between the (110) and (200) peaks, that is, 18° corresponds to the amorphous region.

2.10. Thermogravimetric Analysis (TGA)

The analysis of the phase transitions and decomposition behaviours of the BC films were performed using a Mettler Toledo TGA analyser, Melbourne, Australia. Around 20–30 mg of finely cut dried BC samples were used. The sample's weight was continuously monitored while subjected to a controlled temperature increase from 30 °C to 800 °C.

3. Results

This study demonstrated the successful production of bacterial cellulose (BC) from rice bran and cereal dust pre-treated by SSF using *R. oryzae*, *P. osteratus*, and *R. oligosporus* (Figure 1). SSF pre-treatment enhanced the BC yield compared to untreated substrates, with rice bran fermented by *R. oligosporus* producing the highest yield. The BC produced from SSF-treated substrates showed a comparable nanostructure to that produced in standard HS medium and commercial nata de coco. Both the choice of fungal strain for SSF and the drying method (oven drying vs. freeze drying) significantly influenced the BC nanostructure, fibre diameter, and morphology. The physicochemical properties of BC, including crystallinity, thermal stability, and surface chemistry, were affected by the substrate type and processing conditions. Lastly, a cost analysis revealed that using SSF-treated agro-food wastes could reduce BC media costs by up to 90% compared to the standard HS medium. These results highlight the potential of SSF-treated agro-food wastes as sustainable and cost-effective substrates for BC production. The following subsections will discuss these findings in detail, including the characterisation of BC properties through various analytical techniques.

3.1. Solid-State Fermentation of Rice Bran and Cereal Dust

After 14 days of SSF, extensive fungal growth was observed on all samples (Supplementary Figure S1), with all three strains (*R. oryzae*, *P. osteratus*, and *R. oligosporus*) effectively colonising the rice bran (RB) and cereal dust (CD) substrates. These samples were autoclaved and decontaminated before harvesting the treated substrates. After the decontamination step, the samples were oven-dried and ground into a powder to mimic the substitutes for C and N sources in the HS medium. Visual inspection indicated successful bioconversion of the substrates by the fungal strains, as evidenced by the visible mycelial growth and changes in substrate texture (Supplementary Figure S1). The RB samples exhibited a more compact and darker appearance post-fermentation, while the CD samples showed a lighter, more friable texture. These observations suggest that SSF effectively altered the physicochemical properties of the substrates, improving nutrient accessibility for subsequent bacterial cellulose production. The use of different fungal strains resulted in subtle variations in the fermented substrates, which may influence their efficacy as culture media components.

The SSF process was completed without the addition of harsh chemicals or energy-intensive treatments, relying solely on the native enzyme systems of the fungal strains to modify the substrates. This approach aligns with green chemistry principles and potentially reduces processing costs. Other pre-treatments, such as enzymatic hydrolysis and alkali treatment, have also been investigated to improve the nutritional quality of rice bran. However, these methods are more expensive and often involve the use of harsh chemicals, making them less favourable compared to SSF [27].

3.2. Media Substitution Using SSF-Treated Rice Bran and Cereal Dust

Substituting the carbon and nitrogen sources in the HS medium with SSF-treated rice bran (RB) and cereal dust (CD) significantly enhanced bacterial cellulose (BC) yields compared to the untreated substrates. This was demonstrated through both small-scale and large-scale experiments. Small-scale screening revealed that rice bran fermented with *Rhizopus oryzae* (RB-RO) yielded the highest BC (0.5 g/L), matching the conventional HS medium and surpassing untreated RB (Supplementary Figure S2). This improvement can be attributed to the action of fungal lignocellulolytic enzymes and proteases, which increased the bioavailability of nutrients. Large-scale experiments in 1L Schott bottles confirmed these findings, with rice bran fermented with *Rhizopus oligosporus* (RB-T) yielding the highest BC (1.5 ± 0.6 g/L dry weight), surpassing even the HS medium (Figure 2). This scale-up demonstrated the potential for industrial-scale production using SSF-treated agricultural wastes.

The superior performance in larger vessels aligns with previous studies reporting the influence of surface area to volume ratio on BC yield [25,28–30]. The wet weight BC yields obtained are comparable to or higher than those reported for other pre-treated agro-wastes, such as rice bark (2.4 g/L), oat hulls (2.2 g/L), wheat straw (8.3 g/L), and potato peel waste (4.7 g/L) [10,31,32]. These findings underscore the promise of combining SSF with the biosynthetic capabilities of *Novacetomonas* sp. for cost-effective and sustainable BC production from abundant waste streams.

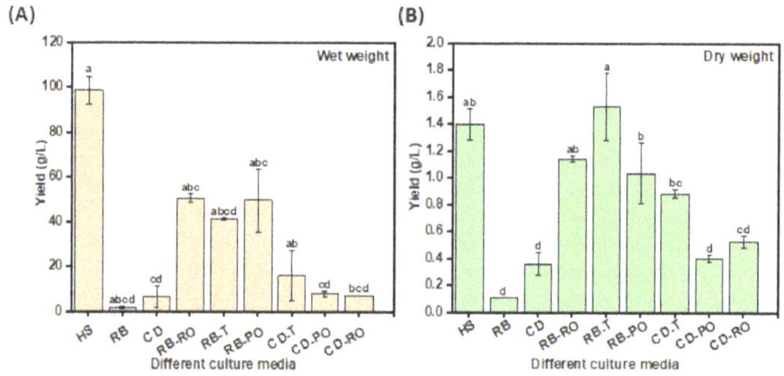

Figure 2. Large-scale BC production yield from untreated and treated waste media formulations. The (**A**) wet and (**B**) dry weight were determined on different waste formulations: Hestrin–Schramm (HS), rice bran (RB), cereal dust (CD), rice bran fermented with *R. oryzae* (RB-RO), rice bran fermented with *R. oligosporus* (RB-T), rice bran fermented with *P. osteratus* (RB-PO), cereal dust fermented with *R. oligosporus* (CD-T), cereal dust fermented with *P. osteratus* (CD-PO), cereal dust fermented with *R. oryzae* (CD-RO). The error bar represents the standard deviation. Different letters represent significant differences.

3.3. Comparative Analyses of the BC Produced from SSF-Treated Rice Bran and Cereal Dust

3.3.1. Scanning Electron Microscopy (SEM)

SEM analysis revealed distinct differences in BC nanostructure based on both the substrate source and drying method. Figure 3 compares the nanostructure of BC produced in HS medium with commercially available nata de coco, while Figure 3 and Supplementary Figure S3 show BC derived from SSF-treated cereal dust and rice bran, respectively.

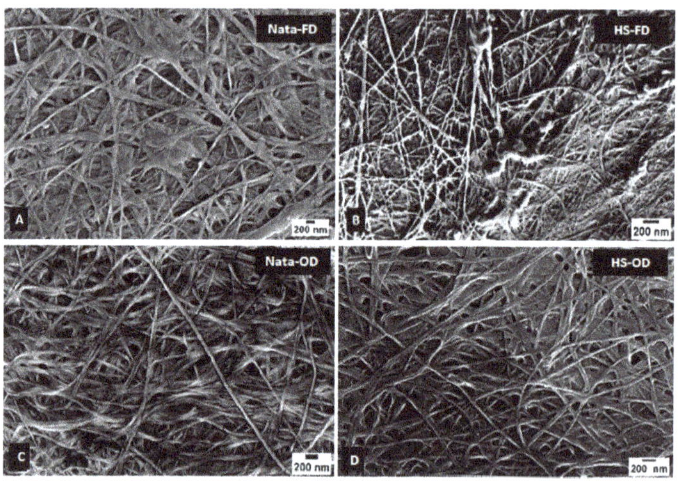

Figure 3. SEM images of commercially available nata de coco cellulose and cellulose produced from *Novacetomonas* sp. in HS culture medium. The (**A**,**C**) freeze-(FD) and oven-dried (OD) nata de coco, (**B**,**D**) freeze-(FD) and oven-dried (OD) cellulose from *Novacetomonas* sp. (scale bar—200 nm).

Across all samples, freeze-dried (FD) BC exhibited a more open, porous structure with visible spaces between fibre bundles, while oven-dried (OD) samples displayed a dense, interconnected network of nanofibers (Figure 3). This trend was consistent for BC produced from HS medium, nata de coco, and SSF-treated agricultural wastes. The effect of the substrate source on the BC nanostructure was evident in the comparison of cereal dust- and rice bran-derived BC (Figure 3 and Supplementary Figure S3). BC from CD fermented with *Pleurotus osteratus* showed a particularly dense and interconnected network, similar to that observed by Mohammadkazemi et al. (2015) in BC produced from mannitol and sucrose media [33]. Samples pre-treated with *Rhizopus oryzae* and *Rhizopus oligosporus* exhibited slightly more varied fibre arrangements.

A fibre diameter analysis (Figure 4 for cereal dust and Supplementary Figure S4 for rice bran) revealed that OD samples consistently produced smaller average fibre diameters (around 30 nm) compared to FD samples (40–50 nm for rice bran, 30–80 nm for cereal dust). This trend aligns with observations by Du et al. (2018), confirming the influence of drying methods on BC nanostructure across various production conditions [34]. Interestingly, the effect of the fungal strain on fibre diameter was less pronounced in the RB samples compared to cereal dust, suggesting that substrate composition may interact with fungal pre-treatment to influence the final BC nanostructure (Figure 5).

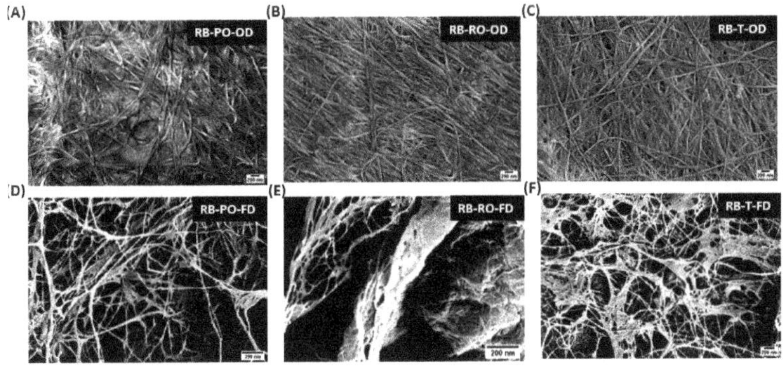

Figure 4. SEM micrographs of BC samples produced from fermented rice bran, using *P. osteratus*, *R. oryzae*, and *R. oligosporus* for SSF. The (**A**) oven-(RB-PO-OD) and (**D**) freeze-dried (RB-PO-FD) samples using *P. osteratus*, the (**B**) oven-(RB-RO-OD) and (**E**) freeze-dried (RB-RO-FD) samples using *R. oryzae*, and the (**C**) oven-(RB-T-OD) and (**F**) freeze-dried (RB-T-FD) samples using *R. oligosporus* (scale bar—200 nm).

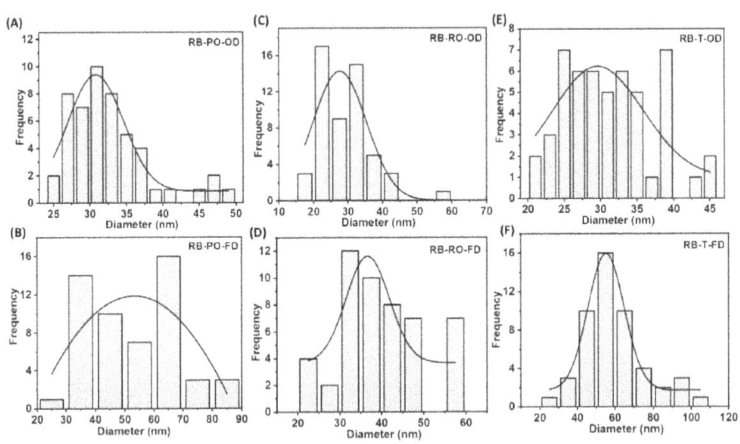

Figure 5. Fibre diameter graphs of BC produced from fermented rice bran, using *P. osteratus*, *R. oryzae*, and *R. oligosporus* for SSF. The (**A**) oven-(RB-PO-OD) and (**B**) freeze-dried (RB-PO-FD) samples using *P. osteratus*, the (**C**) oven-(RB-RO-OD) and (**D**) freeze-dried (RB-RO-FD) samples from *R. oryzae*, and the (**E**) oven-(RB-T-OD) and (**F**) freeze-dried (RB-T-FD) samples using *R. oligosporus*. The curve represents the mean value of the fibre diameter.

These results collectively emphasise the importance of both substrate choice and processing conditions in determining BC nanostructure. The ability to fine-tune fibre dimensions and network morphology through substrate selection, fungal pre-treatment, and drying method offers a versatile toolkit for engineering BC with specific properties suited to various applications [33–39].

3.3.2. Energy-Dispersive X-ray Spectroscopy (EDS) Analysis

EDS analysis revealed that BC produced from SSF-treated rice bran and cereal dust primarily consisted of carbon and oxygen, confirming the preservation of cellulosic composition (Table 1). BC from SSF-treated rice bran showed a higher carbon content (65–69%) compared to cereal dust (53–64%), while cereal dust-derived BC exhibited a higher oxygen content (35–46% vs. 31–35% in rice bran-derived BC). These differences were consistent across oven-dried and freeze-dried samples. The higher oxygen content in cereal dust-derived BC suggests a greater presence of O-containing functional groups, potentially influencing fibre interactions and network compactness. The observed variations in C/O ratios between substrates and fungal treatments indicate that the choice of waste substrate and SSF conditions can influence the BC surface chemistry. These findings align with previous studies showing that BC composition can be affected by bacterial strain, growth conditions, and carbon source [40,41]. Importantly, EDS analysis confirmed that SSF pre-treatment did not introduce detectable contaminants, demonstrating the potential of this method for producing pure BC from agricultural wastes.

Table 1. Elemental composition of C and O in BC produced from fermented rice bran and cereal dust in OD and FD samples.

Samples	Oven-Dried		Freeze-Dried	
	C	O	C	O
Rice bran	68.0%	32.0%	75.1%	24.9%
Rice bran (*Pleurotus osteratus*)	68.7%	31.3%	63.7%	36.3%
Rice bran (*Rhizopus oryzae*)	67.0%	33.0%	57.6%	42.4%
Rice bran (*Rhizopus oligosporus*)	65.2%	34.8%	59.6%	40.4%
Cereal dust	65.4%	34.6%	59.8%	40.2%
Cereal dust (*Pleurotus osteratus*)	64.3%	35.7%	57.1%	42.9%
Cereal dust (*Rhizopus oryzae*)	63.2%	36.8%	53.8%	46.2%
Cereal dust (*Rhizopus oligosporus*)	59.9%	40.1%	53.8%	46.2%

3.3.3. Fourier-Transformed Infrared Spectroscope (FTIR)

FTIR analysis of BC produced from SSF-treated rice bran and cereal dust revealed characteristic cellulose peaks, confirming the preservation of a cellulose type I structure (Figure 5). The key spectral peaks included O-H stretching (3300 cm^{-1}), C-H stretching (2900 cm^{-1}), adsorbed water (1640 cm^{-1}), CH_2 bending (crystallinity band, 1427 cm^{-1}), C-H bending (1350 cm^{-1}), C-O-C stretching (1160–1110 cm^{-1}), and C-O stretching (1060 cm^{-1}). These peaks are consistent with those reported for BC produced from conventional media [42,43]. Some rice bran-derived samples (RB-FD, RB-PO-OD, RB-PO-FD, RB-RO-FD) showed an additional peak around 1700 cm^{-1}, potentially indicating a slight oxidation of surface hydroxyls. In general, the FTIR spectra confirm that BC produced from SSF-treated agricultural wastes maintains the essential chemical structure of pure cellulose, with minor variations that may offer opportunities for tailoring physicochemical properties. It is worth noting that no characteristic lignin peaks were observed in the FTIR spectra, confirming the purity of the bacterial cellulose produced. This is consistent with the known ability of our *Novacetimonas* sp. strain to synthesise pure cellulose. These findings align with previous studies using other waste substrates for BC production [31,44,45] demonstrating the versatility of RB and CD wastes as feedstocks for BC synthesis (Figure 6).

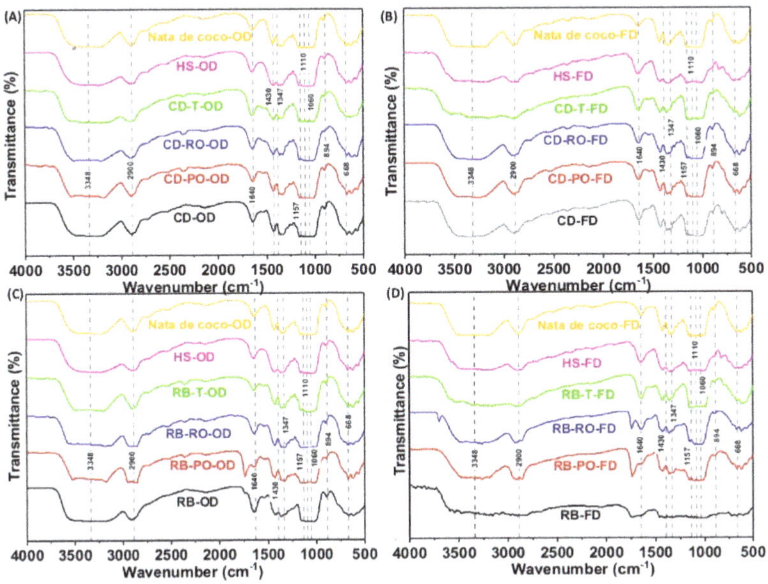

Figure 6. FTIR spectra of oven- and freeze-dried BC synthesised from SSF-treated cereal dust using *P. osteratus*, *R. oryzae*, and *R. oligosporus* during SSF. BC synthesised from SSF-treated cereal dust with (**A**,**B**) *P. osteratus* (CD-PO-OD/FD), *R. oryzae* (CD-RO-OD/FD), and *R. oligosporus* (CD-T-OD/FD). BC synthesised from SSF-treated rice bran with (**C**,**D**) *P. osteratus* (RB-PO-OD/FD), *R. oryzae* (RB-RO-OD/FD), and *R. oligosporus* (RB-T-OD/FD).

3.3.4. X-ray Diffraction (XRD)

XRD patterns of BC produced from SSF-treated RB and CD revealed characteristic peaks at $2\theta = 14.9°$, $16.5°$, and $22.8°$, corresponding to the typical cellulose I structure (Figure 6A–D) [46–48]. These peaks were consistent across all samples, including the nata de coco and HS medium controls, confirming the preservation of BC's crystalline structure (Figure 7). However, some subtle differences were observed between samples. The BC from SSF-treated substrates showed additional broad peaks centred at 30° and between 40 and 50°, not present in the control samples. The freeze-dried rice bran samples exhibited a significant peak at 30°, corresponding to the (122) plane, which was absent in most oven-dried samples. Lastly, the freeze-dried rice bran samples generally showed broader peaks, indicating a lower crystallinity compared to the oven-dried samples. These variations suggest that both the SSF pre-treatment and drying method influence the crystalline structure of the produced BC. The presence of additional peaks and changes in peak intensity indicate potential modifications in cellulose arrangement and crystallinity, which could affect the material properties of the resulting BC. A similar study used heat-pre-treated molasses culture media for BC production and made similar observations in the XRD peaks of the BC films generated from waste, which had peaks identical to the HS medium and showed the characteristic peaks of the crystalline form of cellulose I [1].

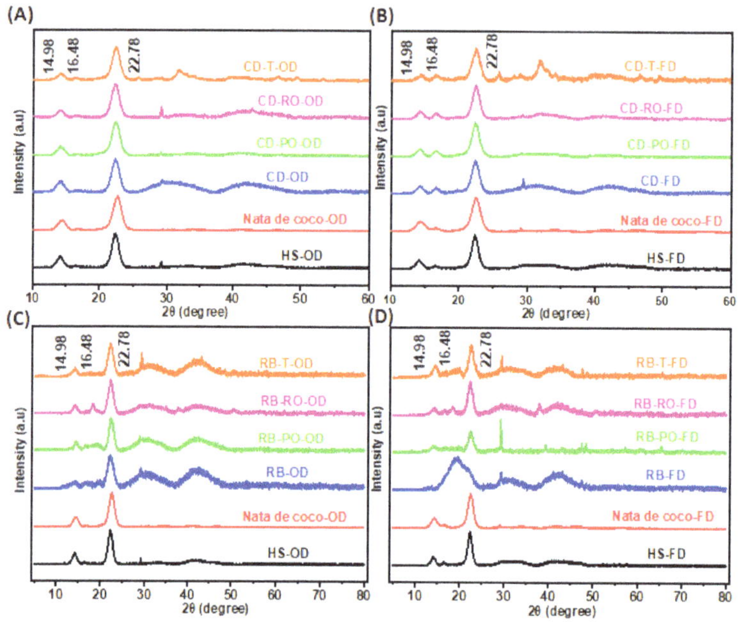

Figure 7. X-ray diffraction patterns of BC synthesised from SSF-treated cereal dust and rice bran. BC samples from (**A**,**B**) cereal dust (CD) and (**C**,**D**) rice bran (RB) using *P. osteratus* (PO-OD/FD), *R. oryzae* (RO-OD/FD), and *R. oligosporus* (T-OD/FD) compared to nata de coco and HS medium, oven-dried (OD) and freeze-dried (FD).

The crystallinity index (CI) of the BC samples is presented in Supplementary Table S3, calculated based on the intensity of peaks at $2\theta = 22.5°$ and $18.5°$, corresponding to the (110) and (200) planes of the crystalline form of cellulose I [1]. It was revealed that the BC from cereal dust showed a high CI (94–99%), with oven-dried samples reaching ~99%. In contrast, the BC from rice bran exhibited a lower CI, particularly in freeze-dried samples (69–79%). In terms of drying, oven-dried samples consistently showed a higher CI than freeze-dried samples for both substrates. Lastly, the SSF treatment with different fungi had a minimal impact on the CI for cereal dust but affected rice bran samples more noticeably. These results align with the XRD patterns, where sharper, more defined peaks correlate with a higher crystallinity. The high CI of cereal dust-derived BC suggests a well-ordered cellulose structure, while the lower CI in rice bran-derived BC, especially freeze-dried samples, indicates increased amorphous regions. The variations in CI between substrates and drying methods demonstrate the potential to modulate BC crystallinity through feedstock selection and processing conditions. Higher crystallinity generally correlates with improved mechanical properties, while lower crystallinity may enhance properties such as swelling and drug-loading capacity. These findings suggest opportunities for tailoring BC properties for specific applications by controlling its crystallinity through waste media substrate choice and processing methods.

3.3.5. Thermogravimetric Analysis (TGA)

The thermogravimetric analysis of BC samples produced from untreated and SSF-treated cereal dust and rice bran revealed distinct thermal degradation profiles (Figure 8). The initial mass loss for most samples occurred around 250 °C, indicating the onset of thermal degradation. The BC from SSF-treated cereal dust (CD-T-OD) showed the highest thermal stability, with only ~60% mass loss at 600 °C (Figure 8A). Additionally, the freeze-dried samples generally exhibited lower thermal stability compared to their oven-dried counterparts. Moreover, the BC from SSF-treated rice bran (RB-T-OD) maintained its structure up to 290 °C, showing improved thermal stability compared to the untreated samples (Figure 8C). In contrast, HS medium-derived BC, particularly freeze-dried (HS-FD), showed the lowest thermal stability, with degradation starting at 150 °C. It was also observed that the SSF treatment generally improved the thermal stability of BC from both RB and CD media compared to the untreated samples and control BC (nata de coco and HS medium). These results suggest that both the substrate type and SSF treatment influence the thermal properties of the resulting BC. The enhanced thermal stability of SSF-treated samples indicates potential improvements in BC structure and composition, which could be beneficial for applications requiring heat resistance.

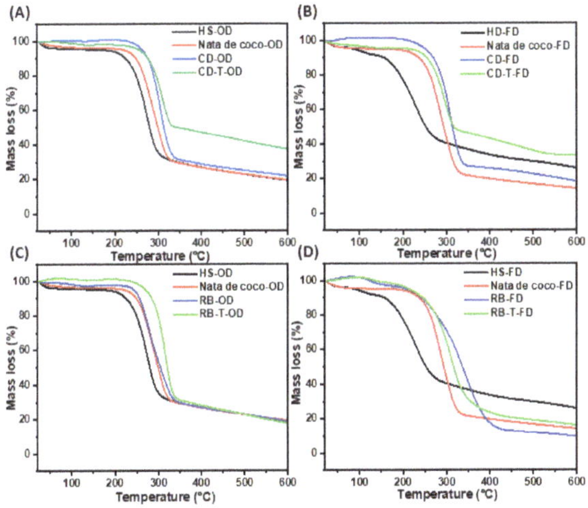

Figure 8. Thermogravimetric analysis of BC samples produced in untreated and treated cereal dust and rice bran using *R. oligosporus* (T) for SSF. (**A**) BC samples produced using cereal dust (CD), oven-(OD) and (**B**) freeze-dried (FD). (**C**) BC samples produced using rice bran (RB), oven-(OD) and (**D**) and freeze-dried (FD).

Derivative thermogravimetry (DTG) analyses revealed distinct thermal degradation profiles for BC samples produced from SSF-treated and untreated cereal dust and rice bran (Figure 9). The SSF-treated cereal dust BC (CD-T-OD) showed an improved thermal stability with a lower mass loss rate (10%) at 300 °C compared to untreated samples (14%) (Figure 9A). The freeze-dried SSF-treated cereal dust BC (CD-T-FD) exhibited the lowest mass loss rate (9.8%) at 290 °C among the cereal dust samples (Figure 9B). The SSF-treated rice bran BC (RB-T-OD) showed a higher mass loss rate (16%) at a higher temperature (320 °C) compared to untreated samples (11% at 280 °C) (Figure 9C). Lastly, the freeze-dried rice bran BC (RB-FD) demonstrated the lowest mass loss rate (7%) at 320 °C among all samples (Figure 9D). These results indicate that SSF treatment generally enhances the thermal stability of BC, with the effect more pronounced in cereal dust samples. The improved thermal properties likely result from increased crystallinity and better-organised cellulose structures, as suggested by XRD analysis (Figure 6 and Table S3). Our findings align with the previous studies, which showed enhanced thermal stability in BC produced from agricultural wastes [33]. The enhanced thermal stability of SSF-treated BC samples suggests potential applications in areas requiring heat-resistant materials, such as biomedical devices or electronics. Furthermore, the ability to modulate thermal properties through substrate selection and processing conditions offers opportunities for tailoring BC for specific applications.

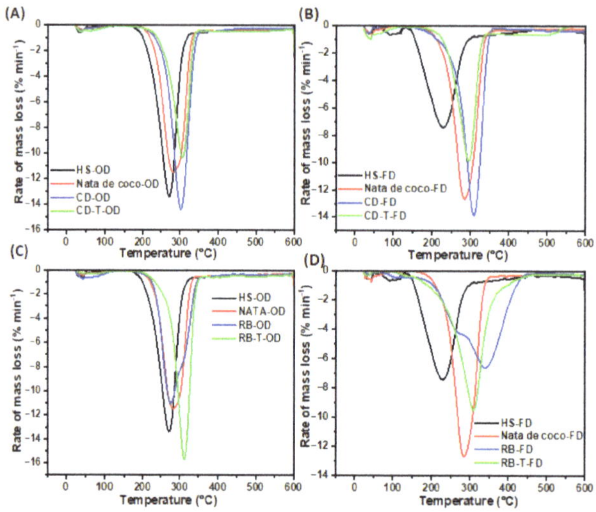

Figure 9. Derivative thermogravimetry (DTG) analysis of BC samples produced in untreated and treated cereal dust and rice bran with using *R. oligosporus* (T) for SSF. (**A**) BC samples produced using cereal dust (CD), oven-(OD) and (**B**) freeze-dried (FD); (**C**) BC samples produced using rice bran (RB), oven-(OD) and (**D**) freeze-dried (FD).

3.3.6. Cost Analysis of Growth Media Input

A detailed comparison of the production costs of bacterial cellulose (BC) based on the input resources used for the culture media has shown significant cost savings when alternative SSF-treated substrates are employed (Table 2). Replacing carbon and nitrogen sources in HS medium with unfermented RB and CD lowers the cost by up to 40%. Meanwhile, using the SSF process to ferment these wastes has not only increased the yield of BC but also reduced the cost of media by around 90%. Table 2 shows the cost analysis of the conventional HS medium with unfermented and fermented RB and CD.

Table 2. Cost analysis of bacterial cellulose produced from HS medium vs. unfermented and fermented rice bran (RB) and cereal dust (CD) in g/L.

		Price		Hestrin-Schramm Media/L		Rice Bran/L		Cereal Dust/L		SSF-Treated RB/L		SSF-Treated CD/L	
Media Composition	**Unit**	**AU $/Unit**	**AU $/g**	**Quantit (g)**	**Cost ($/g)**	**Quantit (g)**	**Cost ($/g)**	**Quantit (g)**	**Cost ($/g)**	**Quantit (g)**	**Cost ($/g)**	**Quantit (g)**	**Cost ($/g)**
Glucose	1 kg	63.5	0.06	20	1.2	0	0	0	0	0	0	0	0
Yeast extract	1 kg	360	0.36	5	1.8	0	0	0	0	0	0	0	0
Peptone	500 g	210	0.42	5	2.1	0	0	0	0	0	0	0	0
Di-sodium hydrogen phosphate	500 g	93.13	0.19	2.7	0.51	2.7	0.51	0.51	0.51	2.7	0.51	2.7	0.51
Citric acid monohydrate	1 kg	185	0.18	1.85	0.33	1.85	1.85	1.85	1.85	1.85	1.85	1.85	1.85
Magnesium sulphate heptahydrate	1 kg	287	0.28	0.5	0.14	0.5	0.15	0.15	0.15	0.5	0.15	0.5	0.15
Ethanol	1 L	117	0.11	100 mL	11	100	11	0	11	0	0	0	0
Rice bran	10 kg	0	0	0	0	76.9	0	0	0	76.9	0	0	0
Cereal Dust	5 kg	0	0	0	0	0	0	58.5	0	0	0	58.5	0
Total cost					**17.08**		**13.51**		**13.51**		**2.51**		**2.51**

The BC (dry weight) produced from SSF-treated RB and CD costs around AUD 2.51 per gram compared to untreated RB and CD, which is AUD 13.51 per gram. In contrast, BC produced using standard HS medium costs around AUD 17.08 per gram. For reference, the approximate cost of commercially available BC in the market is around AUD 29.00 per gram. This substantial decrease in cost is attributed to the dual benefits of using low-cost, abundant waste materials and enhancing the process efficiency of SSF. The cost reduction, coupled with the improved BC yield, suggests that SSF-treated agricultural wastes could be a highly economical and sustainable approach for large-scale BC production. These findings have significant implications for industrial-scale BC production, potentially lowering barriers to entry for BC-based products in various sectors. Moreover, this approach aligns with circular economy principles by valorising agricultural waste streams. Future studies should focus on scaling up this process and conducting a full life-cycle assessment to comprehensively evaluate its environmental and economic impacts.

4. Discussion

This study demonstrates the successful production of bacterial cellulose (BC) from rice bran and cereal dust pre-treated by SSF, representing a significant step towards sustainable and cost-effective BC production. The use of this abundant agricultural waste addresses two critical challenges: the high cost of conventional BC production media and the environmental impact of agro-industrial waste. SSF pre-treatment effectively enhanced the bioavailability of nutrients in rice bran and cereal dust, resulting in improved BC yields compared to untreated substrates. Notably, rice bran fermented with *Rhizopus oligosporus* produced the highest BC yield (1.52 g/L dry weight), surpassing even the standard HS medium. This finding aligns with previous studies showing the potential of SSF to unlock nutrients from complex biomass and extends this approach to BC production [13,19]. On the other hand, it was also shown that the surface area of a vessel is an important parameter, especially in a static cultivation setting, as it impacts the oxygen transport in the medium for BC production [25,30].

A comprehensive physicochemical characterisation revealed that BC produced from SSF-treated substrates maintains the fundamental nanostructure and chemical identity of cellulose, with some variations that could be advantageous for specific applications. SEM analysis showed that both the choice of fungal strain for SSF and the drying method (oven drying vs. freeze drying) significantly influenced the BC nanostructure, fibre diameter, and morphology. These structural differences could be leveraged to tailor BC properties for various applications, from tissue engineering scaffolds to filtration membranes [49]. FTIR and XRD analyses confirmed the preservation of a cellulose I structure in BC from SSF-treated substrates, with some samples showing additional peaks that suggest slight modifications to surface chemistry or crystallinity. These subtle changes could potentially enhance specific properties, such as thermal stability or reactivity, without compromising the essential cellulose structure. The observed variations in crystallinity index between substrates and drying methods demonstrate the potential to modulate BC properties through feedstock selection and processing conditions. On the other hand, thermogravimetric analysis revealed that SSF treatment generally improved the thermal stability of BC from both rice bran and cereal dust media. This enhanced thermal resistance could expand the potential applications of BC to areas requiring heat-stable materials, such as biomedical devices or electronics components [33].

Perhaps most significantly, the cost analysis revealed that using SSF-treated agro-food wastes could reduce BC media costs by up to 90% compared to the standard HS medium. This dramatic cost reduction, coupled with the valorisation of agricultural waste streams, presents a compelling case for the industrial adoption of this approach, aligning with current trends in circular bioeconomy strategies. The potential for significant cost savings while maintaining or improving BC quality could be advantageous for the large-scale production of this versatile biopolymer.

The findings of this study have broad implications for both sustainable material production and waste management in the agro-food industry. By demonstrating the feasibility of using minimally processed agricultural wastes for high-value biopolymer production, this research contributes to the growing body of work on circular bioeconomy strategies. The approach developed here could be extended to other types of agro-industrial wastes, potentially revolutionising the production of bio-based materials while simultaneously addressing waste management challenges. Moreover, the ability to modulate BC properties through substrate selection, SSF conditions, and drying methods offers new avenues for tailoring this material for specific applications. This versatility could expand the use of BC across various sectors, from biomedical applications to advanced materials for electronics and environmental remediation [50]. Hence, this study provides a proof-of-concept for the sustainable valorisation of agro-food wastes into high-value BC biomaterials. These improvements in yield, cost-effectiveness, and potential for property customisation pave the way for scaled-up, environmentally friendly BC production processes.

Future research should focus on optimising SSF conditions for different waste streams, exploring the full range of attainable BC properties, and developing specific applications that leverage the unique characteristics of waste-derived BC. Life-cycle assessments and scaled-up production studies are needed to fully evaluate the environmental and economic impacts of this approach. Hence, future work should include a comprehensive mass balance study of the entire process, from SSF pre-treatment to BC production. This will involve a detailed quantification of all input materials (cereal wastes, fungi, bacteria) and output streams (BC, residual biomass, metabolites) at each stage of the process. Such a study will provide crucial insights into the process efficiency, nutrient utilisation, and potential for further optimisation. Additionally, it will allow for a more accurate assessment of the environmental impact and sustainability of this BC production method.

5. Conclusions

The use of solid-state fermentation with fungi has proven to be highly effective in improving the digestibility and bioavailability of agricultural waste material, which in turn has led to a significant increase in bacterial cellulose production up to double the usual yield. Specific enzymes in this process have led to the breakdown of the complex structures within these waste materials, releasing essential nutrients without damaging the fundamental nanostructure or chemical properties of cellulose. Additionally, the drying process used in this method plays a crucial role in controlling the final structure of BC, allowing the precise customisation of the material's properties. This includes adjusting features such as the width of the nanofibers and the size of the pores, which can be tailored to meet biofunctional needs. A detailed spectroscopic, microscopic, and thermal analysis confirmed the synthesis of signature bacterial cellulose I nanoparticles with commercially comparable properties. Materials characterisation revealed tunability in the thermal stability, surface traits, and nanostructure based on nutritional and drying variations. A cost analysis based on the resources used for the culture medium shows that producing BC from fermented agricultural by-products like rice bran and cereal dust is nearly 90% cheaper than using the traditional Hestrin–Schramm medium. Hence, this research demonstrates a new, integrated approach to converting cereal crop waste into valuable cellulose, offering a solution that is both economically viable and environmentally sustainable, while also allowing for the customisation of the final product's properties.

Supplementary Materials: The following supporting information can be downloaded at: https://www.mdpi.com/article/10.3390/foods13193052/s1, Figure S1: Solid-state fermentation set up for rice bran (RB) and cereal dust (CD) in 5L Erlenmeyer flasks; Figure S2: Small-scale BC production yield from untreated and treated waste media formulations, Hestrin-Schramm (H-S), rice bran (RB), cereal dust (CD), rice bran fermented with *Rhizopus oryzae* (RB-RO), rice bran fermented with *Rhizopus oligosporus* (RB-T), rice bran fermented with *Pleurotus osteratus* (RB-PO), cereal dust fermented with *Rhizopus oligosporus* (CD-T), cereal dust fermented with *Pleurotus osteratus* (CD-PO), cereal dust fermented with *Rhizopus oryzae* (CD-RO) showing (A) wet and (B) dry weight yield (g/L). The error bar represents standard deviation. Different letters represent significant differences; Figure S3: SEM micrographs of BC produced from fermented cereal dust, with *Pleurotus osteratus*, oven-dried (A) (CD-PO-OD), freeze-dried (D) (CD-PO-FD), *Rhizopus oryzae* oven-dried (B) (CD-RO-OD), freeze-dried (E) (CD-RO-FD), *Rhizopus oligosporus* oven dried (C) (CD-T-OD), freeze-dried (F) (CD-T-FD) (scale bar- 200 nm); Figure S4: Fibre diameter graphs of BC produced from fermented cereal dust, with *Pleurotus osteratus*, oven-dried (A) (CD-PO-OD), freeze-dried (D) (CD-PO-FD), *Rhizopus oryzae* oven-dried (B) (CD-RO-OD), freeze-dried (E) (CD-RO-FD), *Rhizopus oligosporus* oven dried (C) (CD-T-OD), freeze-dried (F) (CD-T-FD). The curve represents the mean value of the fibre diameter. Table S1: Total carbohydrate and protein content of wastes used for formulating the waste media for BC production; Table S2: Reformulated media compositions using rice bran and cereal dust (untreated/treated) compared to HS media used for BC production; Table S3: The crystallinity index of the oven and freeze-dried samples of unfermented and fermented rice bran and cereal dust.

Author Contributions: Conceptualization, S.H. and V.B.J.; methodology, S.H., S.D. and V.B.J.; software, S.H.; validation, V.B.J. and S.H.; formal analysis and investigation, S.H.; writing—original draft preparation, S.H.; writing—review and editing, V.B.J., S.D. and H.S. All authors have read and agreed to the published version of the manuscript.

Funding: This research received no external funding.

Institutional Review Board Statement: Not applicable.

Informed Consent Statement: Not applicable.

Data Availability Statement: The raw data supporting the conclusions of this article will be made available by the authors on request. The 16S rRNA data for the molecular identification of the isolate are available at the NCBI as Genbank PP421219.1.

Acknowledgments: We would like to acknowledge Francois Malherbe for his technical support and suggestions. Shriya would like to acknowledge the Swinburne Tuition Fee Scholarship from Swinburne University. The authors would also like to thank SunRice and Rex James Stockfeed Ltd. for the rice bran and cereal dust used in this study.

Conflicts of Interest: The authors declare no conflicts of interest.

References

1. Hussain, Z.; Sajjad, W.; Khan, T.; Wahid, F. Production of bacterial cellulose from industrial wastes: A review. *Cellulose* **2019**, *26*, 2895–2911. [CrossRef]
2. Dungani, R.; Karina, M.; Subyakto; Sulaeman, A.; Hermawan, D.; Hadiyane, A. Agricultural waste fibers towards sustainability and advanced utilization: A review. *Asian J. Plant Sci.* **2016**, *15*, 42–55. [CrossRef]
3. Lahiri, D.; Nag, M.; Dutta, B.; Dey, A.; Sarkar, T.; Pati, S.; Edinur, H.A.; Abdul Kari, Z.; Mohd Noor, N.H.; Ray, R.R. Bacterial Cellulose: Production, Characterization, and Application as Antimicrobial Agent. *Int. J. Mol. Sci.* **2021**, *22*, 12984. [CrossRef]
4. Gullo, M.; La China, S.; Falcone, P.M.; Giudici, P. Biotechnological production of cellulose by acetic acid bacteria: Current state and perspectives. *Appl. Microbiol. Biotechnol.* **2018**, *102*, 6885–6898. [CrossRef]
5. Salihu, R.; Foong, C.Y.; Razak, S.I.A.; Kadir, M.R.A.; Yusof, H.; Nayan, G.H.M. Overview of Inexpensive Production Routes of Bacterial Cellulose and Its Applications in Biomedical Engineering. *Cellul. Chem. Technol.* **2019**, *53*, 1–13. [CrossRef]
6. Kamal, T.; Ul-Islam, M.; Fatima, A.; Ullah, M.W.; Manan, S. Cost-Effective Synthesis of Bacterial Cellulose and Its Applications in the Food and Environmental Sectors. *Gels* **2022**, *8*, 552. [CrossRef] [PubMed]
7. Brunschwiler, C.; Heine, D.; Kappeler, S.; Conde-Petit, B.; Nyström, L. Direct measurement of rice bran lipase activity for inactivation kinetics and storage stability prediction. *J. Cereal Sci.* **2013**, *58*, 272–277. [CrossRef]
8. Narh, C.; Frimpong, C.; Mensah, A.; Wei, Q. Rice Bran, an Alternative Nitrogen Source for Acetobacter xylinum Bacterial Cellulose Synthesis. *Bioresources* **2018**, *13*, 4346–4363. [CrossRef]
9. Galanakis, C.M. Recovery of high added-value components from food wastes: Conventional, emerging technologies and commercialized applications. *Trends Food Sci. Technol.* **2012**, *26*, 68–87. [CrossRef]
10. Urbina, L.; Corcuera, M.Á.; Gabilondo, N.; Eceiza, A.; Retegi, A. A review of bacterial cellulose: Sustainable production from agricultural waste and applications in various fields. *Cellulose* **2021**, *28*, 8229–8253. [CrossRef]
11. Janarny, G.; Gunathilake, K.D.P.P. Changes in rice bran bioactives, their bioactivity, bioaccessibility and bioavailability with solid-state fermentation by Rhizopus oryzae. *Biocatal. Agric. Biotechnol.* **2020**, *23*, 101510. [CrossRef]
12. Ali, H.K.Q.; Zulkali, M.M.D. Utilization of Agro-Residual Ligno-Cellulosic Substances by Using Solid State Fermentation: A Review. *Croat. J. Food Technol. Biotechnol.* **2011**, *6*, 5–12.
13. Kupski, L.; Cipolatti, E.; Rocha, M.D.; Oliveira, M.D.S.; Souza-Soares, L.D.A.; Badiale-Furlong, E. Solid-state fermentation for the enrichment and extraction of proteins and antioxidant compounds in rice bran by Rhizopus oryzae. *Braz. Arch. Biol. Technol.* **2012**, *55*, 937–942. [CrossRef]
14. Ellaiah, P.; Prabhakar, T.; Ramakrishna, B.; Taleb, A.T.; Adinarayana, K. Production of lipase by immobilized cells of Aspergillus niger. *Process Biochem.* **2004**, *39*, 525–528. [CrossRef]
15. Adinarayana, K.; Ellaiah, P.; Prasad, D.S. Purification and partial characterization of thermostable serine alkaline protease from a newly isolated *Bacillus subtilis* PE-11. *AAPS PharmSciTech* **2003**, *4*, E56. [CrossRef] [PubMed]
16. Baysal, Z.; Uyar, F.; Aytekin, Ç. Solid state fermentation for production of α-amylase by a thermotolerant Bacillus subtilis from hot-spring water. *Process Biochem.* **2003**, *38*, 1665–1668. [CrossRef]

17. Sodhi, H.K.; Sharma, K.; Gupta, J.K.; Soni, S.K. Production of a thermostable α-amylase from Bacillus sp. PS-7 by solid state fermentation and its synergistic use in the hydrolysis of malt starch for alcohol production. *Process Biochem.* **2005**, *40*, 525–534. [CrossRef]
18. Ritthibut, N.; Oh, S.-J.; Lim, S.-T. Enhancement of bioactivity of rice bran by solid-state fermentation with Aspergillus strains. *LWT* **2021**, *135*, 110273. [CrossRef]
19. Sukma, A.C.T.; Oktavianty, H.; Sumardiono, S. Optimization of solid-state fermentation condition for crude protein enrichment of rice bran using Rhizopus oryzae in tray bioreactor. *Indones. J. Biotechnol.* **2021**, *26*, 33–40. [CrossRef]
20. Shin, H.-Y.; Kim, S.-M.; Lee, J.H.; Lim, S.-T. Solid-state fermentation of black rice bran with Aspergillus awamori and Aspergillus oryzae: Effects on phenolic acid composition and antioxidant activity of bran extracts. *Food Chem.* **2019**, *272*, 235–241. [CrossRef]
21. Henry, S. Production of Bacterial Cellulose from Agricultural Wastes for Tissue Scaffold Application. Ph.D. Thesis, Swinburne University of Technology, Hawthorn, Australia, 2024.
22. Yuan, Y.; Feng, F.; Chen, L.; Yao, Q.; Chen, K. Directional isolation of ethanol-tolerant acetic acid bacteria from industrial fermented vinegar. *Eur. Food Res. Technol.* **2013**, *236*, 573–578. [CrossRef]
23. Oliveira, M.d.S.; Feddern, V.; Kupski, L.; Cipolatti, E.P.; Badiale-Furlong, E.; de Souza-Soares, L.A. Changes in lipid, fatty acids and phospholipids composition of whole rice bran after solid-state fungal fermentation. *Bioresour. Technol.* **2011**, *102*, 8335–8338. [CrossRef] [PubMed]
24. Borzani, W.; Souza, S.J.d. Mechanism of the film thickness increasing during the bacterial production of cellulose on non-agitated liquid media. *Biotechnol. Lett.* **1995**, *17*, 1271–1272. [CrossRef]
25. Ruka, D.R.; Simon, G.P.; Dean, K.M. Altering the growth conditions of Gluconacetobacter xylinus to maximize the yield of bacterial cellulose. *Carbohydr. Polym.* **2012**, *89*, 613–622. [CrossRef] [PubMed]
26. Nam, S.; French, A.D.; Condon, B.D.; Concha, M. Segal crystallinity index revisited by the simulation of X-ray diffraction patterns of cotton cellulose Iβ and cellulose II. *Carbohydr. Polym.* **2016**, *135*, 1–9. [CrossRef]
27. Kodali, B.; Pogaku, R. Pretreatment studies of rice bran for the effective production of cellulose. *Electron. J. Environ. Agric. Food Chem.* **2006**, *5*, 1253–1264.
28. Wang, J.; Tavakoli, J.; Tang, Y. Bacterial cellulose production, properties and applications with different culture methods—A review. *Carbohydr. Polym.* **2019**, *219*, 63–76. [CrossRef]
29. Holmes, D. Bacterial Cellulose. Master's Thesis, Department of Chemical and Process Engineering University of Canterbury Christchurch, Christchurch, New Zealand, 2004.
30. Aytekin, A.Ö.; Demirbağ, D.D.; Bayrakdar, T. The statistical optimization of bacterial cellulose production via semi-continuous operation mode. *J. Ind. Eng. Chem.* **2016**, *37*, 243–250. [CrossRef]
31. Abdelraof, M.; Hasanin, M.S.; El-Saied, H. Ecofriendly green conversion of potato peel wastes to high productivity bacterial cellulose. *Carbohydr. Polym.* **2019**, *211*, 75–83. [CrossRef]
32. Omarini, A.B.; Labuckas, D.; Zunino, M.P.; Pizzolitto, R.; Fernández-Lahore, M.; Barrionuevo, D.; Zygadlo, J.A. Upgrading the Nutritional Value of Rice Bran by Solid-State Fermentation with Pleurotus sapidus. *Fermentation* **2019**, *5*, 44. [CrossRef]
33. Mohammadkazemi, F.; Azin, M.; Ashori, A. Production of bacterial cellulose using different carbon sources and culture media. *Carbohydr. Polym.* **2015**, *117*, 518–523. [CrossRef] [PubMed]
34. Du, R.; Zhao, F.; Peng, Q.; Zhou, Z.; Han, Y. Production and characterization of bacterial cellulose produced by Gluconacetobacter xylinus isolated from Chinese persimmon vinegar. *Carbohydr. Polym.* **2018**, *194*, 200–207. [CrossRef] [PubMed]
35. Wang, S.-S.; Han, Y.-H.; Ye, Y.-X.; Shi, X.-X.; Xiang, P.; Chen, D.-L.; Li, M. Physicochemical characterization of high-quality bacterial cellulose produced by *Komagataeibacter* sp. strain W1 and identification of the associated genes in bacterial cellulose production. *RSC Adv.* **2017**, *7*, 45145–45155. [CrossRef]
36. Dórame-Miranda, R.F.; Gámez-Meza, N.; Medina-Juárez, L.Á.; Ezquerra-Brauer, J.M.; Ovando-Martínez, M.; Lizardi-Mendoza, J. Bacterial cellulose production by Gluconacetobacter entanii using pecan nutshell as carbon source and its chemical functionalization. *Carbohydr. Polym.* **2019**, *207*, 91–99. [CrossRef]
37. Rani, M.U.; Appaiah, A. Optimization of culture conditions for bacterial cellulose production from Gluconacetobacter hansenii UAC09. *Ann. Microbiol.* **2011**, *61*, 781–787. [CrossRef]
38. Slapšak, N.; Cleenwerck, I.; De Vos, P.; Trček, J. Gluconacetobacter maltaceti sp. nov., a novel vinegar producing acetic acid bacterium. *Syst. Appl. Microbiol.* **2013**, *36*, 17–21. [CrossRef]
39. Bi, J.C.; Liu, S.X.; Li, C.F.; Li, J.; Liu, L.X.; Deng, J.; Yang, Y.C. Morphology and structure characterization of bacterial celluloses produced by different strains in agitated culture. *J. Appl. Microbiol.* **2014**, *117*, 1305–1311. [CrossRef]
40. Kongruang, S. Bacterial Cellulose Production by Acetobacter xylinum Strains from Agricultural Waste Products. In *Biotechnology for Fuels and Chemicals*; Springer: Berlin/Heidelberg, Germany, 2007; pp. 763–774.
41. De Souza, K.C.; Dos Santos, G.R.; Trindade, F.C.; Costa, A.F.D.S.; De Almeida, Y.M.; Sarubbo, L.A.; Vinhas, G.M. Production of bacterial cellulose biopolymers in media containing rice and corn hydrolysate as carbon sources. *Polym. Polym. Compos.* **2021**, *29*, S1466–S1474. [CrossRef]
42. Popescu, C.-M.; Popescu, M.-C.; Singurel, G.; Vasile, C.; Argyropoulos, D.S.; Willfor, S. Spectral Characterization of Eucalyptus Wood. *Appl. Spectrosc.* **2007**, *61*, 1168–1177. [CrossRef] [PubMed]
43. Mendoza, D.J.; Hossain, L.; Browne, C.; Raghuwanshi, V.S.; Simon, G.P.; Garnier, G. Controlling the transparency and rheology of nanocellulose gels with the extent of carboxylation. *Carbohydr. Polym.* **2020**, *245*, 116566. [CrossRef]

44. Castro, C.; Zuluaga, R.; Putaux, J.-L.; Caro, G.; Mondragon, I.; Gañán, P. Structural characterization of bacterial cellulose produced by Gluconacetobacter swingsii sp. from Colombian agroindustrial wastes. *Carbohydr. Polym.* **2011**, *84*, 96–102. [CrossRef]
45. Adebayo-Tayo, B.; Akintunde, M.; Sanusi, J. Effect of different fruit juice media on bacterial cellulose production by Acinetobacter sp. BAN1 and Acetobacter pasteurianus PW1. *J. Adv. Biol. Biotechnol.* **2017**, *14*, 1–9. [CrossRef]
46. Pal, S.; Nisi, R.; Stoppa, M.; Licciulli, A. Silver-Functionalized Bacterial Cellulose as Antibacterial Membrane for Wound-Healing Applications. *ACS Omega* **2017**, *2*, 3632–3639. [CrossRef] [PubMed]
47. Pasaribu, K.M.; Gea, S.; Ilyas, S.; Tamrin, T.; Radecka, I. Characterization of Bacterial Cellulose-Based Wound Dressing in Different Order Impregnation of Chitosan and Collagen. *Biomolecules* **2020**, *10*, 1511. [CrossRef] [PubMed]
48. Ju, X.; Bowden, M.; Brown, E.E.; Zhang, X. An improved X-ray diffraction method for cellulose crystallinity measurement. *Carbohydr. Polym.* **2015**, *123*, 476–481. [CrossRef]
49. Ul-Islam, M.; Ullah, M.W.; Khan, S.; Park, J.K. Production of bacterial cellulose from alternative cheap and waste resources: A step for cost reduction with positive environmental aspects. *Korean J. Chem. Eng.* **2020**, *37*, 925–937. [CrossRef]
50. Azeredo, H.M.C.; Barud, H.; Farinas, C.S.; Vasconcellos, V.M.; Claro, A.M. Bacterial Cellulose as a Raw Material for Food and Food Packaging Applications. *Front. Sustain. Food Syst.* **2019**, *3*, 7. [CrossRef]

foods

MDPI

Protocol

Culinary Home Empowerment for Food Waste Prevention and Minimization: Feasibility and Efficacy Protocol

Brandy-Joe Milliron [1,*], Roni Neff [2], Rachel Sherman [3], DeAndra Forde [1], Lauren Miller [3], Dahlia Stott [1], Alison Mountford [3] and Jonathan M. Deutsch [1,3]

1 Department of Health Sciences, College of Nursing and Health Professions, Drexel University, Philadelphia, PA 19104, USA; dps85@drexel.edu (D.S.); st96d633@drexel.edu (J.M.D.)
2 Department of Environmental Health & Engineering, Bloomberg School of Public Health, Johns Hopkins University, Baltimore, MD 21218, USA
3 Drexel Food Lab, Department of Health Sciences, College of Nursing and Health Professions, Drexel University, Philadelphia, PA 19104, USA; rms548@drexel.edu (R.S.); lem324@drexel.edu (L.M.)
* Correspondence: bm645@drexel.edu

Abstract: The purpose of this research is to evaluate the feasibility, acceptability, and preliminary efficacy of a household food-waste prevention and minimization intervention, titled the Culinary Home Empowerment for Food Waste Prevention and Minimization (CHEF-WPM), which consists of a culinary education video series for home cooks. The specific aims are to (1) assess the effects of the intervention at a population level across process (feasibility, usage, acceptability, satisfaction) and preliminary efficacy (motivation, opportunity, ability) metrics and (2) assess the effects of the intervention at a community level across process (feasibility, usage, acceptability, satisfaction) and preliminary efficacy (motivation, opportunity, ability, household food waste, sustainable dietary practices) metrics. The intervention includes eight modules, each containing three to five brief videos, as well as downloadable recipes and worksheets. The evaluation will explore the effects of the program through two distinct investigations, namely (1) voluntary access to the intervention content in a population-based setting and (2) intensive delivery of the intervention content as part of a remote class in a community setting. Evaluation of the intervention in the population-based setting will use a single-arm, quasi-experimental post-test only study design. All home cooks who access the videos will be invited to answer a five-question post-video survey about acceptability, satisfaction, and potential implementation of the learning. A separate sample of individuals will be recruited to participate in a more in-depth evaluation (pre- and multiple post-test survey). Evaluation of the community-based intervention will use a mixed methods study design. Findings from the two distinct evaluation studies will be jointly discussed and triangulated to support larger conclusions about the intervention's desirability, impact on motivation, opportunity, ability, and food waste, and the potential directions for further improvement.

Keywords: food loss and waste; sustainable development; pedagogy; kitchen literacy; culinary education; public health; community based; consumer

1. Introduction

As the global population grows, the demand for food is projected to increase by over 50% in the next 50 years, assuming current eating patterns remain unchanged [1]. According to the United Nations Environment Programme (UNEP), one-third of all food produced worldwide is lost or wasted annually [2]. Food waste is the largest component of landfills and incineration streams [3], with the majority occurring at the consumer level [2]. In the U.S., over 60% of food waste occurs at the consumption stage, including households and restaurants [3]. This significant amount of wasted food presents opportunities to combat food insecurity, save energy, and address environmental damage and climate change [3].

Food waste in high-income countries is primarily attributed to consumer behavior and policy. The National Academy of Sciences (NAS) has identified 11 consumer-level drivers of food waste based on the Motivation–Opportunity–Ability (MOA) framework [4]. These drivers include consumer knowledge and skills, goals related to food and nutrition, food preferences and diet, psychological distance from food, food environments, and relevant policies and regulations [4].

Block et al.'s four stages to consumer decisions align closely with the NAS's consumer-level drivers to food waste, as each stage is significantly influenced by consumer knowledge, skills, and food environments [5]. For example, in the *pre-acquisition stage*, consumer knowledge regarding the safe handling and culinary use of blemished produce can determine whether it will be purchased and used or discarded. Similarly, in the *acquisition stage*, consumers' ability to accurately calculate purchase amounts helps prevent overbuying food products. During the *consumption stage*, household food environments influence opportunities to avoid food waste by practicing proper food and leftover storage. Finally, in the *disposition stage*, consumer knowledge of food conservation and sustainable practices can reduce a household's food-waste impact [4,6].

1.1. The Consumer Food-Waste Paradox

Despite expressing a dislike for wasting food, consumers often engage in behaviors that result in food waste. To explore this paradox, Le Borgne et al. evaluated consumer concern for food waste and its impact on adopting "waste-prevention routines" [7]. They identified two dimensions of consumer concern: "individual/interpersonal concern", which strongly influenced waste-prevention behaviors, and "global concern", which did not have a significant effect. Their findings suggest that effective strategies to reduce household food waste should focus on enhancing individual/interpersonal concerns through economic concerns, food involvement, and food education. Similarly, Romani et al. emphasized the importance of "routine food management" behaviors in preventing food waste and highlighted the potential of targeted educational programs to promote sustainable food practices [8]. Their research indicates that daily food practices (shopping, cooking, and eating) and key behaviors (over-purchasing, poor meal-planning, and improper food storage) significantly contribute to household food waste and should be the focus of educational interventions.

1.2. Theoretical Underpinnings of Household Food-Waste Prevention

The Theory of Planned Behavior (TPB) [9] and Triandis' Theory of Interpersonal Behavior (TIB) [10] have been used to study consumer and household food-waste behaviors. TPB posits that behavior is driven by *attitudes* (e.g., a belief that wasting food is bad for the environment), *subjective norms* (e.g., societal expectations about food waste), and *perceived behavioral control* (e.g., self-efficacy in being able to reduce food waste). Triandis' TIB expands on TPB by integrating *habit* (e.g., habitual over-purchasing of food), *emotions* (e.g., guilt associated with wasting food), and *facilitating conditions* (e.g., availability of food storage facilities). By integrating these theories into a comprehensive framework, interventions can address both the rational and habitual aspects of food-waste behavior.

The Motivation–Opportunity–Ability (MOA) Model has been viewed as an integration of the Theory of Planned Behavior with Triandis' Theory of Interpersonal Behavior [11]. The MOA Model offers a targeted framework for understanding and influencing consumer behavior by considering *motivation*, an individual's desire or willingness to engage in a behavior (e.g., household food-waste prevention); *opportunity*, external circumstances that enable or inhibit the behavior (e.g., access to tools and resources that make it easier to reduce food waste, such as meal planning tools, food storage containers, kitchen utensils); and *ability*, an individual's capacity to perform the behavior (e.g., skills and knowledge to reduce food waste in the kitchen) (Figure 1) [11]. Adoption of the target behavior builds self-efficacy and reinforces ability and motivation, creating a positive feedback loop. A signature feature of the MOA Model is its integration of motivational, habitual, and contextual factors into a single model that can be used to explain pro-environmental behavior [11].

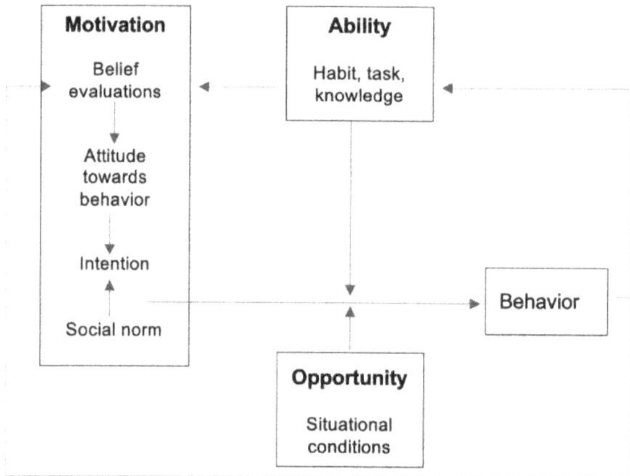

Figure 1. The Motivation–Opportunity–Ability Model.

1.3. Cooking Education as a Tool for Reducing Household Food Waste

Cooking education is a crucial tool for reducing household food waste. It enhances individuals' food agency—the ability to purchase and prepare food within their social, physical, and economic environments—empowering them to achieve their food-related goals [12]. Studies have shown that cooking interventions promote good nutrition and sustainable food practices [13–16]. Countries like Finland and Australia have incorporated home economics into educational curricula to mitigate food waste and enhance household sustainability [17]. Home economics educates consumers to be thrifty and mindful with food, make informed decisions about resources, and understand the environmental impact of their actions [18].

1.4. Specific Aims

While the NAS study provides a host of suggestions to reduce household food waste [4], translating the aforementioned findings into practice at the population and community levels is beyond the scope of most studies. Hence, the protocol described herein tests an approach to translating promising scientific findings into consumer-facing food-waste prevention and minimization lessons. The purpose of this protocol paper is to detail the design and methodology of a household food-waste prevention and minimization intervention titled Culinary Home Empowerment for Food Waste Prevention and Minimization (CHEF-WPM). Consistent with a protocol paper, this paper was submitted before data collection and, as such, does not present results or statistical analysis. Follow-on papers after data collection will do so [19]. The intervention consists of a culinary education video series for home cooks. The aims of the study are to:

Aim 1: assess the effects of CHEF-WPM at a population level across process (feasibility, usage, acceptability, satisfaction) and preliminary efficacy (motivation, opportunity, ability) metrics;

Aim 2: assess the effects of CHEF-WPM at a community level across process (feasibility, usage, acceptability, satisfaction), preliminary efficacy (motivation, opportunity, ability, household food waste), and exploratory (sustainable dietary behaviors) metrics.

It is valuable to study outcomes from both online exposures to the videos and outcomes from a more intensively implemented community-level delivery to gain insights into the effects of different types of deployment. Such insights could shape future motivation–opportunity–ability-oriented efforts using these or other videos.

2. Methods

The evaluation will explore the effects of the program through two distinct investigations, namely (1) when the intervention content is accessed voluntarily in a population-based setting (population CHEF-WPM) and (2) when the intervention content is accessed through intensive delivery as part of a remote class in a community setting (community CHEF-WPM). The population CHEF-WPM investigation will include two types of evaluations, and the community CHEF-WPM will include one type of evaluation. The subsequent sections in this report, as applicable, will be divided into the following three sections for clarity: *Population Intervention 1*, *Population Intervention 2*, and *Community Intervention*. This protocol is reported in accordance with SPIRIT (Standard Protocol Items: Recommendations for Interventional Trials) guidance. A completed SPIRIT checklist is available in Additional File 1. Additionally, Table 1 displays an overview of the objectives and methods for the three distinct evaluation components of the CHEF-WPM intervention.

Table 1. Overview of CHEF-WPM methodology.

Specific Aim	Setting	Study Design	Sample Size	Exposure	Outcomes
Aim 1: Assess the effects of CHEF-WPM at a population level across process and preliminary efficacy metrics.	Population intervention 1	Single-arm post-test only study design.	$N \geq 1068$	All home cooks who access the videos will be invited to answer a brief post-video survey.	Satisfaction, engagement, and comprehension.
	Population intervention 2	Single arm, quasi-experimental pre- and multiple post-test study design (3- and 6-month follow-up).	$N = 500$	Three specific food waste videos that are embedded in surveys.	Feasibility, usage, acceptability, satisfaction, motivation, opportunity, and ability.
Aim 2: Assess the effects of CHEF-WPM at a community level across process, preliminary efficacy, and exploratory metrics	Community intervention	Mixed method evaluation with baseline, post-8-week intervention, and 6-month follow-up assessments. Qualitative data will be collected after 8-week intervention	$N = 60$	All eight modules of the CHEF-WPM intervention will be presented in sequence, over an eight-week period, facilitated by a chef-instructor.	Feasibility, usage, acceptability, satisfaction, motivation, opportunity, ability, household food waste, and sustainable dietary behaviors.

2.1. Intervention

The overall goal of the CHEF-WPM intervention is to increase consumers' motivation, perceived opportunity, and ability to reduce food waste using home-based culinary techniques. The intervention will include eight modules, each containing three to five brief (4–8 min) videos, as well as downloadable recipes and worksheets. CHEF-WPM is designed to be used *a la carte* or as a self-paced online course.

The eight module topics include:

1. My whole plate: Meal Planning & Kitchen Management with a 'No Food Waste' Lens;
2. Veggies: Roots to Stem;
3. Fruit: Fresh or Bruised;
4. Meat- & Plant-Based Proteins;
5. Fish & Seafood: Sear It, Don't Fear It;
6. Bread, Grains, Snacks, Sweets;
7. Cooking Liquids & Dairy;
8. Leftovers Are My Jam.

Each module focuses on improving the *motivation*, *opportunity*, and *ability* to reduce food waste using home-based culinary techniques. The modules will highlight *motivations*, including saving money, avoiding waste, confidence, setting an example for children, and improving nutrition, pleasure, and environmental benefits. The modules will also focus on perceived *opportunity* and illustrate ways to shift contextual factors, such as time and schedule, kitchen arrangement and food storage, and technologies, among other aspects. *Ability* will be addressed via sharing techniques, recipes, and tricks to prevent and minimize food waste at home and problem-solving strategies for challenges to food waste prevention. Adapted from the work of ölander and ThØgersen [11], Figure 2 displays the conceptual framework guiding the CHEF-WPM program.

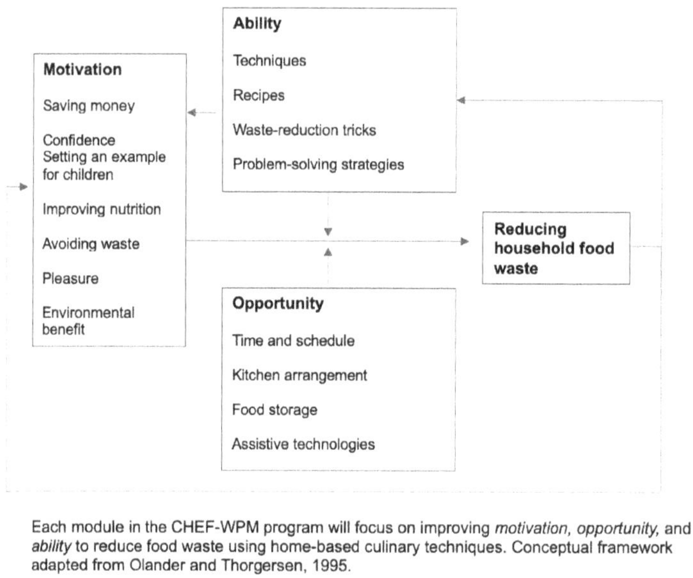

Figure 2. Conceptual framework of CHEF-WPM program [11].

The content will be developed and filmed at Drexel's College of Nursing and Health Professions' Food and Hospitality Management Space, and in the kitchen studio of the food-waste educator. The developed content will be hosted on a freely accessible website, Kajabi (http://kajabi.com; accessed on 1 July 2024). The CHEF-WPM program will be built and delivered to key audiences as an online course on Kajabi. Kajabi is an "all-in-one" platform that we will use to build our program, engage with learners (participants), and collect feasibility and preliminary efficacy metrics. The videos will also be disseminated via social media (i.e., X [FKA Twitter], Instagram, TikTok) and through partnerships with food-security organizations, retailers, nutrition education programs, and others.

2.2. *Population Intervention 1*

2.2.1. Overview

The evaluation of Population Intervention 1 will use a single-arm post-test only study design. All home cooks who access the videos will be invited to answer a brief post-video survey about acceptability and satisfaction (n = 1068).

2.2.2. Study Population and Recruitment

Potential participants will be 18 years or older, responsible for at least half of their household food preparation, prepare meals at home at least twice each week, have access to the internet, and understand English. The following groups will be excluded: adults unable to consent, individuals who are not yet adults (infants, children, and teenagers), and prisoners. Participants will be recruited through Drexel Food Lab networks and in partnership with community and national organizations. Recruitment methods will include digital and paper advertisements, email advertisements, QR codes, and other methods, such as shelf-tags at local libraries. QR codes and email links will direct individuals to the CHEF-WPM videos. At the end of each video, the viewers will have the opportunity to "opt-in" to complete a short survey to assess satisfaction, engagement, and comprehension. Our goal is to recruit a minimum of 1068 individuals to participate in that brief survey, providing 95% power to detect an effect size of 0.3. However, we will collect and analyze data from all of the participants who complete the 5-question survey. In other words, *we will not cease recruitment* once we obtain a sample size of 1068 quality answers.

2.2.3. Data Collection and Outcomes

As described, all home cooks who access the videos will be invited to answer a brief post-video survey about acceptability and satisfaction. At the end of each video, viewers will have the opportunity to "opt-in" to complete a short survey to assess satisfaction, engagement, and comprehension. If they agree, they will be redirected to a study information sheet and screening questionnaire. If they fit the study's inclusion criteria, they will proceed to a brief survey via Qualtrics that will assess satisfaction, engagement, and comprehension of the video content. The brief survey will take less than 2 min to complete.

2.3. *Population Intervention 2*

2.3.1. Overview

The second type of evaluation in the population-based setting will use a single arm, quasi-experimental pre- and multiple post-test study design. A sample (n = 500) of participants will be recruited to participate in a ~10 min survey to evaluate three specific videos across process (feasibility, satisfaction, engagement, comprehension) and preliminary outcome (motivation, opportunity, ability) metrics. The participants will be recontacted again at 3 and 6 months to complete follow-up surveys about longer-term outcomes.

2.3.2. Study Population and Recruitment

The inclusion and exclusion criteria are the same for both population-level interventions. Potential participants should be 18 years or older, responsible for at least half of their household food preparation, prepare meals at home at least twice each week, have access to the internet, and understand English. The following groups will be excluded: adults unable to consent, individuals who are not yet adults (infants, children, and teenagers), and prisoners.

We will recruit participants for this evaluation by utilizing ResearchMatch and Build Clinical. ResearchMatch is a website that connects potential participants with researchers by emailing individuals with a recruitment message that explains the basic details of the research study. If individuals are interested in participating in the study, they will be directed to a screening questionnaire. Build Clinical will run an online advertisement campaign on online and social media websites, such as Facebook, Instagram, and X (formerly Twitter). When individuals press on these digital advertisements, they will be redirected to a landing page where they can learn more about the study and fill out a screening questionnaire.

If the interested individual from either platform is eligible for the study, a research assistant will call the individual and further explain the goals and procedures of the intervention. After this phone call, the research assistant will send the individual an email with a link to the information sheet for exempt research. Proceeding to the survey will serve as a proxy for written consent. Participants will be offered a USD 10 incentive for completion of the pre- and post-video surveys, and an additional USD 5 incentive for completion of each follow-up assessment (3 and 6 months; USD 10 for completion of both; USD 20 for completion of the entire study).

A sample size of 500 is sufficient to provide 95% power to detect an effect size of 0.43. Participants will be purposively sampled by age, race/ethnicity, gender, and income to optimize the diversity of our sample. Our goal is for our sample to mirror the US census demographic composition of the most recent census (Gender—Male: 49.6%, Female: 50.4%, Non-binary: natural fallout; Race—White: 76%, Black/AA: 14%, Asian or Pacific Islander: 6%, American Indian/Alaskan Native/Other: 2%; Hispanic Ethnicity—Hispanic: 19%, Non-Hispanic: 59%).

2.3.3. Data Collection and Outcomes

A structured survey (~10-min pre-post survey) will evaluate specific educational videos across process (feasibility, usage, acceptability, and satisfaction) and preliminary outcome (motivation, opportunity, and ability) metrics. The enrolled participants will also complete 3- and 6-month follow-up surveys. The structured survey will ask about demographic characteristics and will obtain baseline information about motivation, opportunity, and ability as they relate to food-waste prevention. The survey will also contain three embedded food-waste prevention videos for the participants to view. After viewing each video, the participants will be asked to answer questions about the content to assess video satisfaction, engagement, recall, and comprehension of information, in addition to questions about motivation, opportunity, and ability. The 3- and 6-month follow-up surveys will primarily focus on motivation, opportunity, and ability to reduce household food waste.

2.3.4. Data Analyses for Population Interventions 1 and 2

Descriptive characteristics of our study sample will be calculated as means ± standard deviations, frequencies, chi-square contingency analyses, and standardized residuals, unless otherwise noted. Changes in motivation, opportunity, and ability will be analyzed as continuous or categorical variables. Mixed factorial analyses of covariance (ANCOVA) with repeated measures will be used to determine whether mean differences exist in participant satisfaction, engagement, and comprehension after controlling for the influence of socio-demographic- (age, race and ethnicity, marital status, education, and income) and household-related characteristics measured at baseline. Differences will be estimated for all categorical variables using McNemar's exact significance probability testing for paired nominal data. All quantitative statistical analyses will be accomplished using IBM SPSS Statistics.

2.4. *Community Intervention*

2.4.1. Overview

All eight modules of the CHEF-WPM intervention will be presented in sequence over an eight-week period, facilitated by a chef–instructor with assistance from culinary students. The course will include synchronous discussions and group viewing of the CHEF-WPM video series hosted on Kajabi. Evaluation of the remote community-based intervention will use a mixed methods evaluation study design. The objectives of this mixed methods evaluation are to assess (1) intervention implementation (fidelity), satisfaction, engagement, and comprehension; (2) intervention outcomes (motivation, opportunity, ability, and household food waste); and (3) exploratory outcome (sustainable dietary behaviors). A maximum of 60 participants will be recruited in partnership with local community organizations for quantitative data-collection activities. Quantitative data collection will occur pre- and post-8-week intervention and at 6 months follow-up. A purposive sample of 20 participants will be recruited to participate in in-depth interviews at 8 weeks post-intervention. Food waste will be measured via self-report methods among the participants (n = 60) at pre- and post-intervention and 6-month follow-up.

2.4.2. Study Population and Recruitment

Potential participants will be 18 years or older, responsible for at least half of their household food preparation, prepare meals at home at least twice each week, have access to the internet, and understand English. The following groups will be excluded: adults unable to consent, individuals who are not yet adults (infants, children, and teenagers), and prisoners. Individuals who express interest in participating in the 8-week CHEF-WMP program will be screened for eligibility via telephone or a brief Qualtrics survey.

A convenience sample of 60 participants will be recruited for the community-level intervention, drawn primarily from a lower-income population in Philadelphia (3 cohorts of 20 individuals per cohort). This population includes a higher percentage of Black, Indigenous, and people of color than the US population. For the qualitative evaluation (in-depth interviews), a purposive sample (n = 20) of intervention participants will be recruited. Purposive sampling helps ensure sample diversity across age, income, educational attainment, household size, and baseline pre-disposition for food waste prevention. Participants will be recruited through university networks and in partnership with community organizations. Participants will be recruited via phone calls and/or email, digital and paper advertisements, and QR codes. These organizations together generally attract participants reflecting the city's diverse urban demographics, namely 44% Black, 13% Latino, 7% Asian; 46,000 median household income; and 20% poverty rate.

QR codes and email links will direct individuals to a Qualtrics screening questionnaire. If they qualify, they will proceed to a study information sheet that describes participation in detail. If they are still interested after learning more about participation, they will be asked to submit their contact information and a research assistant will contact them with further details about participation. Participants will be compensated a total of USD 150 for completing the evaluation activities (quantitative measures) and an additional USD 40 for completing an interview.

2.4.3. Data Collection and Outcomes

The participants will complete a pre-intervention survey and measure their household food waste before beginning the 8-week intervention. At the completion of the intervention, participants will complete a post-intervention survey and again measure their household food waste. A purposive sample of participants, six from each cohort, will be recruited to complete in-depth interviews at the end of the intervention. We will follow up with all the participants 6-months after the intervention, where they will complete a 6-month post-intervention survey and measure their household food waste. These measures are described below.

For the *pre- and post-intervention surveys*, after viewing the study information page for exempt research, the participants will proceed to a structured survey that will ask about demographic characteristics and will obtain baseline information about motivation, opportunity, and ability as they relate to food-waste prevention. To the extent possible, the evaluation questions were replicated from existing surveys and tailored to this intervention, including the specific content of each media asset. After completing the pre-intervention survey, the participants will be provided a scale and food-waste diary and given a brief training by a study researcher on measuring their home food waste. They will keep a one-week food-waste diary, as described below. The participants will be asked to complete a brief satisfaction survey after each program session.

The post-intervention assessment will occur at the end of the 8-week program. A survey that includes the baseline *survey* questions and additional post-intervention questions will be sent to participants during the one-week period after their last intervention session. In this post-intervention survey, the questions will assess program satisfaction, engagement, recall, and comprehension of information, participants' level of implementation of steps recommended in the videos, further engagement with and shares of videos, reactions to the intervention, and intention to continue engaging in food-waste prevention activities. Participants will again receive instructions on how to measure their food waste for one week. The 6-month follow-up assessment will include a similar survey and food-waste measurement.

For *household food waste*, the participants will measure their household food waste for one week via self-report before and after participating in the remote community-based class. The participants will be given a Taylor compact digital scale and trained on how to measure their food waste. Our team will create a training video and instructions for how to measure food waste at home and will provide visual reminders for key aspects of the measurement, such as what to classify as "avoidable" vs. "unavoidable". The participants will be given a diary form and instructed that, before they or a household member discards any household food waste to any destination including the sink and whether it is trim from meat or produce, plate waste, prepared leftovers, spoiled food, etc., they should weigh the food, describe what it is (fruit, vegetable, cooked grain, etc.), and indicate its type (trim from cooking, spoiled food, leftovers, plate waste, or other), as well as their impression of whether the food waste was unavoidable (example, fish innards) or avoidable (example, spoiled milk from overbuying) through a simple coding system, taking a minute or less. This requires discussion and buy-in from household members. The participants will again be asked to participate in the food-waste assessment at the 6-month follow-up.

For *sustainable dietary practices*, at pre- and post-intervention, and the 6-month follow-up, the participants will be asked to complete a brief dietary assessment to measure the overall sustainability of their dietary practices. This assessment will be conducted using Diet ID via their metric called Dietary Impacts on Environmental Measures (DIEM). The DIEM metric helps us understand the environmental footprint of participants' diets. Diet ID is a novel pattern-recognition-based method that facilitates fast, user-friendly, and scalable dietary assessment [20]. Participants will be prompted with two photos of diets and asked which of the two is more like what they usually eat. Once they have chosen, the participants will be prompted with another two photos and asked the same questions. Participants will answer questions about their gender, height, weight, age, dietary restrictions/allergies, and general physical activity level (low, moderate, high) before completing the dietary assessment. No identifiable information will be collected by Diet ID.

For *qualitative data collection*, a purposive sample (n = 20) of intervention participants will be recruited to maximize diversity across age, income, educational attainment, household size, and baseline pre-disposition for food-waste prevention. They will participate in in-depth interviews after intervention completion to evoke priorities, values, and wisdom related to food-waste prevention; reactions to the intervention; and effects on motivations, opportunities, and abilities, as well as barriers and facilitators for continued engagement in food-waste prevention behaviors, and potential co-benefits for other outcomes as well as undesired tradeoffs and spillovers. Interviews will be led by an experienced qualitative researcher, will be conducted on HIPPA-compliant Zoom, will last approximately 30–45 min, and will be audio-recorded and transcribed verbatim.

2.4.4. Data Analyses

Descriptive characteristics of our study sample will be calculated as means ± standard deviations, frequencies, chi-square contingency analyses, and standardized residuals, unless otherwise noted. Changes in food-waste prevention behaviors, discards, motivation, opportunity ability, potential economic and nutrition-related savings, and DIEM score will be analyzed as continuous or categorical variables. Mixed factorial analyses of covariance (ANCOVA) with repeated measures will be used to determine whether mean differences exist in participant satisfaction, engagement, comprehension, changes in knowledge and planned behaviors, and food-waste variables based on socio-demographic (age, race and ethnicity, marital status, education, and income) and household-related characteristics measured at baseline. Differences will be estimated for all categorical variables using McNemar's exact significance probability testing for paired nominal data. All quantitative statistical analyses will be accomplished using IBM SPSS Statistics. As the purpose of proof-of-concept studies is to determine whether the intervention merits more costly testing using a randomized controlled trial design, small sample sizes are acceptable because clinical, not statistical, benefit is sought.

For qualitative data, the interviews will be analyzed using content analysis based on a deductive–inductive approach. First, interview segments related to the drivers of food-waste prevention (motivation, opportunity, and ability) will be identified. For each driver, themes will be identified inductively.

In this study, once the quantitative and qualitative data are analyzed separately, the data will then be merged together to have a greater understanding of the effects of the intervention. This will allow discongruence and congruence between both sets of data to emerge. We will report the mixed methods results with joint displays.

2.4.5. Community Advisory Board

A community advisory board has been recruited from our community partners to provide input into the design, content, guest-chef representation, and recipes of the videos. The advisory board will meet quarterly via Zoom and in-person (hybrid model), and each member will be paid an incentive for their participation.

2.5. Withdrawal

The study team does not anticipate circumstances under which participants would need to be withdrawn from the research without their consent. Participants may elect to leave the study at any time with no penalty. If a withdrawn participant completed the study survey, their data may still be included in our analysis. Additionally, if participants withdraw from the study before the end of data collection, they will not be compensated.

3. Potential Benefits and Risks of Study Participation

As with any study involving the collection of data, there is a risk of harm resulting from a breach of privacy and/or confidentiality. If information linking participants to their responses is inadvertently disclosed, we do not anticipate any significant harm to result to participants. However, we will still secure the information we collect (as outlined in the data-protection section) to minimize the probability of this risk. We will also maintain feedback and discussion data in separate folders without individual identifiers. Once the research study is complete and individually identifiable information is no longer necessary, the key linking individuals to their responses will be destroyed, further reducing the magnitude of this risk.

During the interviews, it is possible that some activities or questions will lead to momentary embarrassment, discomfort, or unease. The probability and magnitude of discomfort are not expected to be greater than participants would ordinarily encounter in their everyday lives and will be disclosed during the informed consent process. We do not anticipate any costs to participants for taking part in this study. We do not expect any of our procedures will carry unforeseen risks to participants or individuals not participating in the research.

There may be no direct benefit to participants of this research study, though they may value the opportunity to participate in the class. The information collected may benefit society by developing a novel intervention to enhance participants' motivation, opportunity, and ability to reduce consumer food waste. There are minimal risks associated with this study. The probability and magnitude of discomfort are not expected to be greater than participants would ordinarily encounter in their everyday lives and will be disclosed during the informed consent process. The study procedures, particularly the food-waste tracking, require an investment of time and effort.

4. Data Monitoring

The principal investigators will monitor and review the study progress, participant safety, and the accuracy and security of the emerging data. All information collected in this study will be kept confidential as required by law. Names will not be attached to the study participants and a master list, and the data elements and survey responses will be kept in a secure and encrypted database, which can only be accessed by research team members. Due to the low-risk nature of the intervention, a Data and Safety Monitoring Board has not been formally established. All adverse events will be reported to the principal investigators and reviewed by the study team to determine whether the study should be discontinued due to participant safety. Quality assurance procedures will be implemented, including rigorous attention to manual data cleaning, generation of monthly query reports, and ranges and data validation checks.

5. Ethics and Dissemination

5.1. Consent

The consent process for each of the three types of evaluations will be the same. The participants will complete a screening questionnaire to determine study eligibility. If eligible, they will be directed to a study information sheet for exempt research. After reading the information sheet for exempt research, proceeding to the survey will serve as a proxy for informed consent. In the information sheet, the research team's contact information will be provided to the participant. Participants are encouraged to contact the research team if there are any questions and will be given the opportunity to discuss the study with their surrogates or think about it prior to agreeing to participate. They may withdraw consent at any time throughout the study. A copy of the information sheet will be emailed to participants for their records.

5.2. Confidentiality

Participant confidentiality is strictly held in trust by the investigators, study staff, and the sponsor and their agents. The study protocol, documentation, data, and all other information generated will be held in strict confidence. No information concerning the study or the data will be released to any unauthorized third party without prior written approval of the sponsor. The study monitor or other authorized representatives of the sponsor may inspect all study documents and records required to be maintained by the investigator.

Confidentiality will be protected by including only information needed to assess study outcomes, minimizing to the fullest extent possible the collection of any information that could directly identify participants, and maintaining all study information in a secure manner. To ensure participant privacy and confidentiality, a unique study identifier will be assigned to each participant. The data and records will be kept locked and secured, with any computer data password protected. Protected health information (PHI) will not be collected in this study. Therefore, we are requesting a waiver of HIPAA authorization. To ensure privacy, all interviews will be conducted via HIPAA-compliant Zoom technologies. Links to Qualtrics surveys will be encrypted. To protect the confidentiality of participant responses, the data will be de-identified and unique identification numbers will be assigned to participants.

5.3. Dissemination

We will disseminate the study findings through conference presentations and publications in peer-reviewed international journals. Only aggregate data without individually identifiable information will be published. As this study is a feasibility study, the results will be used to inform and develop a larger, adequately powered, randomized controlled trial, if appropriate.

6. Results

This research investigates a highly scalable solution to strengthen consumers' motivation, opportunity, and ability to reduce food waste at the locus where most wasted food occurs by adapting professional culinary education interventions to the home kitchen. This interdisciplinary project combines knowledge from the fields of culinary arts and science, education, consumer behavior, nutrition, environmental science, and public health to provide low-friction interventions and flexible access to content that will resonate with consumers. Rather than teaching home cooks through an overbearing and expensive didactic approach, this intervention empowers home cooks to prevent and minimize food waste in a way that is convenient, accessible, culturally appropriate, and scalable. This intervention offers home cooks the opportunity for professional-level culinary education in a way that saves money, utilizes cultural flavors, provides nutrition background information, and builds confidence, all while reducing food waste.

This research also includes multiple innovative features, addressing critical research gaps and pioneering new approaches:

1. *Cooks as consumer educators*, reflecting the NAS report highlighting the need to use such influencers [4];
2. *Bite-size videos*, adapting food waste education to a social media-friendly modality that many people use to learn about cooking today. This allows us to create or support new or existing influencers, whose reach and appeal are hard to match through more traditional research interventions;
3. *Innovative application and assessment* of motivation, opportunity, and ability (MOA) metrics [11];
4. *Innovation in a longer timeframe of follow-up* and in comparing and contrasting the same intervention in both community and population settings;
5. *Direct measurement of waste.*

7. Discussion

Wasted food is the largest component of the landfill and incineration stream [3], and most of the waste generated remains at the consumer level [2]. Hence, the purpose of this research protocol is to evaluate the feasibility, acceptability, and preliminary efficacy of a household food-waste prevention and minimization intervention, which consists of a culinary education video series for home cooks. The specific aims of this protocol are to (1) assess the effects of the intervention at a population level across process (feasibility, usage, acceptability, and satisfaction) and preliminary efficacy (motivation, opportunity, and ability) metrics and (2) assess the effects of the intervention at a community level across process (feasibility, usage, acceptability, and satisfaction) and preliminary efficacy and exploratory (motivation, opportunity, ability, household food waste, and sustainable dietary practices) metrics.

The evaluation seeks to understand intervention effects both when accessed voluntarily in a population-based setting, and via a more intensive delivery as part of a program in a community setting. Findings from the two distinct evaluation studies will be jointly discussed and triangulated to support larger conclusions about the intervention's desirability, impact on motivation, opportunity, ability, and food waste, and the potential directions for further improvement. The intervention's benefits may be both short and long term because the skills and attitudes presented can shape ongoing kitchen practices and norms.

Final Considerations

Consistent with the protocol paper format, this paper does not present data but rather the methodology (protocol) that will be used to investigate the stated aims. We anticipate that the data analysis will help identify aspects of the intervention that are promising to prevent and minimize wasted food on a larger scale and will inform policy and funding for program interventions. Future publications will present the data and analysis from this protocol, and future protocols will test additional promising interventions, as suggested by our analysis in this study. If successful, the intervention will provide an affordable, scalable, and highly adaptable component that could be integrated into a national food-waste prevention campaign and other communication activities.

Author Contributions: Conceptualization: J.M.D., B.-J.M. and R.N.; Writing—original draft preparation: B.-J.M.; Funding acquisition J.M.D., R.N., B.-J.M. and R.S.; Writing—review and editing: B.-J.M., J.M.D., R.N., D.S., D.F., R.S., L.M. and A.M. All authors have read and agreed to the published version of the manuscript.

Funding: This study is funded by the United States Environmental Protection Agency, Science to Achieve Results (STAR) Program, Grant Number 84053401.

Institutional Review Board Statement: This research protocol was approved by the Institutional Review Boards at Drexel University under exempt status (IRB approval number: 2307010017) and Johns Hop-kins University (IRB approval number: 28247) as not human subject research.

Informed Consent Statement: The consent process for each of the three types of evaluations will be the same. The participants will complete a screening questionnaire to determine study eligibility. If eligible, they will be directed to a study information sheet for exempt research. After reading the information sheet for exempt research, proceeding to the survey will serve as a proxy for informed consent. In the information sheet, the research team's contact information will be provided to the participant. Participants are encouraged to contact the research team if there are any questions and will be given the opportunity to discuss the study with their surrogates or think about it prior to agreeing to participate. They may withdraw consent at any time throughout the study. A copy of the information sheet will be emailed to participants for their records.

Data Availability Statement: The original contributions presented in the study are included in the article, further inquiries can be directed to the corresponding author.

Conflicts of Interest: The authors declare that the research was conducted in the absence of any commercial or financial relationships that could be construed as a potential conflict of interest.

References

1. Yahia, E. (Ed.) *Preventing Food Losses and Waste to Achieve Food Security and Sustainability*; Burleigh Dodds Science Publishing: London, UK, 2019. [CrossRef]
2. UN Environment Programme. UNEP Food Waste Index Report 2021. Available online: https://www.unep.org/resources/report/unep-food-waste-index-report-2021 (accessed on 1 August 2024).
3. US EPA. From Farm to Kitchen: The Environmental Impacts of U.S. Food Waste. Available online: https://www.epa.gov/land-research/farm-kitchen-environmental-impacts-us-food-waste (accessed on 1 August 2024).
4. Committee on a Systems Approach to Reducing Consumer Food Waste; Board on Environmental Change and Society; Food and Nutrition Board; Division of Behavioral and Social Sciences and Education; Health and Medicine Division; National Academies of Sciences, Engineering, and Medicine. *A National Strategy to Reduce Food Waste at the Consumer Level*; Schneeman, B.O., Oria, M., Eds.; National Academies Press: Washington, DC, USA, 2020; p. 25876. [CrossRef]
5. Block, L.G.; Keller, P.A.; Vallen, B.; Williamson, S.; Birau, M.M.; Grinstein, A.; Haws, K.L.; La Barge, M.C.; Lamberton, C.; Moore, E.S.; et al. The Squander Sequence: Understanding Food Waste at Each Stage of the Consumer Decision-Making Process. *J. Public Policy Mark.* **2016**, *35*, 292–304. [CrossRef]
6. van Geffen, L.; van Herpen, E.; van Trijp, H. Household Food Waste—How to Avoid It? An Integrative Review. In *Food Waste Management: Solving the Wicked Problem*; Närvänen, E., Mesiranta, N., Mattila, M., Heikkinen, A., Eds.; Springer International Publishing: Cham, Switzerland, 2020; pp. 27–55. [CrossRef]
7. Le Borgne, G.; Sirieix, L.; Valette-Florence, P.; Costa, S. Adopting Waste-Prevention Routines: The Role of Consumer Concern for Food Waste. *Appetite* **2021**, *163*, 105188. [CrossRef] [PubMed]
8. Romani, S.; Grappi, S.; Bagozzi, R.P.; Barone, A.M. Domestic Food Practices: A Study of Food Management Behaviors and the Role of Food Preparation Planning in Reducing Waste. *Appetite* **2018**, *121*, 215–227. [CrossRef] [PubMed]
9. Ajzen, I. The Theory of Planned Behavior. *Organ. Behav. Hum. Decis. Process.* **1991**, *50*, 179–211. [CrossRef]
10. Triandis, H.C. Subjective Culture and Interpersonal Relations Across Cultures. *Ann. N. Y. Acad. Sci.* **1977**, *285*, 418–434. [CrossRef]
11. Ölander, F.; Thøgersen, J. Understanding of Consumer Behaviour as a Prerequisite for Environmental Protection. *J. Consum. Policy* **1995**, *18*, 345–385. [CrossRef]
12. Wolfson, J.A.; Lahne, J.; Raj, M.; Insolera, N.; Lavelle, F.; Dean, M. Food Agency in the United States: Associations with Cooking Behavior and Dietary Intake. *Nutrients* **2020**, *12*, 877. [CrossRef] [PubMed]
13. Hersch, D.; Perdue, L.; Ambroz, T.; Boucher, J.L. The Impact of Cooking Classes on Food-Related Preferences, Attitudes, and Behaviors of School-Aged Children: A Systematic Review of the Evidence, 2003–2014. *Prev. Chronic Dis.* **2014**, *11*, E193. [CrossRef] [PubMed]
14. Dyen, M.; Sirieix, L. How Does a Local Initiative Contribute to Social Inclusion and Promote Sustainable Food Practices? Focus on the Example of Social Cooking Workshops. *Int. J. Consum. Stud.* **2016**, *40*, 685–694. [CrossRef]

15. Lindgren, E.; Harris, F.; Dangour, A.D.; Gasparatos, A.; Hiramatsu, M.; Javadi, F.; Loken, B.; Murakami, T.; Scheelbeek, P.; Haines, A. Sustainable Food Systems—A Health Perspective. *Sustain. Sci.* **2018**, *13*, 1505–1517. [CrossRef] [PubMed]
16. Murad, M.; Alford, A.-M.; Davis, K. Farm to Future: A Virtual Summer Nutrition Culinary Camp for Kids. *J. Nutr. Educ. Behav.* **2021**, *53*, 445–448. [CrossRef] [PubMed]
17. Finnish Home Economics Teachers Enabling Sustainability and Consumer Skills for Young People. Available online: https://helda.helsinki.fi/items/6fa0f84b-d3fe-461b-91ff-2252fe4797f4 (accessed on 1 August 2024).
18. Renwick, K. Why Home Economics Classes Still Matter. 2018. Available online: https://policyoptions.irpp.org/magazines/june-2018/why-home-economics-classes-still-matter/ (accessed on 1 August 2024).
19. Cameli, M.; Novo, G.; Tusa, M.; Mandoli, G.E.; Corrado, G.; Benedetto, F.; Antonini-Canterin, F.; Citro, R. How to Write a Research Protocol: Tips and Tricks. *J. Cardiovasc. Echogr.* **2018**, *28*, 151–153. [CrossRef] [PubMed]
20. Bernstein, A.M.; Rhee, L.Q.; Njike, V.Y.; Katz, D.L. Dietary Assessment by Pattern Recognition: A Comparative Analysis. *Curr. Dev. Nutr.* **2023**, *7*, 101999. [CrossRef] [PubMed]

MDPI AG
Grosspeteranlage 5
4052 Basel
Switzerland
Tel.: +41 61 683 77 34

Foods Editorial Office
E-mail: foods@mdpi.com
www.mdpi.com/journal/foods

www.ingramcontent.com/pod-product-compliance
Lightning Source LLC
Chambersburg PA
CBHW041018270926
51850CB00024B/1384

* 9 7 8 3 7 2 5 8 7 2 4 0 4 *